12/31

YALE AGRARIAN SERIES

James C. Scott, series editor

The Agrarian Studies Series at Yale University Press seeks to publish outstanding and original interdisciplinary work on agriculture and rural society—for any period, in any location. Works of daring that question existing paradigms and fill abstract categories with the lived experience of rural people are especially encouraged.
—James C. Scott, *Series Editor*

James C. Scott, *Seeing Like a State: How Certain Schemes to Improve the Human Condition Have Failed*

Steve Striffler, *Chicken: The Dangerous Transformation of America's Favorite Food*

Alissa Hamilton, *Squeezed: What You Don't Know About Orange Juice*

Bill Winders, *The Politics of Food Supply: U.S. Agricultural Policy in the World Economy*

James C. Scott, *The Art of Not Being Governed: An Anarchist History of Upland Southeast Asia*

Benjamin R. Cohen, *Notes from the Ground: Science, Soil, and Society in the American Countryside*

Parker Shipton, *Credit Between Cultures: Farmers, Financiers, and Misunderstanding in Africa*

Paul Sillitoe, *From Land to Mouth: The Agricultural "Economy" of the Wola of the New Guinea Highlands*

Sara M. Gregg, *Managing the Mountains: Land Use Planning, the New Deal, and the Creation of a Federal Landscape in Appalachia*

Michael R. Dove, *The Banana Tree at the Gate: A History of Marginal Peoples and Global Markets in Borneo*

Patrick Barron, Rachael Diprose, and Michael Woolcock, *Contesting Development: Participatory Projects and Local Conflict Dynamics in Indonesia*

Edwin C. Hagenstein, Sara M. Gregg, and Brian Donahue, eds., *American Georgics: Writings on Farming, Culture, and the Land*

Timothy Pachirat, *Every Twelve Seconds: Industrialized Slaughter and the Politics of Sight*

Andrew Sluyter, *Black Ranching Frontiers: African Cattle Herders of the Atlantic World, 1500–1900*

Brian Gareau, *From Precaution to Profit: Contemporary Challenges to Environmental Protection in the Montreal Protocol*

Kuntala Lahiri-Dutt and Gopa Samanta, *Dancing with the River: People and Life on the Chars of South Asia*

Alon Tal, *All the Trees of the Forest: Israel's Woodlands from the Bible to the Present*

Drew A. Swanson, *A Golden Weed: Tobacco and Environment in the Piedmont South*

For a complete list of titles in the Yale Agrarian Studies Series, visit www.yalebooks.com.

DREW A. SWANSON

A Golden Weed

Tobacco and Environment in
the Piedmont South

Yale UNIVERSITY PRESS

NEW HAVEN AND LONDON

Yale University Press books may be purchased in quantity for educational, business, or promotional use. For information, please e-mail sales.press@yale.edu (U.S. office) or sales@yaleup.co.uk (U.K. office).

Set in Scala type by Westchester Book Group.

Printed in the United States of America.

Library of Congress Cataloging-in-Publication Data

Swanson, Drew A., 1979–
    A golden weed : tobacco and environment in the Piedmont South / Drew A. Swanson.
        pages cm. — (Yale agrarian series)
    Includes bibliographical references and index.
    ISBN 978-0-300-19116-5 (clothbound : alk. paper)
    1. Tobacco—Piedmont (U.S. : Region)—History.   2. Tobacco—Southern States—
History   I. Title.   II. Title: Tobacco and environment in the Piedmont South.
    SB273.S93 2014
    633.7'10975—dc23   2013050495

A catalogue record for this book is available from the British Library.

This paper meets the requirements of ANSI/NISO Z39.48–1992 (Permanence of Paper).

10 9 8 7 6 5 4 3 2 1

For Margaret, Ethan, and Eugene

CONTENTS

Preface   ix

Acknowledgments   xiii

Introduction: How Did Such a Poor Land Make Some People So Rich, and How Did They So Quickly Become Poor Again?   1

1   On the Back of Tobacco: Sowing the Seeds of a Tobacco Culture   16

2   Let There Be Bright: The Birth of Yellow Tobacco Culture   46

3   Bright Leaf, Bright Prospects: Making Peace with the Idea of Yellow Tobacco   82

4   Tobacco Goes to War   119

5   Fire in the Fields: Reconstructing Labor and Land Following the Civil War   147

6   A Barren and Fruitful Land   182

7   The Decline of the Border   216

Epilogue: A New Deal for Old Land?   246

Appendix: Antebellum Tobacco Prices   263

Notes   267

Index   339

THE FARMER HOLDS a rather schizophrenic place in American history, serving as both hero and villain at various points. On one hand, our histories portray farmers as sources of boundless energy and enterprise, pushing the nation westward, driving economic growth, fueling an almost limitless productivity, and expounding upon the American dream. On the other hand, the farmer often plays the part of historical anchor, a rather greedy dullard who exploits both land and labor in a never-ending pursuit of profit. This farmer is old-fashioned, conservative, opposed to genuine reform or progress, and especially fearful of cities and industry. Rather than making a living in cooperation with the land, this farmer abuses landscapes and then moves on to fresh ones, driving national expansion through carelessness and waste rather than industriousness.

The tobacco grower has long served as the prototype of the latter farmer; few historians have found anything good to say about tobacco farmers as agriculturalists. Recall your American history survey. Tobacco growers appear first as get-rich-quick schemers at Jamestown, so tobacco-crazed that they plant the weed in the streets and fail to raise corn for their own larders; they reappear as aristocratic Chesapeake planters abusing their slave labor even as they demand independence from Britain; and they serve as handy examples of plantation owners who had depleted their eastern lands by the early nineteenth century, only to move west en masse, replicating the plantation system and its ills in such places as Mississippi and Texas. When tobacco farmers garner mention in the post–Civil War era, it is usually to serve as dusty relics of an older culture struggling (ineffectually) to survive in impoverished rural corners of the New South, like Robert Penn Warren's characters in *Night Rider*. Finally, from the New Deal on, tobacco farmers become part of historians' descriptions of one of the twentieth century's great villains, big tobacco companies.

This book approaches tobacco farmers with a basic question in mind: What sort of farmers were they, really? It looks at a century of tobacco cultivation in a place, the Virginia–North Carolina Piedmont border, where tobacco was especially important, and where the crop proved unusually problematic. Between the early nineteenth century and the early twentieth century, this region, known locally as the Southside, transformed from a relatively productive agrarian landscape into a district of extreme environmental problems—severely eroded hills, sandchoked streams, deforested ridges, and impoverished farmers. A particular form of agricultural practice, centered on the development and expansion of the bright (also known as yellow) form of tobacco, deserved much of the blame. The details of this place and this period defy traditional historical explanations, however; it was not simple greed or conservatism that created the region's problems, though these forces were present. Rather, the local environment, the particularities of tobacco plants, and human conceptions of the natural world and people's place in it combined in ways that proved tragic.

This book finds that tobacco farmers in the Southside consistently sought what they believed to be the best use of local landscapes, based on their intimate knowledge of local environments. They spoke in a language of permanence and stability (what we might be tempted to call sustainability today) rather than exploitation, and they thought long and hard about their futures in this place. But for all this concern about permanence, Southside farmers failed to escape the grasp of their present. Local environments provided real limits to plans; they pushed back when farmers pushed too hard. Southern social structures and racial attitudes entangled all management and use of the environment and thus complicated all attempts at agricultural stability. And as is the case with almost all schemes to develop and manage environments, incomplete understandings of how nature worked—the ecology of the place—undermined farmers' best-laid plans.

This book is part of a growing body of literature on the environmental history of the American South. As such it takes agriculture seriously as the primary method of interaction between people and the landscapes in which they lived, complementing impressive recent scholarly work on rice and cotton in the region. As a self-consciously southern history, this study accepts that the region's history of race influenced all, and it works to weave that story into the rise and decline of tobacco's prospects as well. Within the historiography of tobacco, this book also fills a bit of a lacuna. It explores the tobacco landscape in transition, a place no longer the site of the colonial export staple but not yet fully the captive of the big tobacco companies.

In the following pages, to the extent possible, this is a history looking from the tobacco field outward, exploring farmers' views of the world. It is an effort to take their claims and their environmental understandings as seriously as they did and to recover a world that has been all but lost. In this attempt we find a new history of nineteenth-century southern agriculture, no less tragic for its fresh details but important nonetheless if we wish to understand how the modern world came to be.

ACKNOWLEDGMENTS

THIS BOOK THREATENED to become a gully of sorts at times, to suck me into its crumbling banks and dump me somewhere downstream. That it did not wash me away is thanks to the wonderful people who support me. No one has done more to bring this book to fruition, or to shape my scholarly life, than Paul Sutter. He patiently encouraged me to ask bigger questions and to provide more thoughtful answers. Paul has that rarest of talents: the ability to compliment without creating complaisance and to criticize without fostering resentment. I have never met a historian more dedicated to his craft or his students. Other scholars contributed their insightful comments over the course of the last few years. Kathleen Clark, Shane Hamilton, John Inscoe, and Allan Kulikoff all read complete drafts and offered comments. John in particular put up with many a long, rambling conversation as I parsed out various portions of this work in his office. Conversations with fellow graduate students and colleagues shaped and strengthened the book. Sarah Frohardt-Lane, Jim Giesen, Amanda Hagood, Philip Herrington, Mark Hersey, Tim Johnson, Leonard Lanier, Michele Lansdown, Chris Manganiello, Devan McGranahan, Sarah Milov, Barton Myers, Kathi Nehls, Tom Okie, Jesse Pope, Tim Silver, Hayden Smith, Mart Stewart, Claire Strom, Levi Van Sant, and Bert Way influenced my thinking, as did the writing of Evan Bennett and Barbara Hahn. At Millsaps, I profited from an unusually engaged and supportive department, where David Davis, Amy Forbes, Bob McIlvaine, Andrew Paxman, and Stephanie Rolph all expressed an interest in this project and aided it in various ways. Bill Storey deserves special mention for serving as mentor and sounding board as this project moved from dissertation to book. Two anonymous readers for Yale University Press also pushed me to make the book's arguments clearer and to hone my comments throughout the text.

Archivists at the University of Georgia's Hargrett Rare Book and Manuscript Collection, the Small Special Collections Library at the University of Virginia, the Virginia Historical Society, the University of North Carolina's Southern Historical Collection, the Georgia Historical Society, Duke University's Special Collections, the Southeastern Branch of the National Archives and Records Administration, and clerks at the Caswell, Halifax, and Pittsylvania courthouses made research a pleasure. Librarians at the University of Georgia, Delta State University, and Millsaps also provided yeoman's service unearthing rare pamphlets, newspapers, and books on my behalf. The Willson Center for Humanities and Arts, the Southern Historical Collection at the University of North Carolina at Chapel Hill, the Franklin College of Arts and Sciences at the University of Georgia, the UGA History Department, the Virginia Historical Society, and the Wormsloe Institute for Environmental History all furnished funding at critical moments in this project. The Associated Colleges of the South, the Andrew W. Mellon Foundation, and Millsaps College deserve special mention for providing a postdoctoral fellowship at a welcoming academic home with a reduced teaching load where much of this writing took place.

Editorial tasks and production can be exercises in tedium. That they proceeded so smoothly with this book is due to the gifts of executive editor Jean Thomson Black at Yale University Press, series editor Jim Scott, and editorial assistants Sara Hoover and Samantha Ostrowski. Production editor Laura Jones Dooley oversaw the transformation of manuscript to book. Copy editor Eliza Childs graciously remedied my worst grammatical deficiencies. They held my hand as I stumbled through the revision and publication process, creating a better book than I ever could have on my own.

Saving the best for last, I owe more than I can ever repay to my understanding—and patient—family and friends. I was lucky in that my parents and grandparents valued work on paper and work on the land equally and taught me much about doing both. I met my wife, Margaret, in graduate school; as long as she has known me we have carried "the tobacco book" as a part of our lives. I am happy to officially declare that we are on to the next step, but this one was possible only with her love and support.

Introduction

Their life is yellow tobacco.

—JOHN OTT, in Robert Ragland, *Major Ragland's Instructions How to Grow and Cure Tobacco, Especially Fine Yellow* (1885)

AS A CHILD I lived near Virginia's Chesapeake Bay, and I vividly remember road trips to visit my grandparents, who lived on the western edge of the commonwealth's Piedmont. The last hour of our journey wound through rolling hills dotted with spare farmsteads, small fields of tobacco, and mobile homes. The tall, square log barns once used to cure tobacco—before propane barns made them obsolete—were everywhere, overgrown in fields, peeking around modern structures, and moldering in the woods. Although I loved visiting the region, I could imagine few places more "southern" in rural character, poverty, gullied land, and reliance on a single crop. This sleepy, agrarian countryside conjured up images of a land that time, and progress, forgot (fig. 1).

Then, when I was ten years old, my family moved close to my grandparents, and I found myself living in this landscape. At some point in my history classes in my new school I learned that the region's prospects had not always been so bleak; I saw pictures of great jostling crowds of tobacco wagons lining the streets of local towns to unload crops; I read of vast plantations as large as any in the South; and I listened to stories of the frenetic energy of tobacco auction warehouses, where farmers, industrialists, and auctioneers gathered to make tremendous profits on piles of golden leaf. What I saw in the landscape that surrounded me and what I learned of the region's past left me with a question that became

FIGURE 1: A contemporary Piedmont tobacco field along the Virginia and North Carolina border. Courtesy Anna Crawley.

the root of this book: how did such a poor land make some people so rich, and how did they so quickly become poor again?

Another experience that proved formative to this book—sharpening this question—came a few years later from a teenage encounter with a gully, a chasm in the red clay, cutting through a scraggily stand of Piedmont pines. Walking through the woods on my grandparents' (now my parents') farm during hunting season, I came across a rift in the earth, half covered with a mantle of fallen leaves. Almost ten feet deep in places, the gully began as a muddy, root-entangled hole high in the hollow and grew in depth and width as it carved a course to the small creek below me. I had seen many gullies of course, on this farm and others in the surrounding community and adjoining counties, but this fall day, with the leaves swirling around me and a few raindrops promising to fill the gully with storm water later in the afternoon, I was struck with the forlorn nature of this particular scar on the land, and I wondered about its story.

All gullies, like all broader landscapes, have histories. There is nothing extraordinarily perceptive in this observation, but it is easy to forget nonetheless. In the Virginia Piedmont (the belt of rolling hills between the state's flat Tidewater and its western mountains) gullies rarely formed in the woods of geologic accord; forests, after all, are efficient at holding soil in place. They were almost always the result of careless farming on fragile soils. And in the Piedmont since its Euro- and African-American settlement in the eighteenth century, farming has almost always meant tobacco cultivation. This particular gully led me to ponder why would a farmer tend land in such a self-destructive manner? Coming from a farm family, I was reluctant to chalk the gully and its brethren up to simple ignorance or greed; I have met enough old farmers to know that they are reasonably intelligent and modest people—at least as much so as the average person. In short, I do not doubt that farmers of a century ago had their flaws, but I refuse to believe they were entirely governed by them.

Most American history classes offer an answer to the question of gullies in the mid-South, in what has become a virtual catechism: tobacco was an exhausting crop, and the drive for wealth in a region with cheap land and expensive labor pushed planters to use up their land and move on to fresh ground, leaving erosion and soil exhaustion in their wake. In this narrative, tobacco was yet another form of exploitative proto-capitalist (or quasi-feudal, depending on one's take on slavery) plantation agriculture. This thread of land exploitation connected Jamestown to the South Carolina Sea Islands and to the antebellum Texas cotton frontier. And it still courses through the soils of South Georgia onion fields and Mississippi Delta soybean farms. But for all its venerable durability, this explanation felt incomplete. Could people really be so short-sighted? Was greed truly such an overwhelming force?

To begin to answer these questions about wealth and poverty, land use and abuse in the tobacco belt (and from inferences drawn there, to the nation more broadly), I have become convinced that we must take a careful look at the region during the nineteenth century. Beginning around 1840, people started experimenting with making "bright" tobacco along the border of the Virginia and North Carolina Piedmont (a crop also known in the nineteenth century as "yellow" or "light" tobacco or, more recently, "flue-cured" tobacco).[1] Regional farmers had cultivated tobacco since the area's mid-eighteenth-century colonization, but bright leaf was a new form of the old standby, and its culture eventually transformed how people used the land as well as how they thought about tobacco and themselves. The new crop brought amazing benefits; it grew well on the poorest of soils, it sold for astonishingly high prices, and it required relatively little in the way of new farm machinery, buildings, or social structures. The crop also stimulated a manufacturing industry that transformed a raw agricultural product into chewing tobacco, and later cigarettes, destined for regional and national markets. Bright leaf made a poor region rich, or so goes the legend. But the crop also carried consequences. It eventually led to massive soil erosion, deforestation, and indebtedness, and it continued a decades-old reliance on a single cash crop to the exclusion of farm and economic diversity. Within a few decades bright leaf transformed from savior to destroyer. And it was this transformation that helped create the sleepy southern region that so impressed my youthful imagination, as it changed forested hillsides into gullied slopes that eventually reverted to forest once again.

This is a story of the rise of a crop culture, but it is also a tale of the decline of the environment that accompanied this form of tobacco cultivation, from severe erosion to deforestation to insect infestations. It is a miraculous story; some of the poorest land in the South produced, for a time, one of its most lucrative crops. In a transformation akin to alchemy, sandy, weak earth grew tobacco crops that made select farmers rich in a single season. Yet almost as quickly this poor earth melted away in summer thunderstorms, flowing through gullies and in vast silty sheets into the region's creeks and rivers on its way to the Atlantic. Growers understood these dangers, yet somehow they believed that their cultivation methods made agronomic sense. This book seeks to understand not only the material relationships that connected crop, land, and people but also the mental calculations and justifications that accompanied tobacco farming. Why were people willing to destroy the land they lived on in order to raise bright leaf? How did they justify their actions? What general lessons might agricultural and environmental historians draw from bright tobacco's tale?

In exploring these questions, this book focuses on the birthplace of a bright tobacco culture that would eventually cover substantial swaths of the South—three contiguous Piedmont counties, Halifax and Pittsylvania in Virginia, and Caswell in North Carolina—from early experiments with bright leaf around 1840 until the nadir of local farm fortunes at the end of the nineteenth century (fig. 2). The decision to root this study in these three counties was grounded in environmental and agricultural considerations. Farmers, like agricultural pests, storms, floods, and droughts, rarely limited their actions based on county or state lines, and the emergence of bright leaf culture took place with little regard to these political boundaries. Despite this ecological reality, there are good reasons to focus on the three counties in question. Planters in portions of Pittsylvania, Halifax, and Caswell experimented with producing bright tobacco before the Civil War, and these were the only three counties in the nation to do so to any significant extent. Although bright tobacco cultivation spread southeastward following the war, especially by the 1880s, the three-county region (which locals often referred to as the Southside or the Border) remained one of the densest concentrations of bright leaf farms until the Great Depression.[2] In addition, planters and farmers with the environmental and agricultural knowledge that fueled this expansion remained centered around Danville (in southern Pittsylvania County), even as these bright leaf ambassadors spread techniques that originated on the sandy uplands of the Dan and Staunton rivers to other portions of the American South. Many of the most vocal bright leaf advocates and boosters came from these three counties, and the auction system that grew to dominate bright tobacco sales was developed in Danville. In short, these three counties were the first to produce bright leaf, the largest producers over a substantial period of time, the most important in terms of creating and spreading the crop culture, and the first to suffer the consequences as the dangers of bright leaf monoculture became impossible to ignore near the end of the nineteenth century.

An examination largely limited to three counties also presents an opportunity to examine a crop culture on an intimate level. Only through detailed examination of particular crop cultures in particular places can we begin to understand the environmental logic (or illogic) of specific agricultural systems. These details reside at the heart of rural and agrarian history. As historian Orville Burton declares: "We cannot take that material foundation [crops and environments] for granted as self-explanatory. Rather we have to come to terms with its complexity before we can explain culture, to see the frame into which culture fits, to gauge

FIGURE 2: Map of Halifax and Pittsylvania counties, Virginia, and Caswell County, North Carolina. Map by the author.

words and thoughts within the material context from which they were generated."[3] Agricultural history that seeks to examine specific places, then, is as reliant on knowledge of the land as is good agricultural practice, for as farmer and philosopher Wendell Berry has observed, "The land is too various in its kinds, climates, conditions, declivities, aspects, and histories to conform to any generalized understanding or to prosper under generalized treatment."[4] This book avoids a synthetic history of tobacco culture—these exist in abundance—in search of the relationship between farmers and landscape that shaped both a place and a people.

This does not mean that the pages that follow ignore bright tobacco's ties to the broader region, nation, and world. There are points in this story that wander beyond the borders of Caswell, Halifax, and Pittsylvania. For example, the narrative ventures across the Henry County line into the barns of the Leatherwood Valley, follows bright leaf as it makes its way down the Roanoke River or over winding country roads to market, looks at chewing tobacco marketing across the nation, and briefly explores the global reach of the cigarette. These excursions are important parts of bright leaf's story, and though I seek first and foremost an intimate look at farmers and their land, I attempt to remain cognizant of the connections between producer, middleman, and consumer that contributed mightily to the formation of a regional agriculture.

The chronological boundaries of the study also demand a brief discussion. The beginning point of the story is relatively easy to justify: around 1840 a handful of farmers in these three counties began to raise and cure bright leaf with some consistent success. The spread of traditional tobacco culture into the region forms a brief backstory, but this work's true argument begins with the farmers who successfully developed and touted bright leaf. The ending point circa 1900 requires a little more justification. In some ways, the Great Depression, with the identification of the three counties as agricultural and—though the label was not contemporary—environmental problem areas by the Soil Conservation Service (SCS) and the 1933 tobacco quota system imposed by the federal Agricultural Adjustment Act, seems a natural denouement. While I briefly tackle New Deal responses to tobacco growers' problems in the epilogue, the conditions that led to their troubles were firmly in place by the last decade of the nineteenth century. Regional farmers were already experiencing severe soil erosion, the iron tentacles of fertilizer debt, slumping bright tobacco prices, and few obvious alternatives by 1900; SCS officials merely identified and labeled an ongoing condition. Ending the story once the cycle of poverty was firmly in place permits greater attention to what led to the crisis in the first place.

There are shelves of books on tobacco in the southeastern United States, and many of them devote the bulk of their attention to Virginia or North Carolina, yet there are surprisingly few studies focused on bright tobacco as a crop rather than as an industrial input, that is, the raw material for cigarette companies.[5] The bulk of tobacco studies focus either on the emergence of southern tobacco agriculture and its attendant societal structures in the Tidewater of colonial Virginia and Maryland,

or on the development and growth of the tobacco manufacturing industry, epitomized by the expansion of "Big Tobacco" in the twentieth century.[6] The few books that have focused on tobacco culture in the Virginia
and North Carolina Piedmont have generally examined the region's agriculture through the lens of post–Civil War labor struggles. These historians of Reconstruction and its aftermath have explored the transition
from slave to free labor in the tobacco belt and whites' subsequent efforts
to define and racialize agricultural work, though none have paid serious
attention to the particular role of tobacco varieties or the importance of
specific environmental conditions in shaping labor, race, and economic
relations, an omission this book works to remedy.[7] Frederick Siegel's history of Danville, Virginia, does explore the importance of bright leaf in
the changing social structures of the town, but his study ends at the conclusion of the Civil War, at the very moment when bright tobacco cultivation dramatically increased, and his analysis is geographically restricted
to the town and the surrounding portions of Pittsylvania County.[8] Even
the few histories sensitive to the agriculture revolutions wrought by
shifts from dark tobacco cultivation to bright are little concerned with the
effects of the resulting agriculture on its supporting environments.[9] In
total, the most comprehensive history of the subject remains Nannie May
Tilley's *Bright-Tobacco Industry,* now more than sixty years old (1949).

What none of these studies do is examine tobacco as a medium
through which people understood and interacted with the environment.
Tobacco histories typically treat the crop as a commodity, an avocation,
part of an expanding market system, the byproduct of institutional
structures, or a consumer good. Missing is a consideration of tobacco as
a living, evolving plant connecting people to place. With the exception
of a growing body of literature on rice cultivation, southern agricultural
history in general has followed this pattern of omission.[10] The story of
tobacco suggests that southern crops—from cotton to corn—were not
monolithic entities. Tobacco was defined to an extent by regional labor
systems, market prices, transportation networks, and various private
and public institutions, but it was the also the product of specific places.
Each field of tobacco varied, and every tobacco farmer gained certain
understandings of soil, climate, and what it meant to work the land from
that stand of plants. A tobacco grower's success was in large part the
outgrowth of combining knowledge about a crop in general and a piece

of earth in particular. These variations were so significant because farmers drew their understandings of agricultural practice, land and labor management, even their self-worth as farmers from their actions on a particular plot of ground. But it is also crucial to keep in mind that farmers were rarely what we might term rational economic actors; while they surely made some calculations concerning the value of particular crops and practices, the costs and benefits of agriculture often went "underdetermined" to a varying extent from farm to farm.[11] Agriculture, even within staple crop systems such as those that characterized the American South during the nineteenth century, was no featureless abstract thing, no simple equation, and bright tobacco culture provides an example of the variance present within all farming, regional or otherwise.

The relationships between Southside residents, bright tobacco, and the environment were firmly rooted in racially defined (and racist) social and cultural systems, and thus part of the following book is an attempt to reconstruct a hidden "black" history of bright leaf. Environmental history has generally underrepresented African Americans as agricultural and environmental actors in the American South (especially following Emancipation).[12] Slave understandings of soil, cultivation methods, and tobacco biology proved important in the development of the new crop and its manufacturing. These slave inputs were perhaps less formative than the transference of agricultural knowledge that took place in southern tidal rice culture (though that story is by no means settled), but they were substantial nonetheless.[13] Following Emancipation, freedpeople remained tobacco cultivators, but racial struggles over social position, landownership, and political power shaped white perceptions of black agroecological knowledge and African Americans' access to land. White landowners characterized freedpeople as lazy and dumb, and they used bright leaf's stringent cultivation requirements as an excuse to circumscribe black independence. Whites also pointed to the erosion that accompanied tobacco production as proof that African Americans were poor stewards of the land (ignoring similar problems on white-managed farms). Eventually, white farmers created a mythology of bright leaf's emergence that praised influential white planters, written tobacco instructional guides, and systematic book farming, while ignoring black contributions. Bright tobacco thus shaped not only the form of Southside labor but also the ideology that undergirded it.[14]

A number of southern historians have demonstrated that particular agricultural systems shaped slavery and the emancipation experience, which differed in Deep South cotton districts, Louisiana's sugar lands, the Lowcountry, and Piedmont tobacco fields.[15] This book seeks to make the next step, arguing that the environmental foundations of a crop system played an important role in race relations. Planters' successes mating soil and seed types contributed to a thriving slave economy in the Southside even as slave populations in other tobacco districts declined, bright leaf's environmental demands encouraged white landowners to resist independent black labor following Emancipation, and the postwar literature on agricultural techniques and the crop's history were charged with racial language derived from white conceptions of the environment and its degradation. This is not an argument for environmental determinism, for despite U. B. Phillips's infamous declaration, southern culture was not a direct product of the weather (or soil, or particular diseases).[16] Rather, this study is an assertion that culture—in this case, an agriculture—rests upon an environmental foundation that limits the possible and shapes what is likely. In the case of the Southside, a culture of tobacco production and slavery met an environment particularly conducive to producing a high-quality form of the crop, but only at great risk to the health of the land and its people.

Whether black or white, the Southside population was largely rural and agricultural during the nineteenth century. Thus this book seeks to examine relationships between southern people and the environment through their most common means of interaction well into the twentieth century: agriculture. Recently, a number of scholars have emphasized that southern relationships with the environment have been peculiarly rooted in agricultural landscapes.[17] According to historian Mart Stewart, the discussions of wilderness and preservation characteristic of western environmental history should take a backseat to agricultural landscapes in our study of the Southeast, since "the environmental history of the South has been an agrarian one."[18] If, as Benjamin Cohen argues, "any discussion of the environmental sensibilities of the early American Republic must not just *make reference* to agriculture, but should be embedded *within* it," then in the South the statement applies to the entire sweep of the nineteenth and much of the twentieth centuries.[19] Accepting that human relationships with nature in the South often took place through agriculture means dealing with anthropogenic land-scapes in more thoughtful ways. Declension narratives and prelapsarian ideals of the environment are too simple when attempting to explain human-induced change in a landscape that was itself a human-influenced artifact.[20] The thread of preservation that dominates many western environmental histories—be they histories of efforts to prevent damming wild rivers or damning old-growth forest—is hardly a relevant model for understanding a Piedmont scrub forest that has been through a half dozen cycles of field and long fallow. The task, then, is to take the large body of southern agricultural history and "recast and extend" it "as agro-ecological history."[21]

Traditional interpretations of the regional relationship between people and the land have been highly critical; historians have long de-scribed southern agriculture in terms of decline. According to this model, the staple crop plantation system exploited people and the land, depleting southeastern soils as planters marched south and westward like locusts, leaving the wreckage of slavery and erosion in their wake. This exploitation only accelerated after the Civil War as various forms of tenancy, debt peonage, and production for increasingly competitive world markets exacerbated the South's environmental and social prob-lems.[22] The history of bright tobacco complicates this narrative in some interesting ways. It is true that by 1900 the Southside looked much like the cotton, sugar, and rice belts, with white and black farmers strug-gling to make ends meet as fertilizer prices rose, soil washed away, and market prices declined because of overproduction and the power of a monopsony in the form of the American Tobacco Company. But this impoverishment of land and people was not a linear process. For a sub-stantial stretch of the nineteenth century, bright leaf promised a flour-ishing agricultural economy (at least for the white portion of the regional population). The emergence of bright leaf slowed emigration, boosted land values, and stimulated a regional manufacturing industry during the antebellum era; the crop remained an economic engine during the tumult of the Civil War; and bright leaf continued to thrive during the difficult periods of Reconstruction and the financial crises that followed the war. Bright leaf culture developed and expanded not because of its overwhelming appeal to greed and short-term gain (though these ele-ments were present), but through appeals to agricultural and personal improvement in search of a basic economic and social sustainability. Bright tobacco culture called on farmers to study their land and their crops carefully, to read agricultural literature and employ the newest innovations, and to make their land more valuable through intellectual as well as physical labor. These messages were far from the debasing, exploitative agriculture of regional stereotypes. That these efforts eventually failed was as much a product of the Southside environment and plant biology as it was a result of the destructive nature of staple crop agriculture.

This book's emphasis on farmer knowledge and the artisanal nature of bright tobacco cultivation should not be interpreted as romantic revisionism. In some ways these pages celebrate work on the land through the medium of agriculture as an essential way people once understood their surrounding environment, a moral knowledge expressed most powerfully through the writings of the modern agrarian Wendell Berry and environmental historian Richard White. Yet this study is also a critique of the ways in which such knowledge could fail to result in true conservation, the ways in which a true sense of place sometimes failed to result in "a moral superiority" that might have ensured "that those who live and depend on a place will not harm it."[23] Farmers and planters, their families, sharecroppers, renters, and slaves shared an in-depth understanding of crops, soils, and the broader environment, but these understandings were often far from complete, they occasionally clashed with short- or long-term economic interests, and in other instances they were subsumed by greed and ideology. In short, understanding the environment and appreciating it and desiring its preservation were far from coterminous concepts. Just as the region's nineteenth-century Baptists and Methodists believed that knowledge of God alone was little assurance of salvation, people accrued a knowledge of nature absent any guarantee of good stewardship.

A few caveats follow. It is important to note that "bright tobacco" was never a single, fixed, stationary thing. Bright tobacco was as much process and search as it was a distinct crop, a reality beautifully limned by historian Barbara Hahn.[24] The commodification and definition of tobacco, as with any crop, is a process of providing the illusion of predictability to living, variable plants raised by diverse, opinionated people, in an ever-fluctuating material environment. Farmers' pursuit of quality crops and standard products is continuous and is defined by con-

sumer desires; social, cultural, and governmental institutions; and arbitrary notions of how to make something in the "best" way. Combinations
of these variables and the diversity of crops and production in the American South appear in the historical pursuits of long-staple cotton, the
most vivid shades of indigo dye, and long, unblemished grains of rice.
These variables in even the most commercial of agriculture rarely make
the pages of our histories; most of us are far removed from any practical
understanding of the exigencies of farming, and those of us who retain
connections to the land exist in a world of commodity futures trading
and mechanization relentlessly focused on standardization. With these
considerations in mind, throughout this book the phrase "bright
tobacco" refers to attempts to make the crop as much as a fixed and iden-
tifiable product. Trying to define a particular agricultural commodity
retains some usefulness, however. Farmers believed there was such a
thing as bright tobacco, even when they struggled to define what the
crop should look, taste, and smell like. Growers produced a material
product with real weight and presence, and then sold that product as
bright tobacco despite its variability. And it is a human impulse to sort
things into comprehensible groups. We can be assured that the concept
of bright tobacco mattered, despite the variable nature of the "real" thing,
because people at the time were so convinced that it did.

Although bright tobacco proved a somewhat variable thing, farmers' decisions to attempt to produce it were by no means inevitable.
Much of the landscape where the Piedmont met the Blue Ridge along
the Virginia–North Carolina line was ideally suited for raising a valuable, yellow form of tobacco when cured by specialized methods, but it
did not demand a particular form of agriculture. The clay and sand hills
produced corn, hemp, and other crops; the landscape was perhaps most
safely and sustainably employed in producing hay and mast-fattened
pigs and cattle; and bright tobacco never completely eliminated these
forms of farming. Culture encouraged and sustained the pursuit of
bright tobacco: consumers motivated by desire and addiction coveted
fashionable and flavorful forms of tobacco; buyers and manufacturers
held particular notions of markets and tobacco types; institutions, such
as governments and commodity markets, shaped how leaf was raised,
manufactured, and sold; and technological systems created feedback
loops that made certain techniques and routines seem only natural. And

yet it would be misguided to push ideas of social construction too far, to portray them as omnipotent. Piedmont nature and tobacco plants remained real, active components of history. Hillside soils provided foundations for production, and all the desire and cultural predilections in the world could not transform clay into sandy loam. Any landscape, when pushed too far, pushes back (to indulge in a little anthropomorphic conceit), most severely in the Piedmont in the form of dramatic soil erosion. The environment—in its material manifestations and in ideas of it both local and distant—was omnipresent, as important for farmers who ignored its limitations as for those who sought to intimately understand it.

A final warning: readers looking for detailed discussions of tobacco's health effects and its medical history will be disappointed. Tobacco as an agent of addiction and disease is comprehensively covered elsewhere.[25] It goes without saying that tobacco has served as one of history's greatest killers, a plant that transformed from novelty and cultural symbol to a widespread addiction, leading to lingering, painful death for millions of people over recent centuries. The addictive power of nicotine and the variable effects of different forms of consumer tobacco do form a part of the following story and are mentioned, but only so far as they help explain the story in the Southside. In a few places, especially the long-term health consequences of the shift to phosphate fertilizers and the accidental effects on addictiveness created by particular forms of tobacco curing, this environmental history delves into relationships between bodies and plants in a bit more detail, but this book largely leaves tobacco's epidemiological history in more capable hands.

The phrase "tobacco culture" appears quite often in these pages. Culture is at its most basic a term of convenience, shorthand for a given people's package of shared intellectual concepts, social interactions, and material practices, and it serves that purpose here. Tobacco was a culture; it was the glue that welded a region—from fields to slave quarters to courthouses to factories—together. Tobacco found its way into countless aspects of everyday life; it became a part of how people spoke, how they acted, how they thought about the world. Making "good" or "bad" tobacco served as an indicator of a person's moral fiber. The lives of farmers in general revolve around seasonal tasks, and tobacco in particular developed its seasonal rhythms, complete with a unique language. The "season" came when spring rains thoroughly wet the earth, leading to the time of "setting out," or transplanting. Knocking low-quality leaves off the bottom of plants became "priming," and the harvest was "pulling" or "saving" time, followed by "housing" leaf in the barns. Tobacco even penetrated and changed the very bodies of some Piedmont residents. Tobacco consumers altered lungs, mouths, esophagi, and brain chemistry as they smoked, chewed, or dipped snuff; some lived to a ripe old age and others, undoubtedly, died of cancer or respiratory illnesses. Field hands absorbed nicotine through their skins on dewy or rainy morning, often suffering from a mild (though exceedingly uncomfortable) poisoning known as tobacco sickness. During curing months farmers and slaves went without regular sleep, their eyes reddened and their skin dried out by constant exposure to smoke from fires in the barns. And one of the most distinctive features of tobacco workers in the Piedmont throughout the nineteenth century was an extra-long thumbnail or two, not for gouging out an opponent's eyes during backwoods brawls as condescending outsiders speculated, but for killing hornworms that fed on tobacco and for pinching the axial flowers from the growing plant during topping. Tobacco was a culture that permeated body as well as soul.

The story of the emergence of bright tobacco is a tale of promise proved hollow. Part of this failure came from the limitations of the land, and part from the incompatibility of certain human desires and the Piedmont landscape. Decisions that seemed to make a certain economic and agricultural sense in the end proved but steps on a path to poverty. The crop at first seemed to offer the Southside's white farmers all that they might desire: it worked well on their poor soil, it brought high prices, and it meshed to a certain extent with their ideas about land and agriculture. Ultimately yellow tobacco would prove the region's downfall, as it wore down land and impoverished farmers by the end of the century, leaving the region with few economic alternatives. The crop's promise proved even more illusory for the region's African Americans. Following slavery, strong tobacco prices hinted at the potential success of freedpeople, if only they could obtain land. These high prices, coupled with white landowners' notions of black agricultural abilities, led to some of the lowest black land ownership rates in the South. Although its details were unique, by the early twentieth century bright tobacco culture had proven another route to the agricultural New South of tenancy, poverty, and ravaged land.

ONE

On the Back of Tobacco

SOWING THE SEEDS OF A TOBACCO CULTURE

AMONG THE FIRST EURO-AMERICANS to write extensively of the Virginia and North Carolina Piedmont was William Byrd II of Westover, the famous (and infamous) chronicler of planter life in the early eighteenth century. Byrd led a survey into the Piedmont interior in 1728, and he journeyed west again in 1732 and 1733 to assess the region's prospects for settlement and real estate speculation. The first expedition set out to firm up the boundary between Virginia and North Carolina; the latter colony claimed the mouth of the Nottoway River as the line, while the Old Dominion, unsurprisingly, declared the border to be fifteen miles farther south. Survey work in the spring of 1728 hacked a path through the pocosins and black water swamps of the Tidewater. Byrd's party returned home for the summer to avoid snakes, humid weather, and noisome mosquitoes and resumed work in late September, pushing into the uplands and eventually past the last English settlements on their way toward the mountains and the approaching winter season.[1]

As Byrd drew within sight of the eastern outposts of the Blue Ridge—White Oak, Turkeycock, and Smith mountains—he entered a landscape essentialized in the modern geographers' label for the physiographic province: the Piedmont. Its rolling hills literally lay at the foot of the mountains. The Dan and Staunton rivers, the southern and northern branches of the Roanoke River, respectively, cut deep valleys, sometimes hundreds of feet lower than the surrounding bluffs, in the shadow of slopes often as steep as any found in the mountains to the west. Smaller rivers and large creeks, including the Hyco, Pigg, Turkeycock, Bannister, Countryline, and Sandy, also carved their own paths through the countryside, with their numerous tributaries dividing the whole into thousands of finger-like ridges separating one stretch of bottomland from another. Most of the landscape lay on a slope, with the exception of riparian bottoms and a few stretches of level land atop the larger ridges. Located to the west of the contemporary English settlement line, the landscape abounded in wildlife; Byrd noted turkeys, rattlesnakes, black bears, beavers, whitetail deer, gray squirrels, fox squirrels, bobcats, buffalo, wolves, cranes, grouse, geese, mussels, mountain lions, passenger pigeons, raccoons, flying squirrels, chipmunks, elk, opossums, snapping turtles, and river otters populating the hills, hollows, and watercourses.[2]

Although the land away from the rivers was rugged, Byrd found much to admire in the river bottoms. He would in fact go on to claim a vast swath of the watershed in a speculative attempt to attract European immigrants to a new land of Eden. Byrd wrote of the land east of modern Danville, Virginia: "All the land we travelled over . . . is exceedingly rich, both on the Virginia side of the line, and that of Carolina. Besides whole forests of canes, that adorn the banks of the river and creeks thereabouts, the fertility of the soil throws out such a quantity of winter grass, that horses and cattle might keep themselves in heart all the cold season without the help of any fodder." Gaining steam, he rhapsodized, "I question not but there are thirty thousand acres at least, lying altogether, as fertile as the lands were said to be about Babylon . . . a colony of one thousand families might, with the help of moderate industry, pass their time very happily there."[3] Byrd envisioned an agricultural empire complete with farmers raising livestock and silkworms, cultivating cotton, hemp, and perhaps even rice, and tending rank orchards.[4] For all this talk of diversified farming, Byrd also noted the land's potential for that great Virginia staple—"that bewitching vegetable"—tobacco, a commodity that appeared as early as the second page of his account.[5]

Byrd's survey helped open the southern Virginia Piedmont and the adjoining reaches of North Carolina to Euro-American settlement, and tobacco shaped these districts from their earliest days. Over the course of the century following the line survey, planters, farmers, and slaves built a rural landscape centered on the cultivation of fire-cured (or dark) tobacco, the traditional Chesapeake crop that had served as a staple since the early days of Jamestown. This landscape was the product of an existing tobacco culture transported from the coastal plain of eastern Virginia and adapted to the Piedmont environment, and it would shape regional farmers' adoption of a new staple beginning in the mid-nineteenth century. In this chapter I will also examine in detail the cultivation of dark tobacco, which laid a foundation for the bright tobacco culture that would sweep the region before the Civil War. Dark tobacco fueled the birth and growth of the Southside, but local environmental constraints left planters and farmers searching for an agricultural alternative by the antebellum era. Bright tobacco would prove to be that replacement.

Despite Byrd's glowing reports, regional growth proceeded slowly at first. In 1738 there were so few Southside residents that the colonial Virginia government offered incentives to induce settlement. A commonwealth statute waived taxes for a period of ten years for any new landowner along the Staunton or Dan rivers (the region's two main waterways), reduced taxes thereafter, and provided for the automatic naturalization of alien settlers.[6] Although the 1738 statute had little immediate effect on the population (or lack thereof) of the Staunton and Dan valleys, speculators, such as Byrd, used the incentives to engross hundreds of thousands of acres of the best bottomlands in the region during the 1730s and 1740s. The real influx of settlers into what would become Caswell, Halifax, and Pittsylvania began around 1750, largely because of declining economic opportunity in the Virginia Tidewater. As eastern populations increased, the best lands were entirely taken by wealthier planters, and a number of small farmers sought their fortunes in the fresh lands of the southern Piedmont.[7] Many of the region's earliest settlers came from Tidewater tobacco counties, but Scots-Irish farmers who moved down the Valley of Virginia from Pennsylvania and into the southwestern Virginia and North Carolina Piedmont composed a substantial portion of post-1750 immigrants. Facing a shortage of the richest river and creek bottomland (patented by the Chesapeake settlers), these small farmers often settled the weaker soils of the region's uplands.[8] With the jaundiced eye of an Englishman toward the Irish, traveler John Stuart described the inhabitants of the Caswell hills around the time of the

Revolution as "chiefly natives of Ireland, most wretchedly ignorant and uncivilized."[9]

Although cultural differences separated these two waves of settlers, both groups relied on the Chesapeake's historic staple as their primary cash crop. The first Euro-American settlers in the three counties selected their farm sites with an eye toward tobacco cultivation. Desiring fertile alluvial soil and easy access to water transportation for heavy tobacco hogsheads (barrels holding cured leaf), almost all of the region's initial residents sought out land along the numerous rivers and creeks that dissected the hilly landscape. Although these watercourses were not suitable to large boats, small bateaux could navigate most local waters even before internal improvements. Peter Wilson, one of the earliest recorded planters in Pittsylvania, selected a narrow spot on a bend of the Dan River sometime prior to 1750 where he could operate a ferry and farm.[10] Such frontiersmen as John Smith, Jr., and Benjamin Clement settled on good bottomlands along the Staunton River, first surveyed in the 1740s, and began raising tobacco soon after carving out their homesteads.[11] In the land that would become Caswell County, south of Byrd's line, the first farmers arrived in the 1750s and 1760s, and almost all settled along the district's watercourses. The Hyco and Dan rivers and Countryline Creek were particularly popular locations.[12] Many of these first settlements in the three counties sat on the same sites as earlier Native-American villages, taking advantage of the open land of old Indian fields and clearings such as the ones found along the Staunton and Pigg rivers and Sycamore and Cherrystone creeks.[13]

These emerging Piedmont tobacco outposts were part of a shift in tobacco production that swept the mid-Atlantic in the middle and late decades of the eighteenth century. As tobacco cultivation grew in the Piedmont, the importance of the crop's culture lessened on the older lands of the Chesapeake. Beginning in the mid-eighteenth century, and accelerating after the Revolution, many Tidewater farmers, especially on the Eastern Shore of Virginia and Maryland and on Virginia's Northern Neck, turned to wheat as an alternative cash crop to tobacco. This shift occurred for a number of reasons. Tobacco cultivation on fresh Piedmont lands increased competition at a time when tobacco prices fluctuated (prices gradually increased but there were periods of severe recession, and especially steep price declines would occur during the French Revolution). Wheat culture involved less hoe and more plow work, which lessened Chesapeake farmers' dependence on slave labor. The Tidewater's connections to Atlantic trade also promoted grain cultivation after 1760: a series of European conflicts increased demand for bread, several poor weather years hampered other American and foreign grain-growing regions, and the rapid growth of the northeastern colonies fueled grain prices. Also contributing to this shift toward grain cultivation, eastern planters ranging from large masters, such as George Washington, to smaller landowners, saw in wheat the potential to diversify their agricultural practices.[14]

This Tidewater shift to wheat cultivation before the American Revolution was dramatic, but it did not entirely displace tobacco cultivation in the region. A few figures reveal the continued importance of the traditional crop on Chesapeake farms. According to Philip Morgan, in 1740 the value of the Tidewater's exported tobacco was fourteen times greater than the value of the region's grain. With the growth of wheat culture, the gap shrank over the ensuing decades, but in 1770, Tidewater tobacco exports were still worth three times as much as grain.[15] Allan Kulikoff calculates that as of 1775, two out of three residents of coastal Virginia and Maryland still raised tobacco on a regular basis.[16] Tobacco remained a crop with widespread appeal in the eastern portion of the colony, but the gradual decline in Tidewater leaf production after the Revolution accelerated Piedmont tobacco cultivation, where fresh land promised larger crops and good profits (at least initially). Tobacco's shift from the Tidewater to the Piedmont was a lengthy transition, but the latter region was the center of American tobacco production by the early decades of the nineteenth century.

Piedmont tobacco growers were thus subject to the forces of national and transatlantic trade and to changes in distant landscapes, but they also had to deal with the material realities of raising tobacco in a new environment. And the Southside environment was quite different from the Tidewater or the Great Valley from which most regional settlers came. The Piedmont province of Virginia and North Carolina is located between the flat Tidewater to the east and the Blue Ridge Mountains to the west. Caswell, Halifax, and Pittsylvania lie along the Piedmont's western edge, a landscape of rolling hills, deep creek and river courses, and the occasional low mountain. Although the region receives moderate precipitation—between forty and fifty inches per year—it is a particularly well-drained landscape thanks to its numerous streams and rivers. As Byrd learned, these waterways dominate regional topography: the Staunton River drains the northern portion of Pittsylvania and Halifax, and the Dan drains the southern stretches of both counties and all of Caswell, before the two rivers combine to form the Roanoke in eastern Halifax. Water figures prominently even in the corners of the region farthest from the major rivers, as hardly a square mile of the three counties lacks a regular stream of some sort.

With the exception of moderately rocky ridges containing Appalachian soil strata, such as Turkeycock, Smith, and White Oak mountains, these watercourses cut through characteristic Piedmont soils. To draw on the language and understandings of modern soil scientists, most regional soils are ultisols, weathered from parent material from the eastern slopes of the Appalachian range. These soils leave much to be desired from an agricultural perspective. Ultisols tend to be low in nutrients and minerals essential to most plant growth, including nitrogen, phosphorus, potassium, and calcium, and they are consistently acidic, posing barriers for species reliant on neutral or basic environments (such as many nitrogen-fixing legumes that are important components of convertible husbandry). These nutrient-poor, acidic conditions were the result of a number of factors: ultisols are quite old and weathered, even from a geological perspective; the Piedmont experiences consistent heat, humidity, and heavy rainfall, which leach nutrients and minerals from upper soil horizons at a rapid rate; and the region missed out on a potential source of soil enrichment in the form of the most recent glaciation, which made it no farther south than Pennsylvania. Regional soils were also very slow to incorporate organic materials, including leaves and pine needles, into their upper horizons. Typical Piedmont soils thus, as one environmental historian observes, "bear a strong resemblance to tropical soils, which tend to be deeply weathered and prone to exhaustion and erosion if not handled with care."[17]

Of course the land along the Virginia and North Carolina border at the foot of the mountains contained its own distinctive mosaic of these Piedmont soils. At a most basic level the region is divided between lowlands and more extensive uplands. Lying along the numerous creeks and rivers, the region's bottomland is quite dark and fertile, rich in organic matter deposited by periodic flooding. For this reason bottoms were typically the first settlement and agricultural sites in the Southside, first worked by Native Americans who were followed by colonial farmers. The upland is much poorer for agricultural purposes. It is composed of sandy loam that modern soil scientists place in the Appling, Cecil, and Durham series. Like other ultisols, this sandy loam is the product of the weathering of ancient bedrock; it is granular in structure, permeable, highly acidic, and it overlays a subsoil of stiff red or yellow clay. These light soils drain well, but they are low in organic matter and thus in nitrogen (even where relatively rich in phosphorus and potassium), and their structure combined with their position on sloping hillsides make them some of the most highly erodible topsoils in the South.[18] Although eighteenth-century farmers did not understand or define the soils they encountered using the same language as today's soil scientists, they quickly identified the basic properties of local landscapes and drew their own mental and physical maps that assigned specific values to particular soils.

The Piedmont, when Euro- and African-Americans began to resettle the land, was a landscape dominated by trees as well as waterways. Noted botanist William Bartram wrote that the region was everywhere covered with "forests of Trees of vast growth where uncultivated except levell, bottoms, where Branches of Rivers take their Rise," part of a "great Forest when in its primitive State of Nature."[19] Ecologists generally classify Piedmont woodlands as mixed hardwood or oak-pine forests, but these general type descriptions obscure as much as they reveal. The woods Byrd traveled and later settlers aimed to farm were quite diverse on the local scale, reflecting variations in soils and microclimate from acre to acre. River cane, sycamores, and shrubs grew in lowlands; tulip trees, grape vines, red oaks, walnut, and beech covered coves and rich, shaded slopes; the poorer soils of ridge tops and high grounds harbored hickories, chinquapins, dogwoods, various pines, blueberries, locusts, and turkey oaks; and the borders of high and low ground nourished persimmons and towering American chestnuts.[20]

Topography, vegetation, and hydrology had a dramatic influence on the region's Euro-American settlement and agriculture. The rich, forested bottomland attracted dark tobacco farmers, and the sandy upland topsoil—which proved poor ground for tobacco brought from the Chesapeake—would eventually produce the best-quality bright leaf in the world. This thin soil, with its lack of organic matter and its hillside and ridge-top situation, was a precarious resource held in place by the region's forest cover. The area's abundant rivers and streams provided water and local transportation networks, but the Roanoke River's drainage into the Albemarle Sound of North Carolina, with only limited access to the Atlantic, rather than into the Chesapeake Bay, retarded early economic growth. Rocky ledges and falls throughout the river system also hindered water transportation, limiting initial water traffic to small boats and canoes. These factors proved an early obstacle to the development of the Southside, but navigational improvements to the river system and the construction of new roads in the late eighteenth and early nineteenth centuries connecting the lower Roanoke to Richmond and Petersburg transformed the waterway into a transportation thoroughfare.[21]

These environmental realities had to compete with the hyperbolic early written accounts of the land. Byrd and other commentators wrote glowingly of the Southside's agricultural potential. Byrd lavished praise on the region's thick forests, replete with game, and believed this abundance a sign of fertility.[22] He recorded flocks of passenger pigeons so large that "whenever they fly from the country, [they] darken the sun for quite a while, and when they want to rest in the woods, they break the branches because of their numbers."[23] He also praised the broad river and creek bottoms thick with stands of river cane.[24] He declared the Dan a "Charming River," and "perfectly clear."[25] Byrd was especially impressed by the Southside's floral and faunal abundance because he saw in these natural resources evidence that the land would support agriculture. A few years after his survey, Byrd attempted to establish a settlement near the junction of the Dan and the Staunton, which he declared a "Newly Discovered Eden." After recording the region's natural advantages, he reminded readers that tobacco was "the foremost crop which one plants in Virginia," linking the compatibility of the historic staple with the new landscape in his sales pitch.[26]

This "new" land was not a howling wilderness, however, and Byrd's optimism regarding the land's agricultural potential came in part from evidence of past Native American farming. Occaneechi and Saura peoples had long lived in the Southside, but beginning in the early 1700s, these groups moved their villages farther from encroaching Euro-American settlements, leaving the Southside as a sort of buffer zone between the English on Virginia's coastal plain and the Appalachian Mountains and central Carolina Piedmont. Sixty years before Byrd's first journey, adventurer John Lederer had traveled through and commented briefly on the branches of the Roanoke, noting of the river bottoms the "rich soyl, capable of producing many commodities," as well as the substantial tracts of land kept clear and cultivated by local Indian towns—but this landscape had changed by the time the survey party rode through it.[27] It is unclear if this movement was the product of disease epidemics or conflict with settlers, to facilitate trade, or some combination of these factors.[28] Everywhere he traveled, Byrd saw remnants of Indian fields and villages. Near the southwestern corner of Pittsylvania, along the Dan River, Byrd discovered abandoned Saura fields that were still open and covered in tall grass, "Truly a Land of Plenty, both for Man and Beast."[29] The Dan and its tributaries were also lined with stands of river cane up to sixteen feet tall, covering ground that had until recently been cleared.[30] Byrd was likewise impressed by the large expanses of old fields near the confluence of the Dan and the Staunton in eastern Halifax County. During the seventeenth century the Occaneechi had inhabited a series of islands where the two rivers met, and Byrd believed these old farms some of the best land in the Piedmont. He described the peach trees and wild hops that grew on the river banks where the Occaneechi village once stood, noted the "brambles, vines and poke bushes" that were filling in the abandoned fields, and rode through clearings where "the grass grew as high as a horse and his rider." Byrd's account of these abandoned Southside fields emphasized the availability of land suited to agriculture, particularly the cultivation of tobacco.[31]

Despite Byrd's glowing descriptions of Southside land, the region's defining agricultural characteristic was that it was not exceptionally fertile. The majority of Southside land, the hills and high ground between watercourses, was, as noted above, acidic, relatively deficient in nitrogen, and low in organic matter. Byrd's assessment was in part a product of his desire to promote and sell land he owned in the region and the fact that his travels by and large followed the rivers, where the richest soil was located. The best land could make fine dark tobacco and was fertile enough to support continuous cultivation of the crop for a number of years, but the weaker uplands made thin, light, poor tobacco and wore out quickly. The Southside as it developed only highlighted the differences between high and low ground, and regional cultivators would deal with these limitations in the decades to come.

With tobacco cultivation came another Tidewater institution: slavery. Slaves made up a substantial portion of the southern Piedmont's early settlers. As a number of historians have pointed out, the institution was not transferred from the Tidewater to the Piedmont whole cloth, but all of its component pieces arrived with the region's initial settlers, who quickly wove a plantation society in the hills, although one that initially held more space for yeoman farmers than the established Chesapeake districts.[32] The Piedmont's first white settlers owned fewer slaves than their counterparts in eastern Virginia or North Carolina, if they held any slaves at all, but the institution grew steadily in the Southside. Out of eighty-three wills recorded in Halifax County between 1752 and 1773, 45 percent (thirty-seven) listed slaves. Most area farmers owned fewer than ten slaves prior to the Revolution, but a few—such as Halifax's Edward Booker with twenty-two and John Smith, Jr., of Pocket Plantation along the Staunton, who owned twenty-three slaves at the onset of the war—built sizable plantations. Bound labor was almost always connected to tobacco production. Estate inventories of farmers who owned slaves also typically included tobacco notes, interest in tobacco crops, or quantities of cured leaf. In just two examples, Paul Chiles left eleven slaves and a little more than 4,500 pounds of tobacco to his inheritors in 1761, and Smith's slaves produced as much as 40,000 pounds of tobacco annually on his Pocket Plantation.[33]

The tobacco produced by Chiles, Smith, and their fellow Piedmont planters largely flowed overseas. From its commercial origins at Jamestown, colonial American tobacco was a crop destined for European markets. By the mid-eighteenth century, Europeans of all classes, ages, and both genders smoked pipes, chewed tobacco, dipped snuff, or lit cigars, and a number of cultures even celebrated tobacco consumption as a healthy or prophylactic habit. The bulk of Virginia and North Carolina tobacco (small amounts were consumed locally) moved down the region's rivers to Chesapeake ports, where it sold to Scottish merchants who carried the leaf to Glasgow or London. From there the weed was dispersed to Great Britain's towns and cities or transshipped to the continent. Small amounts of tobacco even trickled into corners of the world as remote from the Chesapeake as Japan and Uganda. Tobacco taxes supported the English government (to the tune of 300,000 pounds sterling annually at the outbreak of the American Revolution), hauling tobacco kept the Scottish shipping industry afloat, and provisioning tobacco planters and farmers kept Scottish merchants in business. Virginia tobacco (as buyers called all tobacco from the greater Chesapeake region) acquired an international reputation for consistent quality, and the transcontinental demand fueled the growth of the Piedmont.[34]

Consistent with its reliance on tobacco, the Piedmont long remained a rural landscape. The region's largely agricultural economy, with its focus on production for distant markets, did little to promote urban growth before the Revolution. As early as the 1730s, William Byrd laid off the town of Eden along the southern bank of the Dan River, just west of its juncture with the Staunton, in modern Halifax County. Despite his efforts to attract Swiss colonists to the proposed town, Byrd's plan never resulted in a physical settlement.[35] A few hamlets and villages developed, but without exception these small centers languished. Both Peytonsburg in Halifax (1759) and Chatham in Pittsylvania (1767) were founded as courthouse sites but grew little after county divisions led to the relocation of their courthouses to more central locations. (To confuse matters, the first Chatham lost its courthouse to the hamlet of Competition in 1777, whose boosters adopted the name Chatham. The first Chatham then became known as Callands.)[36] Following the Revolution, two towns on the Dan River, Danville in Pittsylvania County and Milton in Caswell, became entrepôts centered on tobacco warehousing and manufacturing, but their growth was slow prior to the 1830s.

Carville Earle and Ronald Hoffman have proposed a convincing theory explaining the general lack of urban development in the early plantation South centered on the needs of agro-environments, which they label "staple theory." Expanding on the work of other scholars, Earle and Hoffman credit the region's shortage of towns and cities to an overwhelming reliance on agricultural staples, such as tobacco and cotton, which demanded little in the way of infrastructure or services.[37] In general agreement with this "staple theory," Charles Farmer believes that a focus on tobacco undercut the growth of towns in the Virginia Southside. Tobacco demanded little but transportation arteries and storage facilities in the form of warehouses, especially during the eighteenth century when the vast majority of leaf was exported to Europe. Even the need for roads in the region was often obviated by the availability of water networks connecting the Piedmont to the Tidewater. If production needs did little to encourage urban growth, regional consumption was hardly more stimulating. Farmer, following Earle and Hoffman, credits the slave/plantation system with the ability to provide for most of its own needs, and the absence of a large middle class meant little demand for artisans and services. Country stores and roving peddlers supplied the majority of luxuries and other purchased goods, and the larger planters dealt with their factors in Richmond and Petersburg for manufactured items such as furniture, silver, and carriages.[38] Country store account books and peddler records from the late eighteenth and early nineteenth centuries reveal that local residents purchased two main classes of goods: basic necessities that were often cheaper to buy than to produce, such as cloth, nails, salt, whiskey, pocket knives, and sugar, and the occasional luxury items, such as silk handkerchiefs, ivory combs, nutmegs, and ribbon.[39] Southside merchants like Scotsman James Glenn typically exchanged finished goods for tobacco rather than cash, which they transported to Petersburg or Richmond en route to European markets.[40] Although local farmers and planters were hardly self-sufficient, there was little in the way of the widespread demand for commercial goods and services necessary for the growth of large towns. This dispersed rural landscape was a product of tobacco culture, but it would also help sustain it. A lack of urban industry presented few alternatives to tobacco cultivation, and poor transportation networks meant that only agricultural crops with a high value to weight ratio (such as dark and later bright tobacco) were cost-effective. Profuse waterways and bottomland soil presented the possibility of tobacco cultivation, which in turn promoted use of waterways, often to the exclusion of other transportation routes, and this reciprocal relationship encouraged local settlers to think of their region as a land "naturally" adapted to tobacco production.

While towns and cities were largely stillborn, as tobacco culture expanded the Southside's rural landscape grew more thickly populated. Although the region was slow to be settled in the early and mid-eighteenth century, the population expanded rapidly after mid-century (table 1.1). By the first federal census in 1790 the three counties held 36,397

Table 1.1.

Combined Free and Slave Populations of Caswell County, North Carolina, and Halifax and Pittsylvania Counties, Virginia, 1790–1860

YEAR	TOTAL POPULATION	FREE POPULATION	SLAVE POPULATION	PERCENT ENSLAVED
1790	36,397	25,114	11,283	31
1800	40,775	25,943	14,832	36.4
1810	51,981	30,956	21,025	40.4
1820	53,777	30,326	23,451	43.6
1830	69,242	37,289	31,953	46.1
1840	67,027	34,129	32,898	49.1
1850	70,027	35,007	35,020	50
1860	74,839	36,247	38,592	51.6

Sources: Compiled from *Fifth Census; Enumeration of the Inhabitants of the United States—1830* (Washington, DC: Duff Green, 1832), 10–13, 84–85, 90–91; *Compendium of the Enumeration of the Inhabitants and Statistics of the United States—Sixth Census* (Washington, DC: Thomas Allen, 1841), 34–35, 42–43; *The Seventh Census of the United States: 1850* (Washington, DC: Robert Armstrong, 1853), 259–260, 307; and *Population of the United States in 1860; Compiled from the Original Returns of the Eighth Census* (Washington, DC: Government Printing Office, 1864), 358, 516–517.

inhabitants, a number that grew by almost 50 percent over the next two decades. As fast as the total population increased, the number of slaves swelled even more rapidly. In 1790 there were already more than 11,000 slaves in Caswell, Halifax, and Pittsylvania, a figure that almost doubled—to 21,025 slaves—by 1810. When residents organized Pittsylvania County in 1767, its white population owned only 271 slaves; by 1800, the slave population had grown to 4,200. The region's free population continued to increase until 1830, at which point economic depression and the opening of rich western lands contributed to the stasis or even decline of the white population in certain districts. This decline failed to affect slave populations, however, as the number of enslaved workers grew steadily until the start of the Civil War.[41] This steady expansion of slavery came at least in part because of the viability of tobacco cultivation on the counties' stronger lands, and bright tobacco's development in the 1840s only solidified this trend. By the end of the antebellum era, the Southside's commitment to slavery was closer to that of the

Deep South than the Chesapeake, where slaves as a percentage of the population were in decline.

Despite the limitations of the Southside environment, planters who were wealthy and keen enough to secure substantial tracts of fertile bottomland, who expanded their slave holdings, and who married into equally well-to-do families managed to parlay dark tobacco planting into lavish wealth. Lederer and Byrd were right: the valleys of the Staunton and Dan could create agricultural empires—at least in limited numbers.

First among the wealthy families of the region during the antebellum period was the Bruces of Halifax County. Of Scottish ancestry, James Bruce moved to Halifax around 1780. He worked as a merchant buying tobacco from county farmers and reselling it on the Richmond and Petersburg markets, shipping the products of river and creek valleys downstream. Bruce also ran a chain of country stores throughout the Southside and began buying thousands of acres of quality tobacco land in Halifax, Charlotte, and Mecklenburg counties, planting in his own right. By some accounts, Bruce became the third millionaire in America by the 1830s, trailing only furrier John Jacob Astor and railroad magnate Stephen Girard. His son James C. Bruce built one of the finest Greek Revival mansions in the South at his plantation, Berry Hill, on the banks of the Dan River, and was worth an estimated $4 million by 1860.[42] The younger Bruce's fabulous wealth and exorbitant lifestyle led one early twentieth-century hagiographer to describe him as "a worthy representative of that noble class of country gentlemen who in the old times gave so much social distinction to Virginia."[43]

The Hairstons of Pittsylvania and Henry counties were equally well off and surpassed even the Bruces in terms of their slave holdings. Like James Bruce, George Hairston was a merchant who settled along the Dan in southwestern Pittsylvania County in the late 1700s, and by the early nineteenth century he had amassed 238,000 acres of land along the river at the base of the Blue Ridge. By the 1850s, George's son Samuel, who planted tobacco in Pittsylvania, Halifax, and Henry counties as well as raised corn on several North Carolina plantations, had created perhaps the largest individual agricultural empire in the history of the antebellum South. Family biographer Henry Wiencek estimates that Samuel and his brothers and sisters controlled as many as 10,000 slaves throughout the South at the outbreak of the Civil War.[44] An 1851 article in the *Richmond Whig and Public Advertiser* declared Hairston the richest man in Virginia and likened the grounds of his home plantation, Oak Hill, west of Danville, to paradise. The author declared, "I have travelled over fifteen States of this Union, and have never seen anything comparable to his yard and garden, except some of them in the Mississippi Delta—and none of them equal it . . . the public grounds [at Washington, DC] were *nearly* as handsome as Samuel Hairston's."[45] Although his estimate was more conservative than that of Wiencek, historian Ulrich B. Phillips credited Hairston as the master of as many as 1,600 slaves in the 1850s and pronounced him "the greatest of the tobacco planters."[46] Samuel's mother, Ruth, owned almost as many slaves. An 1852 inventory of her estate totaled 1,135 slaves in Virginia and North Carolina.[47] In Henry County in 1860, roughly 1 of every 7 slaves—or 704 of 5,018 total slaves—belonged to a Hairston.[48]

Although James C. Bruce and Samuel Hairston were extraordinary examples of planter wealth and power, there were a number of other planters who had parlayed the three counties' limited fertile ground into tobacco riches. In addition to Bruce and Hairston, John Clark, Benjamin Garrett, William Sims, John Coleman, Ethelbert Coleman, John Edmund, J. L. Garland, and W. L. Stamps all owned more than 100 slaves in their respective home counties by 1860. In fact, 566 masters—or almost 18 percent of all slaveholders in the three counties—were "planters," as defined by the ownership of 20 or more slaves (table 1.2). A substantial portion of regional slaves lived and worked on large plantations. A full 19 percent ($n=7,342$) labored on units of 50 or more slaves, and 4 percent ($n=1,378$) resided on plantations with more than 100 enslaved workers, veritable villages of black labor. Slave ownership was more concentrated in Caswell, Halifax, and Pittsylvania than in Virginia or North Carolina as a whole. Only 11.1 percent of Virginia slaveholders owned more than 20 slaves, and the corresponding figure for North Carolina was 11.7 percent.[49]

Although the Bruces, Hairstons, and similar wealthy planters throughout the three counties planted tobacco as a part of their economic activities, most also increasingly relied on other forms of income. Profitable tobacco cultivation was largely limited to the rich bottomlands of the larger rivers and creeks, and most upland soils were productive as tobacco land for only two or three years following clearing. Planters would

Table 1.2.

Slaveholders in Caswell, Halifax, and Pittsylvania Counties, 1860

COUNTY/ STATE	TOTAL SLAVEHOLDERS	SLAVEHOLDERS WITH 20 OR MORE SLAVES	PERCENTAGE WITH 20 OR MORE SLAVES
Caswell	748	142	19
Halifax	1,051	235	22.4
Pittsylvania	1,413	189	13.4
Counties Combined	3,212	566	17.6
North Carolina	34,658	4,065	11.7
Virginia	52,128	5,777	11.1

Source: Compiled from Historical Census Data Browser, 2004, retrieved December 19, 2008, from the University of Virginia, Geospatial and Statistical Data Center, http://fisher.lib.virginia .edu/collections/stats/histcensus/index.html.

then plant corn on the thin soil for a few years before allowing the slopes to grow up in broom sedge and briars, followed by pine, black locust, and persimmon saplings. Large planter diversification seems to have been a response to the limits on traditional Chesapeake tobacco culture imposed by the environmental conditions of the southern Piedmont. Samuel Hairston produced corn, cattle, cotton, and grain as well as tobacco, and his siblings carved out cotton plantations in Mississippi on the backs of slaves purchased with dark tobacco money.[50] Among his many enterprises, James C. Bruce ran nine country stores, several large flour and lumber mills, a plaster plant (producing a popular soil amendment), a cotton manufactory, and a blacksmith shop. He also owned a wagon train and bought and sold real estate on a large scale.[51] Aside from alternative crops and livestock, one of the most common ways for dark tobacco planters along the border to diversify their income was to manufacture finished tobacco products on their plantation or in an adjoining town. Wealthy planters existed throughout the three counties, but that wealth was based on varied economic activities, ownership of the richest bottomlands in the region, and the intermarriage of wealthy families.

Bruce, Hairston, and their fellow large planters were exceptional in terms of numbers, but their influence was far-reaching. They dominated the tiny manufacturing economy of the Southside, and they had control of some of the most fertile stretches of land in the three counties. The bulk of farmers and planters who raised tobacco did not have these advantages. Dark tobacco culture in the Southside existed within certain environmental limits—the acreage of quality bottomland largely circumscribed the dark tobacco crop and encouraged planters and farmers to search for alternative uses of the upland landscape. And the rural, agrarian nature of the landscape and its meager infrastructure centered on transporting tobacco led residents to seek a staple that fit within these existing systems. The Southside environment was just productive enough to support tobacco agriculture as practiced in the eighteenth century, but it was not rich enough to sustain a large population of tobacco farmers over the long term. Despite Byrd's assertions that the Southside was a fertile tobacco country, a century of cultivating the crop, and the wealth dark tobacco produced among a select group of planters, the region's environmental constraints led to a search for a new form of agriculture.

By the mid-nineteenth century tobacco had been entrenched in the southern Piedmont for nearly one hundred years, and the tasks of the tobacco season governed the actions of planters, small farmers, merchants, and slaves alike. Planters often referred to tobacco as a "thirteen month" crop; its cultivation not only demanded work throughout the year, but the care of the previous year's crop often overlapped with beginning a new one.

Creating a plant bed was the first step in making a crop (fig. 3). Unlike corn or cotton, which farmers directly sowed in the field, workers started tobacco in a nursery bed and then transplanted the seedlings to the field. Farmers took great care in selecting the sites for plant beds, locating them on new ground each season. Beginning in December and continuing as late as the end of February, they cut and cleared small patches of woods on south-facing slopes.[52] These plant beds were usually one hundred square meters or smaller, and they did best in the rich lands along small streams or in wet hollows. Many farmers were carefully attuned to the soil type and microclimates that supported rapid seedling growth; they sought locations that grew hazels or alders, as these

FIGURE 3: The first task in any tobacco season was to clear and burn a plot of forestland in preparation for a plant bed for seedlings. This work involved careful selection and preparation of the site, the first of many tobacco tasks that entailed hard work and environmental knowledge. From Edward King, *The Great South; a Record of Journeys in Louisiana, Texas, the Indian Territory, Missouri, Arkansas, Mississippi, Alabama, Georgia, Florida, South Carolina, North Carolina, Kentucky, Tennessee, Virginia, West Virginia, and Maryland* (1875). Delta State University Library.

shrubs thrived on ground well suited to raising tobacco seedlings.[53] After cutting the timber and grubbing out stumps and roots, farmers usually burned the top few inches of the bed's soil in order to kill weed seeds, improve tilth (the soil's texture, especially pertaining to its suitability for planting seeds), and release nutrients for the tobacco seeds. Laborers built a portable fire using a set of hardwood skids with brush piled on top, and, beginning in one corner of the plant bed, they lit the brush and allowed it to burn for an hour or more, long enough to sterilize the first few inches of topsoil. Then they added more brush to the fire and moved the pile to the next section of bed using chains and iron hooks. This process continued until they had burned over the entire plant bed.[54]

Once the bed was burned over thoroughly, the farmer pulverized the soil as fine as possible to prepare it for seeds. Workers always mixed tobacco seeds—at nearly a half million per ounce, they are among the smallest of all agricultural seeds—with a substrate, usually ashes, to facilitate even coverage over the bed. A typical plant bed required only two tablespoons of seed mixed with a gallon of ashes for thorough coverage. Many farmers also fertilized their plant beds with stable manure or guano to facilitate large, early seedlings.[55] Once they had sown the seed and harrowed in the manure, planters covered their beds with hog hair or a thick layer of brush. These coverings served a variety of purposes: they protected the young plants from frost, kept many insects—especially tobacco flies, which chewed holes in the tender foliage—away, and they helped the bed retain moisture during dry spells. Throughout the late winter and the early spring months, farmers kept a close eye on their plant beds; they weeded if grass became a problem, top-dressed the patch with fertilizer if the plants progressed slowly, and added or removed brush based on weather conditions.[56]

Just as farmers selected certain portions of the landscape for plant beds, they also used their knowledge of local environments in their placement of tobacco fields. Most fields were located in or near creek or river bottoms. Bottomland was generally richer than the uplands, it typically retained moisture well, its proximity to water permitted primitive irrigation in emergency circumstances, and it tended to be flat and thus easier to cultivate than the surrounding ridges. Regional topography, laced as it was with small rivers and their numerous tributaries, provided a substantial quantity of bottomland acreage suited to traditional tobacco cultivation.

Transplanting seedlings from the plant bed to the field could take place anytime following the last frost to the middle of July, depending on weather conditions.[57] During late winter or early spring, farmers worked the land several times with cultivating plows, which broke up heavy soil, turned under weeds, and removed roots and rocks. Hoes followed plows, forming small dirt mounds called hills; each hill would become home to one tobacco seedling. Planters who chose to use manure on their tobacco lands usually applied small amounts directly to each hill rather than broadcasting it over the entire field. This practice reflected a shortage both of fertilizer and of labor to haul it. Most planters relied on the natural fertility of new land or the lasting power of bottomland rather than manure, as too much manure tended to result in a harsh leaf that was slow to ripen. As William Tatham, an English expert on Virginia tobacco culture, recorded, "Tobacco which is produced from manured or *cow-penned* land, is only considered, in ordinary, to be a crop of the second quality."[58] The crop's intolerance of heavy manuring worked against farmers' attempts to incorporate tobacco into systems of convertible husbandry, though planters must have found the plant's light demands for manure a blessing rather than a burden, especially given southern farmers' attachment to an extensive form of livestock husbandry.

After farmers prepared their tobacco land, they waited for a good rain (or a "season") to transplant seedlings. A soaking rain loosened the soil and made planting easier, and it provided the moisture necessary for sustaining a young seedling through the shock of transplanting. Following the wet spell, all hands on the farm would draw plants from the beds and carefully move them to the fields in baskets or barrels. At the field they placed one seedling to each hill, using a short wooden peg to make a hole for the plant. Planting was a period of intense labor. The faster hands moved seedlings to the field, the better the plants could take advantage of the "season's" soil moisture, resulting in a higher plant survival rate. Quick work during planting meant a lighter workload over the next few weeks, when the workers would walk over the fields numerous times, replanting hills where the transplants had perished.[59]

Once a farmer was satisfied with the stand, cultivation began, a process that would continue until a few weeks before harvest. Cultivation, undertaken with hoes and horse- or mule-drawn shovel plows, accomplished two purposes: it killed weeds that competed with the growing plants, and if done properly, it continually built up the hills to support the tobacco as the plants grew heavier over the course of the season.[60] As in corn and cotton culture, tobacco cultivation was an exercise in vigilance. Over the course of a season workers cultivated around each plant several times, and even industrious hands could fall behind in the never-ending battle against weeds if unusual weather intervened. Planter William Sims conveyed this anxiety following a wet spell in the summer of 1855, lamenting tobacco grown "quite foul and grassy."[61] Once the tobacco reached full height and was difficult to work around, farmers made one last cultivating pass, building up the hill to its final height, a process commonly referred to as "laying by."[62]

Once the plants began to form flowers at the apex of the stalk, the tobacco was ready to be "topped." During topping, hands broke out the flowering structure and small apical leaves, and often knocked several leaves off the bottom of the stalk as well (a process referred to as "priming," as it left only prime leaves on the plant). The purpose of topping and priming was to force the plant to direct its energy into growing larger and heavier leaves rather than reproductive structures. Whereas un-topped tobacco grew tall and spindly, topped tobacco filled out short and squat. Each farmer had his own formula, but topped and primed tobacco plants generally retained eight to twelve leaves, depending on soil and weather conditions. Like many other tobacco tasks, farmers considered topping highly skilled work. They expected a topping hand to carefully examine each plant and determine the optimal number of leaves to leave based on appearance, season, soil type, and plant variety.[63] Topping added quality to the crop, but it also created a problem. Breaking out the flowering structure encouraged tobacco to form axial branches, called suckers, where the leaves meet the stalk, in the plant's attempt to create a new flower. Hands had to repeatedly walk over the crop after topping, removing any suckers that might redirect vegetative growth away from the leaves.[64]

Just as tedious as suckering was the task of removing the hornworms that infested the crop throughout the summer. Tobacco hornworms are the larvae of the Carolina sphinx moth (*Manduca sexta*), and a heavy infestation of worms could consume a field of tobacco in a period of just days if not combatted. The most common method of fighting worms was to remove the finger-sized larvae and any unhatched eggs by hand, which workers then crushed between their fingers or under their heels, though some planters tried more innovative methods, from ranging insectivorous guineas and turkeys in their fields to having slave children with torches and paddles work the edges of the fields at dusk, swatting as many adult moths as possible before the insects could lay eggs.[65] Worming was often a gender- or age-specific task; planters typically relegated the work to women and children. Nancy Williams, who was a slave on a Caswell plantation, remembered worming as a child, and also what happened when she missed a few worms one day (in the racially charged dialect of ex-slave narratives): "Purty soon old Masser come 'long, dough, an' see dat I done been missin' some of dem terbaccy worms. Picked up a hand full of worms, he did, an' stuffed 'em inter my mouth; Lordy know how many of dem shiny things I done swallered, but I sho' picked 'em off careful arter dat."[66] Worming began as soon as the larvae appeared in the spring and continued through the harvest.

Once the tobacco leaves grew heavy enough and began to yellow in the early fall, they were ready for harvest. Determining exactly when to harvest was one of the most difficult and crucial decisions a tobacco farmer made; harvest before the plants were completely ripe and the tobacco cured poorly and spoiled or brought a poor price at market, wait too long and the entire crop might be destroyed by an early frost. As former slave Gabe Hunt recalled, "[You] got to pick dem leaves what's jus' startin to brown. Pick 'em too soon dey don't cure, an' you pick 'em too late dey bitters [brittle]."[67] Although tobacco culture manuals tried to provide advice on how to determine the perfect moment for harvest, they generally seemed to agree that only experience allowed farmers to accurately gauge when they should harvest their crop. In a typical passage, Tatham declared, "Much practice is requisite to form a judicious discernment concerning the state and progress of the ripening leaf."[68] For farmers who planted early and experienced favorable summer weather, tobacco could ripen as early as the end of August, but a late season that delayed planting could force hands to battle the coming of frost in October to get the crop safely in the barn.[69]

During harvest workers cut each stalk off near the ground with a long tobacco knife and then generally split the stalk for half its length to facilitate curing. Growers cured cut tobacco one of several ways. They could leave the crop in the field for a short amount of time to wilt, then hang it on outdoor racks to cure in the sun; they could transport the tobacco to log barns, where it would be dried out over a period of several days using small wood fires built on the barn floor; or they could cure the leaf using a combination of these methods.[70] In both sun-curing and fire-curing, workers affixed the split tobacco stalks to wooden sticks that hung in the outdoor racks or on tiers built in the log barns. Green tobacco is 80 to 90 percent water by weight, and the object of these curing methods was to remove enough moisture from the plant so that the tobacco did not rot or mold, while preventing the leaves from becoming so dry and brittle that they crumbled to dust at the touch. Tobacco dry enough for storage but remaining pliable was described as in "case." Fire curing took several days, depending on ripeness and conditions, and outdoor curing could take as long as several weeks. Once growers had sufficiently cured their tobacco, they placed it in a storage shed (if sun-cured) or left the plants hanging in the curing barn until they had time to prepare the crop for market.[71]

Although both sun and fire curing remained in practice in the Southside into the antebellum period, the use of fires was the predominant curing method employed by most farmers by the 1840s. Barns enjoyed several benefits over outdoor curing: fires made for faster curing, barns provided more protection against the elements during the curing process, and if barn space was not needed for an additional cure, the leaf could remain in the curing barn until ready to be handled further. The typical tobacco barn in the three counties was built with rough pine logs and clay chinking and was sixteen or twenty feet square. The structures had packed dirt floors and one or two small doors, and they were generally tall enough to accommodate five or more tiers of poles which supported the hanging sticks of tobacco. Poles were typically spaced about four feet apart horizontally, with each ascending tier roughly three feet above the lower one. Farmers built the small wood curing fires directly on the floor below the first tier of poles, and the smoke exited the barn through vents located along the ridge of the roof or in the eaves. Farmers often erected their barns on the edge of the tobacco field rather than near the farmyard in order to facilitate the movement of tobacco from the field to the barn.[72]

Fire curing in barns presented many advantages, but it also posed certain risks. Open flames in a wooden barn filled with resinous plant matter were obvious fire hazards, threatening at any time to burn up both the structure and the year's crop. This omnipresent threat of fire meant farmers usually watched their barns diligently throughout the curing process, often staying inside their smoky confines for hours on end to regulate the fires. A few years after the Civil War, when fire curing still remained popular, a *Danville Register* article described the hazards of tending the curing fires. "They [farmers] have to watch their barns day and night, and get pretty well smoked, so that from the loss of sleep and red eyes and hard work, some of them are nearly as crazy as bed bugs."[73]

Once the crop was cured, it generally remained hanging in the barns until a lull in farm activity over the winter months. During cold snaps and poor weather, farmhands worked to prepare the tobacco for market. First, they removed the plants from the wooden sticks and stripped the individual leaves from the stalks. The process went best during damp weather when the leaves were in "order," pliable enough to withstand stripping without crumbling. Once stripped, the ordered leaves were graded into classes. Although these grades were somewhat arbitrary and not legally binding, farmers and merchants generally referred to the best grade as leaf tobacco, followed in descending quality by fillers, lugs, and substandard tobacco (which went by various names, from "trashy" to "foul"). Farmers strove to ensure that each hogshead of tobacco was of consistent quality, as inspectors usually graded hogsheads based on the poorest-quality leaves they observed when sampling. After hands sorted the tobacco into grades, they tied the leaves into small bundles referred to as "hands." The typical hand was composed of four or five leaves with their tails and stems aligned, tied together at the stem end with another leaf. Farmers carefully stacked these hands in radial layers inside a wooden hogshead, and the whole mass was compressed (or "prized") down with a lever or screw press until the hogshead was full. Just as the process of bringing tobacco to "case" was difficult, prizing tobacco in proper order was a tricky task. If farmers prized damp tobacco it would mold and spoil, if too dry, the leaf might crumble into dust when the hogshead was unpacked.[74]

Virginia law regulated the size of hogsheads, but over time improvements in prizing presses and a slight enlargement of the legal dimensions allowed hogsheads to grow heavier and heavier. Tatham also noted that many tobacco inspectors "winked" at planters who used hogsheads slightly larger than the legal dimensions. In 1822 Peter Minor estimated that the average Virginia hogshead weighed 1,350 pounds, up from an average of roughly 1,000 pounds around the turn of the century, and many enterprising planters hoped to pack as much as 1,500 pounds into the fifty-six inch-tall by thirty-six inch-wide container.[75]

By the 1840s and 1850s, well-packed hogsheads weighed almost 2,000 pounds, a staggering sum made more manageable by improved roads and a move to railroad freight in the latter decade.[76] As haulers and warehouses usually assessed fees for handling or inspecting tobacco by the container rather than by weight, farmers had a strong economic incentive to increase the amount of tobacco in each hogshead.[77]

Once prized, the leaf was ready for transport to one of Virginia's or North Carolina's inspection sites, state-regulated stations where officials weighed the hogsheads and certified their quality. No tobacco could be exported without undergoing inspection. Tobacco buyers also gathered at these inspection stations, where they purchased inspected hogshead for transport to manufacturers in Richmond, Petersburg, the northeast United States, or Europe. Most tobacco moved across the countryside in one of three ways: by wagon, by "rolling," or by small boats known as bateaux. Farmers close to an inspection station carried most of their tobacco in ox- or horse-drawn wagons that could hold two hogsheads apiece. Transporting over longer distances often entailed "rolling" the tobacco to market (fig. 4). In this method, they drove short wooden pegs into the center of the hogshead's top and bottom lids, and these pegs fitted into a pair of wooden poles, creating a rig that could be pulled by a horse. To reduce the possibility of damage to the tobacco inside the hogshead, farmers usually affixed several bands of tough hickory around the body of the hogshead to act as wheel rims. This method of transport was particularly popular among smaller farmers as it was cheap and easy, and the farmer could simply discard the poles and ride his horse back from market once he sold his tobacco.[78]

Farmers who lived on local rivers had another option. They could send their tobacco down the Staunton, Dan, Pigg, Hyco, or Bannister on bateaux. Bateaux were narrow, flat-bottomed boats usually around fifty feet in length, but occasionally as long as eighty, which could haul five to twelve hogsheads with a two- or three-man crew. Improvements along the Roanoke drainage, including channels cut through the rapids and dredging the shallows, made the once almost impassible waters navigable during the early 1800s. Bateaux pilots plying these waters were often experienced slaves entrusted to work independently for months at a time. During the winter and spring, the bateaux plied the local rivers, with their pilots making arrangements with planters to pick up large

FIGURE 4: Before the advent of railroads, transporting tobacco was often a long and arduous process in the Southside. Here a farmer rolls a hogshead along a rutted Piedmont road on his way to market. From Edward King, *The Great South; a Record of Journeys in Louisiana, Texas, the Indian Territory, Missouri, Arkansas, Mississippi, Alabama, Georgia, Florida, South Carolina, North Carolina, Kentucky, Tennessee, Virginia, West Virginia, and Maryland* (1875). Delta State University Library.

crops or simply beaching their boats at popular landing spots and blowing a horn to announce they were available for work. At several sites along the Staunton and Dan merchants built short-term storage warehouses for the hogsheads of small farmers and planters without river access. Moses Gilliam's warehouse on the Staunton in Halifax was representative. In 1842, Gilliam stored a total of ninety-six hogsheads for twenty-four farmers, from one hogshead for small growers, such as George Spencer and Eden Tally, to eight hogsheads for Dabney Raglin. Alternately, small tobacco producers could roll their hogsheads to the homes of large planters, such as James Bruce, who charged a fee to store the crops and ship the produce of a number of growers en masse. Once the bateaux were loaded, the river men would float local tobacco down the streams and rivers that combined to form the Roanoke to Weldon or Gaston, North Carolina. From there, most hogsheads traveled overland to Petersburg or Richmond, though a few continued on to the small ports of the Albemarle Sound or north through the Dismal Swamp Canal to Norfolk.[79] After 1833, a railroad connected the falls of the Roanoke to Petersburg, largely displacing the wagon trade.[80]

By the 1840s, the Southside was largely a settled, agricultural landscape, where the cycles of the tobacco season formed an endless chain of work that blurred the years even as they emphasized the seasons. This countryside relied heavily on the cultivation of tobacco with slave labor. Caswell, Halifax, and Pittsylvania were rural counties, with only a few centers that could be considered towns: Danville led the way with a few thousand residents in 1840, Milton was less than half as large, and Chatham and Halifax Courthouse each held around three hundred residents.[81] A ride through any of the three counties would have revealed similar scenes. Wealthy planters monopolized the river and larger creek bottoms, where their substantial frame or brick plantation houses, flanked by rows of small slave houses, watched over dozens or even hundreds of slaves who labored in the tobacco and corn fields. Among the most numerous structures were the log tobacco barns clustered around the edges of fields, vertical structures that emitted long plumes of smoke during the curing season each fall. At certain points during the year, bateaux plied the waterways, laden with hogsheads bound for Danville, Milton, or headed down the Roanoke watershed on their way to fall-line markets, and wagoners and farmers rolling hogsheads made their way along the counties' dirt roads. Moving from the rich lowlands to the rolling hills that dominated the Piedmont uplands, our rider was more likely to see broad forested tracks and smaller farms of log houses and less productive fields. Free-ranging livestock roamed the woods and roadsides, and the smaller farms had fewer or no slaves, though many still produced some tobacco.

Although the Southside remained a rural place, its environment had changed a great deal since William Byrd and his party had surveyed the state line in the 1720s. The buffalo and black bears were gone, deer populations dwindled due to over-hunting, and large portions of the thick woods that once covered the region's uplands had been cleared for fields, building material, and firewood. Of course not all wildlife had disappeared. Species such as rabbits and quail, which relied on edge habitats and disturbance, thrived as farms spread across the counties. James Bruce and his hunting companions killed several hundred quail apiece each winter during the late 1830s and early 1840s; in just one day of hunting in 1838, the small party shot forty-four rabbits along Halifax field edges.[82] Likewise, George Jeffreys of Caswell found rabbits abundant on his farm in the 1840s but noted that foxes were "too *scarce*," as "there are too many fences—the excitement is too great."[83] Many of the fish that had attracted settlers to the Dan and the Staunton in the colonial period remained in the rivers in the mid-nineteenth century. Robert Withers remembered heavy runs of Atlantic salmon, suckers, and black bass in the Staunton before the Civil War, and the fugitive slave Henry Goings recalled the schools of anadromous sturgeon that still swam up the Roanoke to its forks. These "sturgeon in large quantities," along with rockfish and shad, made the river "a splendid place for carrying on fishing to a very heavy extent."[84] As late as 1858, Daniel Merritt commented on the wild abundance periodically on display along the border. After a fall with "a fine red oak mast," Merritt marveled at the "thousands & millions of wild pigeons in the land."[85] Despite these remnants of natural abundance, regional life was centered on agriculture, or more specifically, on a single agricultural crop.

Certain environmental conditions stimulated the spread of tobacco culture throughout the Southside. The geography and natural resources of the region helped shape its agriculture and society to a great degree. These environmental influences often took the form of limits rather than encouragements. The region's abundance of animal life and vegetation that so awed early Euro-American explorers and settlers proved to cover relatively infertile upland soil: ground that included most of the southern Piedmont and that promised little in the way of agricultural productivity. Farmers forced to raise crops on small patches of productive bottomland turned to the high-value crop with which they had experience: dark tobacco. Early settlers' decisions to grow tobacco were also influenced by the Piedmont's climate (the season was too short to grow cotton reliably), water transportation routes, and the availability of wood for building barns and curing fires. Like the regional shortage of rich soils, the direction of the Roanoke River's drainage influenced Southside agriculture. Historians of the region have described the Roanoke's outlet into the Albemarle Sound of North Carolina, with its narrow and dangerous outlets to the Atlantic, rather than into the Chesapeake Bay, as a retarding factor in the development of the Southside.[86] Although these critiques are true to a certain extent, the cost and inconvenience of the region's major water route also meant that farmers had to focus on high-value crops. Tobacco was one of the few agricultural staples that could still turn a profit after a journey to market that involved rolling, floating, and hauling a hogshead—which could weigh almost a ton—well over a hundred miles. Like mountain farmers who converted their heavy corn into whiskey, a value-added product, Southside tobacco growers focused their agricultural efforts on a crop that it paid to transport to distant markets.

Despite these environmental and geographical conditions conducive to a particular form of agriculture, the Southside was not destined to become a tobacco kingdom. Tobacco's evolution in the region also relied on the culture that settlers brought to the southern Piedmont. At least some of these first pioneers were enmeshed in a tobacco economy that linked North American agricultural landscapes to European consumers through Scottish merchants and British ports, a system developed in the Tidewater over the previous century. Small farmers and large planters settled along the border with knowledge of and experience in tobacco cultivation—they knew the crop made money and they knew how to raise it. Previous settlement in the central and northern Virginia Piedmont also proved that tobacco farming west of the Tidewater paid. When farmers looked at their new land with an eye toward its promise, they thought naturally of where and how tobacco might fit in the landscape. The power of this cultural perception was evident in other regions of frontier Virginia. Planters and farmers in districts with rich soil, such as the Shenandoah Valley, also cultivated tobacco briefly during the early settlement period, before turning to more diversified farming that took full advantage of the valley's fertility.[87] Tobacco proved a much more durable staple in the Southside because this cultural affinity for tobacco agriculture encountered an environment of limited resources. These same cultural experiences and environmental realities contributed to the regional shift to bright tobacco cultivation that would so shape life in the Southside for the next century.

Beginning in the 1840s, the search for a new version of tobacco culture would begin to challenge the traditional crop for agricultural and social prominence. But the existing culture of tobacco, which governed or influenced labor relations, daily and seasonal routines, transportation infrastructure, politics, and conceptions of landscape and even self-worth, would prove a crucial framework for the changes to come. The crop was central to the worlds of the poorest "dirt farmer," tens of thousands of slaves walking the fields, and rural patricians Bruce and Hairston in their brick mansions. The southern Piedmont, simply put, was a place where tobacco was life and life was tobacco.

TWO

Let There Be Bright

THE BIRTH OF YELLOW TOBACCO CULTURE

LIKE ALL SIGNIFICANT CULTURAL TRANSITIONS, the development of bright tobacco has a creation myth that makes the crop seem at once revelatory and natural. Yellow tobacco's genesis involves the Caswell planter Abisha Slade, who grew tobacco in the Blanch community on a sandy ridge a few miles southeast of Danville, and his young slave, Stephen. According to the story, in the fall of 1839 eighteen-year-old Stephen was tending the traditional wood fires that were drying a barn of tobacco. At some point during the curing process he fell asleep and the unattended fires died. Upon awakening, a panicked Stephen rushed to the plantation's blacksmith forge and carried several loads of charcoal to the dormant barn. The resulting coal fires heated the barn well above the typical curing temperatures for several hours before cooling off, and the tobacco that resulted from this high-temperature—yet relatively smoke-free—curing turned out to be a bright yellow color. Stephen's master Abisha carried this charcoal-cured crop to the Danville market, where he sold the leaf to a manufacturer for $40 per hundredweight, and a tobacco expert was born. Abisha tinkered with the use of charcoal over the following years, perfecting the system, and he traveled throughout Caswell and the southern reaches of Halifax and Pittsylvania, often accompanied by Stephen, spreading his coal-firing instructions to planters who sought a reliable formula for curing yellow tobacco. Abisha's

curing work, combined with new seed varieties and careful soil selection, created a new crop culture in the Southside.[1]

The Abisha and Stephen story appeared in its most influential guise
in 1949, in historian Nannie May Tilley's magisterial study of the development of yellow tobacco, *The Bright-Tobacco Industry*. Following Tilley's
account, almost all subsequent histories of flue-cured tobacco, from
county volumes to scholarly tomes, have included the Stephen Slade
story, codifying the tale through repetition.[2] Memorials to Stephen even
include a roadside historical marker, that arbitrator of all legitimate southern history, commemorating the "discovery" of bright tobacco in the
Slade barn. Before Tilley's account, the story had received little circulation. Prior histories did acknowledge Abisha Slade and his brothers
Thomas, Elias, and William as early bright tobacco experts, but they made
no mention of Stephen and they generally dated Abisha's influence on
regional tobacco cultivation to the early 1850s rather than 1839.[3] Historian Barbara Hahn has recently shown that Stephen's story is most likely
apocryphal. Although Abisha Slade and his brothers figured prominently
in antebellum and early postwar accounts of bright leaf development,
Stephen and the blacksmith forge apparently made their first appearance
in an 1886 newspaper article in the *Pittsylvania Tribune*, followed by a
similar story a month later in the agricultural journal *Progressive Farmer*,
accounts upon which Tilley relied in her study.[4] Aside from these brief
references, until the publication of *The Bright-Tobacco Industry*, tobacco
historians, including some of the Slades' acquaintances, made little mention of Stephen as the accidental originator of charcoal curing.

Although its veracity may be in doubt, the "Stephen myth" emerged
in the 1880s because it so neatly captured the coalescence of a white understanding of the birth of a crop and economy. The *Progressive Farmer*
account brought together several key elements of the development of
bright tobacco culture in an overt fashion. According to the article, a
"Captain John Lee" approached an older freedman at a Danville tobacco
auction in the spring of 1886 and "gave a cheer" which "was taken up by
the crowd." A reporter inquired as to the reason for the excitement and
was informed that the freedman was "Stephen Slade . . . the first man to
cure bright tobacco." The reporter spoke with Slade, who told him that
the curing was in fact an accident; after the application of charcoal, the
tobacco " 'kept on yallowin' and yallowin' tell it got clear up.' " Stephen

followed this explanation, according to the reporter, with a declaration of his loyalty to the Democratic Party and his fondness for his former master and the social status quo antebellum: "I wish he [Abisha] was alive today and I was his slave."[5] The article referred to Abisha Slade as "Elisha Slade" but confirmed 1839 as the year of the fortuitous accident.[6]

Whether or not the Stephen Slade story is true, the tale has proven so durable because it captures several central truths of early bright tobacco culture. Stephen's status as a slave represented a vital aspect of early bright tobacco experimentation: most early experts were slave own-ers who employed black hands in the management of their tobacco crops. It is also telling that Stephen's discovery was an accident. Indeed, the story of his inadvertent nap mirrors popular antebellum accounts of the "lazy slave." This representation allowed white tobacco growers to credit Stephen with the discovery of the importance of charcoal while minimizing the significance of black expertise in the process of produc-ing quality tobacco, a transformation that made the story entertaining rather than threatening for local planters. And though Stephen may have discovered the key to successful curing, according to the tale it took Abisha's intelligence to master and systematize the process. As was the case with other antebellum southern crops, white "expertise" worked to hide black contributions to agricultural knowledge.[7] Thus when Captain Lee and his fellow auction-goers cheered Stephen, their acclaim was perhaps more for their memory of the Old South and the rise of a new regional cash crop—and their mastery over this arrangement—than it was for the ingenuity of a former slave. Stephen capped his fond tribute to the old order by asserting his continued loyalty to his deceased master and the antebellum political world, an assertion that the white crowd must have found comforting in the racial tension of the 1880s South.[8]

Stephen's tale is a classic creation story. As David Nye has pointed out, "Nineteenth-century Americans repeatedly told themselves stories about the mastery and control of nature through technology in which radical transformations of the landscape were normal developments." These stories were selective; they singled "out particular objects while deemphasizing or even deleting others."[9] Stephen's tale presents these themes. Charcoal curing became the force that transformed dark tobacco into bright, and black experience raising and curing tobacco became a happy accident rooted in a slave's carelessness. The story's focus on curing minimized the importance of seed and soil selection in creating quality bright leaf, and Stephen's nap hid generations of slave understandings of tobacco and its cultivation in the Southside environment. The story's acceptance likely owed a great deal to its simplicity, for as Nye argues, the popularity of period creation stories "arose from their apparent ability to explain historical events and fuse them with cultural values."[10] The article in the *Progressive Farmer* attached a date, place, and names to amorphous agricultural labors, environmental conditions, and social systems.

The farmers and warehousemen gathered around Stephen that day sought a creation story because bright tobacco, already an emerging economic force during the antebellum period, had become the literal lifeblood of the region's countryside and towns following the Civil War. The elements of a bright tobacco culture that came to dominate Southside farming entailed subtle changes to the traditional routines of dark tobacco. Farmers selected varieties of seed for certain color and taste characteristics; they planted these seeds on new portions of the Piedmont landscape; and, like the Slades, they experimented with new methods of curing their tobacco. While none of these changes seemed dramatic departures from the tobacco culture that had existed in the Southside since the mid-1700s, collectively these practices would alter both land and people over the following decades. Within a generation of Abisha and Stephen Slade, a Virginia writer could without exaggeration declare of regional cultivators, "Their life is yellow tobacco."[11]

To explain this agricultural addiction, we must examine the region around Danville, Virginia, in the years leading up to the Civil War. Around 1840, planters and farmers along the border of Virginia and North Carolina began developing a new form of tobacco: bright leaf. This crop was characterized by a fine-textured, mild-tasting leaf that ripened to a yellowish green color in the field and cured golden yellow. From early experimentation on a few farms, the new crop techniques spread across Caswell, Halifax, and Pittsylvania counties (as well as into the edges of adjoining counties) during the twenty years leading up to the Civil War. Bright tobacco brought exceptionally high prices at local and regional markets thanks to its attractive color and mild flavor, and the crop and its associated manufacturing enterprises stimulated an economic revival in the Southside. Underlying this agricultural transformation was an alteration of farmers' perceptions of tobacco as a crop and of the regional environment. In the search for lighter tobacco and greater profits, old ways of farming and patterns of managing tobacco fell by the wayside, to be replaced by an agronomy that was simultaneously more intensive and extensive.

American attempts to produce "fancy"-colored or yellow tobacco predated the advent of Southside bright leaf by more than a century. All North American domesticated tobacco is *Nicotiana tabacum,* a native species of the nightshade family. *Tabacum* is an extremely plastic plant, expressing quite diverse characteristics on different soil types; its phenotype (or outward appearance) is easily transformed through environmental interaction. Soon after tobacco cultivation began in the New World, growers worked to shape *tabacum* to suit their desires, and a number of varieties or strains of tobacco were the result.[12]

The pursuit of distinctive tobacco often veered toward the production of light-colored leaf, though it is not entirely clear why this was the case. The demand for pale cured tobacco was likely part biological and part consumer fetish. Tobacco that grew light in the field (and thus often cured a lighter color) generally contained lower levels of tar and nicotine, and hence was easier to consume in quantity than stronger, harsher leaf. At the same time it also was often sweeter than darker tobaccos, satisfying users' sweet teeth. At the same time, there seems to have been a perception that darker tobacco contained more impurities, natural or otherwise. Yellow tobacco was also more exclusive and costly, being hard to produce, and thus became a luxury good and status symbol. In these regards, consumer attraction to yellow tobacco seems akin to the popularity of white sugar described so adroitly by anthropologist Sidney Mintz, in *Sweetness and Power.* White sugar satisfied consumer demand for high-calorie, energy-rich food, while appealing to notions of purity and exclusivity.[13] Whatever intrinsic attraction light tobacco held, mid-Atlantic tobacco planters recognized the consumer demand for "yellow" tobacco and had long sought to grow light-colored leaf, whether through seed selection, the exploitation of soil particularities, or curing methods.

As early as the last decades of the seventeenth century, Virginia planters along the Rappahannock and the York rivers grew a particularly flavorful variety of tobacco, known as sweet-scented, that may have been lighter in color than the traditional Oronoko. Growers claimed that this sweet-scented tobacco was the product of light, sandy soils.[14] Through curing or cultivation experiments, planters throughout the Tidewater and Piedmont occasionally produced small quantities of tobacco that were lighter in color and of a milder flavor than the majority of dark tobacco, and these fancy lots sold for high prices at regional markets. By the first decades of the nineteenth century, Maryland "kite-foot" tobacco had gained renown as a light-colored, smooth-tasting leaf. Growers along the western shore of the Chesapeake Bay perfected soil selection and curing techniques that reliably produced kite-foot, and carved out a niche in international markets.[15] Following the War of 1812, the export market—especially in France—grew particularly interested in fancy grades of Virginia and Maryland tobacco (leaf grown in North Carolina was invariably included in the umbrella term "Virginia tobacco"). Farmers in pockets across the Virginia Piedmont labored to discover curing methods that might produce lighter-colored leaf, and they cultivated fanciful descriptions of their produce. Planters in the late 1810s and 1820s described their tobacco as "pie-bald," "calico," "green streak," "straw," "fawn," and "hickory-leaf color," among other names, attempting to draw the interest of fancy tobacco buyers.[16] These attempts to produce new forms of tobacco highlight the historic interest in light-colored leaf, but they also demonstrate the scattered and inconsistent nature of tobacco cultivation and curing. Growers had a difficult time reproducing seed lines, finding soils suitable to light tobacco, and mastering a reliable curing method. Southside farmers would be the first to consistently combine all three elements in their search for yellow tobacco, though a great deal of trial and error was involved.

The first, and perhaps least discussed, element necessary for producing bright tobacco was the proper seed. As with any crop, farmers had long employed a form of intentional selection when preserving tobacco seed for the following year. As the crop matured and topping time rolled around, planters selected a number of the hardiest plants to go to seed. The aim for cultivators who desired to make "fancy" tobacco was to select plants that matured to a light color and fine texture in the field, plants most likely to cure a yellow color. (To make a complicated relationship simple, tobacco that looked yellow when harvested had the best chance of leaving the curing barn yellow—dark green leaves had little chance of curing to a light color). Growers selected plants that exhibited likely characteristics in their specific farm environments; those "having the finest fibre and texture, and that ripened yellow on the hill."[17] Over time, this annual selection process led to hundreds of tobacco varieties (or landraces) adapted to particular soil conditions. Promising seed lines circulated within communities, and some of these sought-after varietals acquired their own distinct names. Caswell's George Jeffreys acquired "one and all" seed from a fellow planter and bought a "Pryor" variety (most likely "Yellow Pryor") from a plantation across the Virginia line.[18] Other named varieties that produced yellow leaves in Jeffreys's neighborhood included "Daniel Jones" and "White Stem."[19] In a postwar seed catalogue, Halifax's Robert Ragland described "Yellow Oronoko" and Yellow Pryor as the classic antebellum bright leaf varieties, and he traced their evolution to farmer selections in the 1840s.[20]

Planters were adamant about the folly of attempting to produce bright tobacco without careful seed selection. Conscientious growers mated seed type to soil type, doing so through field trials. Samuel Shelton, a Henry County planter, wrote that "there is a very great difference in crops, managed in the same way, arising from the different kinds of tobacco planted." He went on to illustrate that, when planted on similar land, narrow and broad leaf Oronoko varieties produced very different ripe tobacco. On relatively weak land the former produced "a finer and sweeter article than any other kind," whereas the latter turned out "decidedly coarser than the other" in the same fields. Rather, broad leaf Oronoko excelled in turning out large, dark tobacco on rich bottomlands.[21] Over years of trial and error, planters interested in growing bright tobacco learned—as a postwar commentator declared—that "no one but a novice would sow [dark tobacco varieties] for the fine manufacturing grade, or choose the fine varieties for heavy coarse stock."[22] While various *tabacum* varieties are genetically quite similar, selective breeding created surprisingly diverse plants when coupled with local environments and cultivation practices.[23]

Farmer's seed selection was an important factor in growing yellow tobacco, but the best seed failed to produce a high-value crop without appropriate soil. The best soil types for producing smooth-tasting, light-colored tobacco, as described by modern soil scientists, were Appling, Cecil, Durham, and Granville sandy loam. These soil series all shared a similar structure. Their topsoil was a light-colored sand, typically a few inches to a foot deep, underlain by a stiff, bright red or yellow clay pocketed with disintegrating rock. This topsoil contained little organic material and was notably deficient in nitrogen. Water percolated rapidly through this sandy layer and then flowed horizontally across the surface of the denser clay; the soil across both horizons was moderately to strongly acidic.[24] These sandy soils did not uniformly blanket the three counties. Soil types appropriate for growing bright tobacco covered roughly one-third of Caswell, Halifax, and Pittsylvania counties, primarily the high ridges that lay between the numerous creeks and rivers. Pockets of sandy soil snaked through richer bottomland loam, weak soil often lay side-by-side with more fertile ground, and most farms included a variety of soil types.[25] As a postbellum tobacco historian described, "Fine, bright tobacco land may be separated by only a few feet from a heavier clay soil, which will produce only a heavy manufacturing or export leaf."[26] These soil types were not unique to the three counties; belts of this sandy loam extended east and south into the corn and cotton region of North Carolina. But the hills along the Dan, Bannister, Staunton, and Hyco rivers did contain the most abundant sampling of these soils among traditional tobacco-growing regions.

Even though bright leaf varieties thrived on "thin ridge-land," like all other agricultural crops they required soil nutrients, including phosphorus, potassium, and nitrogen.[27] Indeed, yellow tobacco varieties were heavy feeders of phosphorus and potassium, but they needed much less nitrogen than dark tobacco.[28] Soils with a great deal of nitrogen produced tobacco with a harsh, heavy taste, as tobacco plants converted excesses of the nutrient into a number of nitrogenous compounds—the most prominent of which was an alkaloid, nicotine—that made for "biting, strong Tobaccos."[29] Contemporary agriculturalists understood the toxic and stimulating, but not the addictive, properties of nicotine, but they incorrectly believed that the substance was created during the curing process rather than in plant growth.[30] And though regional farmers did not define soils in the same manner as later soil scientists, through working the earth they quickly came to recognize the drainage characteristics, acidity, and fertility of local land.

High concentrations of nitrogen also hindered the proper ripening and curing of bright tobacco. Planters found that curing worked most effectively on perfectly ripe tobacco, plants that were just beginning the process of senescence (the gradual deterioration of an annual plant prior to its death). Soils that contained a great deal of nitrogen grew tobacco with larger, heavier leaves and thicker stalks, traits that encouraged plant vitality. These robust plants continued to grow long into the fall harvest season and were slow to ripen, and thus farmers found them difficult to cure properly before the onset of frost, which killed the crop. Drainage was also an essential element of quality tobacco land. Tobacco was a crop that was notoriously intolerant of soggy ground, and plants subjected to standing water quickly wilted and drowned. The relative lack of organic matter in sandy Southside soils, which contributed to low nitrogen levels, also led to the ground's rapid water percolation.[31]

Although antebellum farmers lacked modern understanding of soil nutrients, nicotine production, and plant biology, they quickly learned what soil types produced quality tobacco, and sought effective ways to select fields. As soil selection was so vital to producing a quality crop, local planters and farmers became quite adept at identifying appropriate tobacco lands. A shortage of quality bottomland meant that dark tobacco growers had long experimented with the poorer ridge soils, and though these lands were relatively unproductive for most crops, sections of sandy land remained in grain and tobacco cultivation throughout the early nineteenth century, especially on marginal farms. Where the ridges were not in cultivated fields, the scrubby pine and hardwood forest served as wood lots and livestock range. Planters and farmers turned to this experience with regional soils when they sought the best land for raising yellow tobacco. A key indicator of sandy loam was the vegetative cover. Newly cleared ridge land almost always made good tobacco, but old fields on the weak soil also produced a fine leaf, and growers sought grounds covered in such scrubby successional species as dogwoods, sourwoods, chinquapins, and hickory saplings.[32] Enterprising planters also discovered that even old fields too weak to regrow woodland could make quality bright leaf, and they placed ground with sparse weed cover or broomstraw—that historic indicator of exhausted southern land—into cultivation once again.[33]

For many struggling tobacco farmers, bright tobacco's reliance on poor soil appeared all but providential. Overnight, ground that had been the least productive stretches of regional farms, suitable only for scratching out a few bushels of corn or ranging lean hogs, became the most valuable agricultural acreage in the Piedmont.[34] As tobacco experts from the late nineteenth century marveled in retrospection, "This porous, spongy, sandy earth, destitute of humus, and incapable of growing any crop without the most abundant application of manures, became the corner stone of a new agriculture."[35] It was a crop culture that seemed to defy normal agricultural strictures, turning poor land into valuable ground and old fields into assets. As one farmer noted, "a peculiarity of these lands, and one which greatly enhances their permanent value, is that, after being worn down by continuous and exhaustive cropping, and then turned out" within a few years they could recover enough to "produce the finest cutters and smokers, which always command high prices."[36]

But poor ridge soils and carefully selected seed alone were not sufficient to produce high-quality yellow tobacco. Perhaps the key element in bright tobacco production was the development of reliable curing methods. Between 1840 and the end of the Civil War, bright tobacco growers gradually moved from the open wood fires and air curing favored by dark tobacco growers to the use of charcoal fires and then metal flue systems (fig. 5). Both new techniques heated tobacco to higher temperatures than open wood fires could, and both triggered biochemical changes in the leaf that brought out its yellow color. These curing technologies fixed tobacco characteristics developed through seed and soil selection, and as they were perfected they transformed yellow tobacco from a lucky occurrence to a repeatable agricultural staple. Bright tobacco thus brought together nature and culture. Soil types created by millions of years of geological forces met curing methods shaped by human desires and generations of experience cultivating tobacco, with seeds—themselves the products of *tabacum*'s evolutionary biology and human selection—as the medium. In the confluence of these forces, yellow tobacco emerged as the product of a particular place and a particular people.

The use of flues to cure tobacco predated the bright tobacco boom by at least four decades. By the early years of the nineteenth century Virginia planters were experimenting with pipes that could carry heated air through the tobacco barn without the accompanying smoke, which tended to give the leaf a darker color and stronger flavor. J. Robinson of Charlotte County (1809), Peter Minor of Albemarle County (1822), Thomas Gay of Goochland County (1824), and Edmund Pendleton of

FIGURE 5: Piedmont tobacco barns contained several log tiers to hold the curing tobacco above a heat source, whether it be wood fires, charcoal, or flues. In this image the log tiers running the length of the barn are evident, as are the remains of two fire vents on either side of the door. Courtesy Library of Congress.

Louisa County (1826) all wrote of curing systems that utilized stone flues to carry heat throughout the tobacco barn.[37] In 1828, Halifax's Davis Tuck developed an improved flue system with a stone and iron firebox that opened to the outside of the barn, a series of sheet iron flues to carry heat inside the barn, and a chimney that vented the smoke. By sawing a few holes and installing these flues, farmers could convert their existing log tobacco barns into flue-cured barns. The farmer could stoke the fire without entering the barn, and Tuck advised that growers use a primitive thermometer visible through a small glass window to regulate barn temperatures. In all important respects, Tuck's 1831 patent described "an improved flue [that] was almost identical with the style" used throughout the bright tobacco belt well into the 1960s.[38]

Despite Tuck's forward-thinking flue system, regional planters were slow to adopt the flue as tobacco-curing technology due to the brief popularity of an intermediate curing method: charcoal fires. Popularized by Abisha Slade, charcoal curing combined traditional fire-curing techniques—open fires on the dirt floor of a barn—with an improved yet common heating source. Although Slade spread the charcoal-curing method across the region, experiments with the fuel likely dated to at least the 1820s, as regional planters dabbled with curing techniques. By 1861, planter Samuel Shelton could declare (with some exaggeration) that the use of cordwood in curing tobacco "has been totally abandoned in all the fine tobacco-growing sections." Indeed, he stated, "to cure tobacco with chunks [of wood] is a disgrace to the tobacco-making community."[39]

Whether done by flues or charcoal, curing tobacco to a yellow color both solidified and altered existing relationships between tobacco planters and the Southside environment. Both curing methods used wood, the traditional barn fuel, but they did so in new ways. Both also placed a layer of simple technology between burning wood and curing leaf; in the case of charcoal it was the work of the brazier's rick, in that of flue curing, a firebox and a set of metal flues. Flue and charcoal curing required growers to heat their barns to higher temperatures for longer periods of time, using greater quantities of wood, and the process demanded that farmers pay careful attention to their barns throughout the process. Unlike wood fire curing, which essentially dried the leaf to slow decomposition, flue- and charcoal-curing methods relied on complicated formulas to develop and fix yellow leaf color.

The new curing methods proved much more demanding in terms of the attention and care a grower exhibited while tobacco was in the barn. Bright leaf experts developed elaborate curing formulas that called for steadily increasing temperatures, often at the rate of only a few degrees per hour, and they demanded a curer's constant vigilance for days on end. The following small excerpt from the instructions of "R. J. S.," a Caswell farmer, are representative of the complexity of contemporary curing guides: "Continue this heat [90°] for 12 hours, then raise it 10° and in 12 hours 10° more, the next 12 hours 5° (115°) at this stage apply charcoal for 6 hours, by which time the tobacco will be through the process of fermentation, and as yellow as it will be. Then pass up to 120°, which in 12 hours will cure the leaf, after which go up 10° every 2 or 3 hours, until you reach 150–60 or any higher heat within bounds of safety from burning, and continue until the stem and stalk are cured."[40] The writer went on to note that his method demanded that "the strictest watch must be kept day and night."[41] Guides that furnished such exacting instructions must have challenged farmers, who worked with charcoal of varying quality, often without thermometers. Success under these systems and conditions would have been spotty at best. The point of these machinations, as with many other versions of curing, was to provoke changes in the appearance, taste, and stability of tobacco, especially to make light-colored leaves yellower, and to make mild leaves sweeter.

As the above instructions intimate, curing continued biochemical changes that had started in the field. As mature tobacco ripened on the stalk, chlorophyll (green pigment) in the leaves slowly degraded. The act of curing tobacco, like that of aging beef, was an exercise in controlled decomposition. The gradually increasing heat of the tobacco barn accelerated this chlorophyll degradation, revealing the yellow carotinoid pigments that existed in the leaves and which were normally masked by chlorophyll, and—if the leaf was harvested at the correct level of maturity and had grown in soil with an appropriate chemical composition—this process resulted in a bright yellow leaf. To provide a common analogy, this sequence of pigment unmasking was similar to the process by which deciduous tree leaves change color in the autumn. The heating process also converted plant starches into sugars and proteins into amino acids, transformations that imparted flavor and aroma to the curing tobacco.

Yellowing and starch and protein conversion occurred simultaneously but independently, and, by happy coincidence for tobacco growers, peak sugar content and the point of high chlorophyll degradation coincided in most bright tobacco varieties. These chemical reactions solidified connections consumers made between yellow color and sweet flavor. Increasing the heat after yellow color was revealed fixed the appearance. This high heat also destroyed certain enzymes that if left active would slowly brown the cured leaves. Once curing fixed the color, only prolonged scorching heat, rot, or burning could alter the tobacco's appearance. Thus curing acted to accentuate certain biological qualities in the tobacco leaf; it could not "make" tobacco yellow or flavorful, but it could reveal colors and tastes that existed in the plant matter at the time it was harvested.[42] As with soil selection, antebellum planters did not understand the biochemistry of curing in the same terms that we do today, but they quickly developed a practical knowledge of the effects of various temperatures and rates of heating on particular tobacco varieties.

In *Making Tobacco Bright,* Hahn credits institutional structures with shaping producer and consumer conceptions of tobacco, and the cultural technology of curing with the physical act of yellowing tobacco. Hahn's insights regarding the power of institutions are particularly perceptive—tobacco definitions did owe a great deal to inspection practices, auction grades, taxation, and federal variety descriptions—but her claim for the predominant role of curing in the production of bright tobacco obscures the vital importance of soil and seed types in the creation of a new tobacco form.[43] Curing could only reveal yellow pigment created by biological processes that originated in particular soils, and it could convert to sugar only the starches that existed in a given tobacco variety. As a biochemist explains in a recent text, yellow tobacco is made on the soil rather than in the barn: the "potential quality of tobacco is determined in the field . . . harvested leaves with poor quality characteristics are not improved during curing, even when curing procedures are ideal."[44] It was no accident that bright tobacco in its modern form first appeared along the sandy ridges flanking the Dan River and its tributaries. Planters had experimented with flue curing for several decades across Virginia, but it was only in the Southside that flues (and charcoal) met soil and seed types appropriate for reliable bright leaf production. Thus bright tobacco is perhaps best described as a process rather than strictly as a crop, yet it was a process that was firmly rooted in local environmental conditions and within the biological constraints of seeds.

Efforts to grow bright tobacco changed how farmers thought about seeds, soils, and curing, but many elements of dark tobacco cultivation were left relatively unmodified. In general, bright leaf called for an intensification of cultivation methods rather than outright alterations to seasonal activities. Guides to raising yellower tobacco advised planters to plow their fields more frequently than normal, as the aeration of the soil increased percolation in addition to killing competing weeds. This plowing advice also called for the exclusive use of shallow shovel plows in cultivation to prevent mixing the stronger (and less porous) clay subsoil with the thin, sandy topsoil. Topping also drew experts' attention. Bright tobacco with too many or too few leaves could ripen too early or too late, or it could develop leaves either too thin or too heavy. This was true for dark tobacco as well, but it was even more critical for yellow tobacco as coal and flue curing were more delicate processes than wood firing. Most critical was harvesting the crop at the peak of ripeness; green tobacco refused to cure properly, and overripe tobacco had no body and was almost worthless. Only tobacco that had just begun the process of senescence in the field would reveal its underlying yellow color in the barn, and only a trained eye could determine exactly when to cut the crop.[45]

All of these changes were a matter of degree rather than type. Treatises on bright tobacco stressed that only expert farmers could produce quality yellow tobacco whereas any grower was capable of raising dark tobacco. Thus farmers who turned out bright leaf were "good" farmers by default, agrarians who intimately understood soils, plants, and weather, an expertise directly reflected in crop quality. As one contemporary writer advised of bright tobacco cultivation, "a great deal depends on the management, and . . . the finest crop on the hill may be butchered and ruined by bad management."[46] With bright leaf cultivation, then, came a value judgment about the cultivator as well as the crop, and many farmers were anxious to prove that they were capable of mastering fine tobacco. These instructions, and the agricultural critique implicit in them, appealed to farmers' pride as well as their pocketbooks. Successful growers such as Abisha Slade were acknowledged artisans as well as agriculturalists, and they commanded respect at agricultural meetings, on court days, and in local warehouses.

Although tobacco growers prided themselves on their ability to make fine yellow tobacco, the crop on many plantations also relied on the agricultural and environmental expertise of slaves. Planter language actively obscured these relationships. Most planters wrote and spoke as if they did not trust overseers, much less slaves, to make fine-quality tobacco. As one Caswell planter wrote to another, "I have never seen many Overseers that I thought knew how to manage Tobo. well. it [sic] is seldom where they have the entire management that they get good prices."[47]

Frederick Law Olmsted, while traveling in the eastern tobacco belt, recorded similar sentiment regarding slaves. A planter confided in Olmsted that he made only common tobacco, as the "finer sorts required more painstaking and discretion than it was possible to make a large gang of negroes use. 'You can make a nigger work,' he said, 'but you cannot make him think.' "[48]

As a consequence of this attitude, some planters micromanaged tobacco cultivation in their attempts to control all aspects of crop production. After Emancipation, Henrietta Perry, who had been a slave in southern Pittsylvania County, recalled the close attention her master paid to his slaves' tobacco work. "Us black people had to look arter dat 'baccy lak it was gold. Us women had to pin our dresses up roundst our necks fo' we stepped in dat ole 'baccy fiel', else we'd git a lashin'. Git a lashin' too effen you cut a leaf fo' its ripe." Her owner did not extend the same attention to his other crops. "Marse ain' cared what we do in de wheat an' corn fiel', cause dat warn't nothin' but food for us niggers, but you better not do nothin' to dem 'baccy leaves."[49] Jordan Johnson, a slave in neighboring Campbell County, told a similar tale. He remembered a pregnant woman, Annie, who accidentally cut down a tobacco plant while weeding. Under the overseer's harsh glare, she grew nervous and chopped off another seedling, "Ole overseer lif' up dat rawhide an' beat Annie 'cross de back an shoulders 'till she fell to de groun'."[50]

Despite planters' belief that slaves made poor tobacco hands due to inattention or incompetence, black hands were the true tobacco experts, at least on some of the larger plantations. Stephen Slade's role in the origin of bright tobacco curing may be apocryphal, but thousands of slaves shaped and guided the cultivation of the crop through the antebellum era (as they did other crops across the South).[51] Slaves shepherded the plants from the plant beds to the fields, hoed and hilled the tobacco by hand, culled hornworms from the leaves when they appeared, topped each plant at the appropriate height, and judged when ripe tobacco should be cut. Masters gave instructions—perhaps based on input from experienced hands—but slaves executed them. Gabe Hunt, another former slave, recalled the complex judgments necessary in a number of daily tobacco tasks. When the tobacco harvest came, hands had to carefully judge the ripeness of each plant, and this expertise extended to the mechanics of harvesting the leaves as well. "Got to break 'em [leaves] off clean at de stem an' not twist 'em cause if dey bruised dey spile."[52] Like-wise, topping quality tobacco was an exercise in experience and judg-ment. Hands had to quickly determine the appropriate number of leaves to remain on the docked plant, a calculation that factored in soil strength, plant health, a field's production history, seed variety, and weather con-ditions, among other variables. In a postwar pamphlet, tobacco expert Robert Ragland advised his white readers to seek out former slaves as topping tutors. He wrote, "Young man, if you don't know how [to top], get some old negro to show you."[53]

Processing tobacco once it had been harvested demanded just as much experience and environmental knowledge. Planters may have su-pervised most curing—though the tale of Stephen Slade challenges this assumption—but slaves typically sorted, packed, and prized the cured leaf. Planters desired hogsheads or loose lots of clean, well-sorted (leaves of similar quality packed together) leaves that were moist enough to travel without crumbling but too dry to rot. Each stage of the post-curing labor involved assessments of leaf quality, color, and pliability (fig. 6). Workers had to judge humidity, the tobacco's moisture content, and its general grade according to color and texture, all while factoring in the distance to market and the form of transportation. These calculations relied on experience and on slaves' intimate understanding of tobacco as a living and decaying plant.

Combining all of the elements of bright leaf cultivation to success-fully and reliably produce yellow tobacco was a challenging undertaking that often involved a good deal of trial and error. The efforts of George Jeffreys, a Methodist preacher who also farmed along Hyco Creek in the eastern Caswell, reveal some of the difficulties inherent in early efforts to master bright tobacco.[54] In the mid-1840s, inspired by the cultivation practices of his neighbors, Jeffreys became determined to make "*fine high priced Tobo.*"[55] Likening the complicated cultivation to the chal-lenges of his ministerial labors, he recorded a prayer in his diary "that God will give me a *bright* crop this year."[56] Pleased with his determina-tion to enter the ranks of bright tobacco growers, he resolved to "devote all my attention to its culture & curing and then publish a *pamphlet.*"[57]

Jeffreys planted Little Frederick, One and All, and an unidentified Pryor variety on his plantation, types that he selected based on his ob-servations and the advice of neighbors. He apparently also understood

FIGURE 6: Slave knowledge and expertise was crucial to many parts of antebellum tobacco cultivation, including grading and packing the cured crop. In this illustration, slaves are shown sorting tobacco leaves for prizing in hogsheads. From Edward King, *The Great South; a Record of Journeys in Louisiana, Texas, the Indian Territory, Missouri, Arkansas, Mississippi, Alabama, Georgia, Florida, South Carolina, North Carolina, Kentucky, Tennessee, Virginia, West Virginia, and Maryland* (1875). Delta State University Library.

the vital importance of planting on weak soil, as he moved his corn crop to the plantation's creek bottomlands, and planted tobacco on his upland fields, though those lands were covered in "two important evils, the great nuisances of our farms viz broomsedge & gullies."[58] He also followed cultivation advice that would become bright tobacco gospel over the following decades, relying exclusively on shovel plows to till his tobacco fields. Jeffreys had earlier employed dagon plows, which turned the land over to a deeper depth and slowed erosion, but he worried that mixing topsoil and subsoil would prove detrimental to producing fine tobacco. Like other regional planters, he believed that the sandy, thin topsoil contained the elements essential to growing bright tobacco and that the clay subsoil inhibited proper drainage. Jeffreys recognized the dangers of intensive tobacco cultivation using these methods on erosive lands and planned to combat gullying with a crop of spring oats planted on the fields, in the hopes that the crop would hold the soil together during the region's heavy spring rains. He also emphasized the vital importance of allowing tobacco to mature fully in the field before harvest. He wrote, "As regards *curing* great attention should be paid to the *yellowing* of the Tobo as the first & most important process."[59] In all of these techniques, he anticipated the regional yellow tobacco culture of the 1850s and 1860s.

Whereas Jeffreys understood certain precepts of bright tobacco culture, he was quite confused about other elements necessary to successfully curing yellow leaf. He believed that the region's upland soil was well suited to the growth of light tobacco, but he also penned livestock on a portion of his tobacco ground to enrich the soil, efforts that surely proved counterproductive to producing a fine-bodied leaf with a smooth, sweet flavor. Jeffreys was equally uncertain about the best method of curing his tobacco. He practiced "sunning" his cut tobacco—leaving the plants in windrows in the field to begin the drying process before his slaves transferred them to barns—and believed these curing methods alone could produce bright tobacco.[60] With his plant beds seeded in 1845, he wrote that he intended to "raise plant & cure the Tobo without fire & wood," efforts that surely failed to produce top-quality leaf.[61] Jeffreys's successes and struggles emphasize that the creation of a bright tobacco culture was a process rather than an instantaneous occurrence. Planter such as Jeffreys, or the Slades for that matter, struggled for years to perfect methods of cultivation and curing that led to barns full of bright yellow tobacco, and these early efforts were no doubt marked by more failures than successes.

However sporadic, the efforts of Jeffreys and his fellow bright leaf experimenters found positive reinforcement in regional warehouses and the local press. New agricultural products often experience a "honeymoon" price effect as consumers are willing to pay inflated prices until production increases satisfy the existing demand, but bright tobacco's price advantage over dark tobacco continued throughout the antebellum period, Reconstruction, and well into the twentieth century. Traditional tobacco prices fluctuated throughout the 1840s and 1850s, responding to swelling and shrinking foreign and domestic demand and varying crop size; in general, after the Panic of 1837 prices declined through the mid-1840s, climbed gradually to a peak in 1856, and slid downward again following the Panic of 1857. Despite this variance, for every year during the period except 1856, common tobacco grades brought less than $10 per hundredweight on Virginia markets (and often far less). At the 1844 price nadir, average tobacco brought less than $4 per hundredweight on the Richmond market.[62]

Determining the average value of tobacco sold in the Southside before the Civil War is a challenging task. There are good records of prices paid for bright and dark tobacco on the Danville market from 1869 forward, when yellow tobacco consistently commanded prices twice as high as fire-cured leaf, but there are no corresponding figures for the antebellum and Civil War periods.[63] Even where receipts and accounts are available, planters, agents, and warehousemen often failed to differentiate between bright and dark tobacco in writing; the growers involved knew their own crops and the prices commanded by each type and generally used the generic term "tobacco" to refer to either type, and leaf ran the spectrum from dark to light colored, often with no firm line where one stopped and the other began. Thus it is impossible to give exact figures for each type of tobacco before the war, but we can make some educated deductions. An examination of receipts for well over 1,000,000 pounds of tobacco sold by Caswell, Halifax, and Pittsylvania growers between 1840 and 1865 hint at the profit potential of yellow tobacco. While the average hundredweight of tobacco from this sample sold for $7.04, in line with Richmond prices during the same period, lots designated as bright leaf could sell for several times this figure. Some planters who experimented with making fine tobacco reaped handsome rewards; hundredweights of yellow tobacco raised by William Bailey brought $45.25 (in 1859), Joel Hubbard's tobacco sold for $28 (1858), George Clement garnered $18 (1854), Joseph Totten found buyers at $20 (1856), and Pleasant Womack sold several thousand pounds of bright leaf for between $38 and $48 (1860). At $48 per hundredweight, a farmer who could raise 2,000 pounds of tobacco (a reasonable sum for one hand at the time) might gross $960 in a season, compared to only a little over $140 for a cultivator who garnered the median tobacco price for the same weight of cured dark leaf. This was a substantial sum as a typical farmhand would do well to make $100 per year during this period. These and similar sales suggest that bright leaf brought two to three times as much money as dark tobacco during the antebellum era, and exceptionally yellow leaf—suitable for wrappers—had the potential to sell even higher.[64]

Newspapers, especially in market towns, called special attention to the extremely high, or "lottery," prices paid for select lots of yellow tobacco. A *Milton Chronicle* article advised planters that carefully cured and packed yellow wrappers could bring between $40 and $100 per hundredweight in 1857, and another piece in the same paper the following year informed readers of a local planter who obtained $60 per hundredweight from a Caswell manufacturer.[65] The *Danville Republican* touted similar prices for the best leaf, with "excellent tobacco" garnering prices five to ten times higher than typical dark tobacco.[66] And a *Virginia Echo* advertisement from 1859 pointed to a Halifax planter who received $71 per hundredweight as an example of the limitless potential of bright tobacco.[67] In part, these extraordinary prices served to entice bright tobacco growers to a particular town or warehouse and were not representative of the prices the average farmer could expect for good yellow tobacco; nevertheless, these reports did serve to create a mystique regarding bright leaf's profitability. There was always a chance, however small, that a farmer could make a fortune on one extraordinary crop.

With this promise came a concomitant pressure. Planters who sought to become bright tobacco experts tied their farming ability to their tobacco's quality, as reflected in the prices paid for their crop. Tobacco agents often played on this perceived tie between fine tobacco and expert farming to flatter or pressure planters. When William Bailey of Halifax sold some bright leaf for $45.25 per hundredweight in 1859, his agent wrote "We think you now have a right to brag as well as some of your neighbors."[68] Elisha Barksdale's agent sold his crop for a respectable $15 per hundredweight but admonished Barksdale that he would have to do better if he wanted to raise the best bright leaf as his crop was still "too dark for wrappers."[69] Other agents challenged their suppliers more directly. Caswell's John Garland, who was experimenting with bright leaf production in the late 1840s, received a receipt for tobacco sales from his agent that criticized his crop management. Garland's agent wrote, "You Tobo. was prized too high in order . . . & two thirds of them [the hogsheads] were moulded & funked."[70] These and similar communications between tobacco agents and planters associated tobacco quality with a farmer's knowledge; good cultivators had a right to brag to their neighbors, and planters such as Garland who struggled to master the complicated steps necessary to produce and ship quality tobacco faced the scrutiny of their business partners and, by extension, their community. And to the extent that planters were judged on the quality of their tobacco, they often built their reputations on the unacknowledged expertise of their slaves, whose technical skills and knowledge were obscured by a conspiracy of neglect.

Newspapers' and journals' propensity to publish the names of the best growers highlighted the public nature of successful bright tobacco production. Articles dealing with high prices for fine tobacco invariably listed the growers who commanded such premiums and almost always mentioned their county or community, further tying bright tobacco to particular individuals and landscapes. A typical article appeared in the *Virginia Echo* on August 19, 1859. The Halifax County paper gave the prices obtained that season by seven county bright leaf planters, ranging from Samuel Adams's crop of three hogsheads that brought $20.50 per hundredweight, to M. P. Trible's crop of the same size that brought an average of $44.62 per hundredweight.[71] With such publicity, certain districts—such as Mount Vernon in Halifax, the Caswell ridge land between Milton and Danville, and the Leatherwood Valley along the Pittsylvania-Henry line—developed reputations for producing the "finest manufacturing [tobacco] that grew" anywhere.[72]

The labors of farmers and slaves in the Piedmont environment were instrumental in the creation of bright leaf, but the work in regional tobacco factories was equally important in the emergence of a distinctive crop culture. Bright tobacco production was the result of seed selection, the unique environmental conditions of the Southside, and curing methods developed by enterprising planters and their workers, but the spread of bright leaf cultivation also depended on consumer demand and manufacturer innovation. Consumer interest in mild yet flavorful varieties and cures had long standing, displayed in the popularity of distinctive styles, such as sweet-scented and Maryland kite-foot tobacco, in the eighteenth and early nineteenth centuries. But manufacturers also realized the need to increase the market for high-quality tobacco.[73] Local chewing tobacco manufacturers believed the showy, flavorful tobacco varieties could be popular with the tobacco-consuming public, and they quickly experimented with new products and advertising campaigns touting yellow leaf. By the late 1850s, bright tobacco had become an integral input in local manufacturing enterprises, spurring the rapid growth of Southside tobacco factories. Plugs produced in the three counties quickly developed a national reputation for bright leaf tobacco products, and consumers as far away as California began to identify yellow tobacco with the countryside surrounding Danville. Developments in rural fields spurred industrial growth, which, in turn promoted increases in yellow tobacco cultivation, and tobacco manufacturers worked hand in hand with growers to improve bright leaf supplies.

Tobacco manufacturing in Virginia—and to a lesser extent in North Carolina—had a long history as Civil War storm clouds brewed. There were 252 tobacco manufacturing facilities scattered across Virginia in 1860, with the greatest concentration in the district encompassing Richmond and Petersburg, historic centers of commonwealth industry which had manufactured tobacco products since the colonial era.[74] That year Richmond alone contained more than 50 tobacco factories that employed roughly 3,400 workers.[75] Tobacco manufacturing along the Virginia–North Carolina line dated to at least the 1820s, when one or more small factories began operations in Danville.[76] At the outbreak of the Civil War, there were at least 53 tobacco factories of various sizes in Caswell, Halifax, and Pittsylvania, with the majority concentrated in or near the towns of Danville, Milton, and Yanceyville.[77] Within less than four decades of the founding of its first factory, Danville had grown to be the fifth largest tobacco manufacturing center in the world, trailing only Richmond, New York, Petersburg, and Lynchburg.[78] By 1860, Danville tobacco factories purchased more than 3.5 million pounds of local tobacco annually, employed almost 500 hands, and produced finished goods worth $610,332.[79] In the first ten months of 1854, the company of Sutherlin and Ferrell alone purchased 389,574 pounds of tobacco for $37,574 and sold $52,950.45 worth of manufactured tobacco.[80] These factories specialized in chewing tobacco production, and, by the 1850s they began to focus on incorporating bright leaf into their traditional products. The close proximity of so many tobacco growers and the high quality of Southside yellow tobacco made regional factories unusually profitable on the whole. The rate of return among Danville factories was the highest in the state at 24 percent.[81]

Almost all regional tobacco factory workers were slaves.[82] The use of enslaved men and women dated to the first Danville tobacco factory. Slaves formed the largest regional labor pool, they were subject to their owner's or lessee's control to a greater degree than were free laborers, and, crucially, most slaves had experience and skill handling tobacco. According to a local writer, that first facility produced "boxes and little hooped kegs of plaited and twisted chewing tobacco, as black as the ebony faces of the Negroes who prepared the plugs and worked the presses."[83] Manufacturers often owned a number of their laborers, and they also leased slaves on annual contracts from local planters. By 1860, roughly half of the black workforce was leased. Sutherlin and Ferrell leased a portion of their black workers, typically paying a slave's owner between $100 and $150 per year and promising to clothe, feed, and provide medical care for the slave.[84] Due to the variety of ownership and varied employment patterns, unraveling factory labor forces is quite complicated. By way of example, at the census enumerator's 1860 visit to Sutherlin and Ferrell, founder William T. Sutherlin owned 39 slaves and rented 26 more; his partner Peter Ferrell owned 1 worker; Sutherlin's brother John, who worked for the company, owned an additional 9 slaves; and the company's ownership owned 1 slave collectively. Danville manufacturer A. S. Wyllie owned 42 slaves but occasionally hired factory workers. Beverley Barksdale and Joshua Hightower, who owned a Halifax factory, collectively owned 93 slaves, though they were both active planters as well as manufacturers and likely used only a portion of their workforce in manufacturing. Barksdale and Hightower may also have leased slaves, but the census for the southern district of Halifax did not record leases.[85] Despite these archival thickets and census snares, it is clear that a good number of local slaves worked in tobacco factories.

For these factory slaves, work routines were nearly universal. Almost all local manufactured tobacco during the antebellum period was chewing tobacco, and the transformation of cured tobacco leaf into a manufactured tobacco product was relatively simple. When hogsheads of tobacco arrived at the factory, slaves unpacked the leaf, sorted and graded the tobacco and removed any foreign matter, and spread the leaves out to dry further. Workers accomplished drying in a number of ways, ranging from simple air drying to the use of mechanical tumblers. Once the tobacco was sufficiently dry, laborers remoistened the leaf with just enough water so they could handle the tobacco without having it crumble, and they then removed the thick, tough central stem—or rib—from each leaf. Slaves next soaked the "stemmed" leaves in liquid flavoring agents—most commonly a mixture of rum or licorice and sugar—which also served as preservatives, and once again dried the tobacco. Hands then coated certain products with various spice mixtures. At this point the tobacco had taken on its final flavor, and factory workers turned to molding the product into standardized shapes. Workers hand-formed the tobacco into chewing plugs, a universal term for both plaited ropes ("twists") and small rectangular bars ("lumps"); wrapped the shapes in particularly attractive leaves; and then placed these plugs into metal or wood molds and used mechanical presses that exerted pressure in order to set the shape. Finally, workers packed the finished plugs into wooden boxes for shipment to local merchants and distant markets. In addition to being race specific, this factory labor was typically divided along gender and age lines. Young children ran errands and sorted tobacco, women and older children worked at the monotonous task of stemming, and male slaves usually formed the plugs and operated the presses, tasks that manufacturers considered skilled work.[86]

Although this factory work seems far removed from a slave's agricultural labor on Piedmont tobacco plantations, plug making was in certain ways an extension of field labor. Almost all factory slaves had personal experience cultivating tobacco. Either they had worked in the fields before being leased to manufacturers, or, in many cases, they worked as both field laborers and industrial hands over the course of a single year, moving from field to factory floor as plantation labor slowed in the winter months. Like farm work, factory labor required specific understandings of tobacco biology and environmental conditions. In the days before climate-controlled factory floors, workers who handled moisture sensitive materials, like tobacco or textiles, relied on their experience and understandings of climate and material.[87] Stemmers had to use their sense of touch and vision to determine when appropriate moisture levels in the tobacco permitted stem removal without causing the leaf to crumble, and this experience often came from handling cured tobacco on the farm, where moisture levels were equally critical. The action of judging when tobacco should be taken from the curing barn, sorted, tied into hands, and packed in hogsheads was very similar to the judgment required by stemming. Likewise, selecting plug wrappers entailed practice and an understanding of consumer demand, but it was also akin to the sorting and grading that took place on every plantation and farm before tobacco left for the market, work that became particularly critical as bright tobacco cultivation emphasized quality above all else. Planters and factory bosses may have set the guidelines for handling tobacco, but it was often slave judgments that shepherded the leaf through the curing, packing, and manufacturing regimens. Throughout the manufacturing process, tobacco remained an organic product, subject to shifts in humidity and temperature, and plug production was reliant on careful handling. It seems likely that manufacturers leased local slaves for their factories precisely because these workers had experience with tobacco from the plant bed to the packing house; their environmental and agriculture knowledge was part and parcel with their industrial knowledge.[88]

Charles Dew, in his influential study of the Virginia iron industry, *Bond of Iron*, has argued that slave ironworkers used their industrial knowledge to forge a better lives for themselves within the bounds of early southern industry. Slaves learned industrial skills that made them an invaluable part of iron making and as a consequence extracted better housing, food, and more privileges from their masters. In addition, skilled ironworkers often garnered payment for their work beyond basic requirements. Dew claims that "the industrial skills that" these slaves possessed forced their master "to permit them to behave in ways he never would have allowed his field hands to act."[89] Early bright tobacco manufacturing suggests a different but equally intriguing source of industrial knowledge. Whereas ironworkers acquired their abilities thanks to years of work experience in the forge, tobacco hands brought much of their tacit knowledge of tobacco with them from local fields and barns. Certain manufacturing skills were learned on the factory floor—pressing plugs, flavoring the mix, and proper stemming—yet slaves came to manufacturing with a knowledge of tobacco that supported much of this work.

The tobacco historian Joseph Robert argued that tobacco factory work was a particularly mild form of slavery, in which hands, like the iron workers Dew described, often received bonuses for meeting certain production levels, and "the exploitative evils otherwise inherent in any form of absenteeism" were kept at a minimal level by manufacturers' desires to keep their skilled workforce content.[90] Despite Robert's claims, it seems likely that factory slaves suffered from many of the same indignities and hardships common among agricultural laborers, as their planter masters and their industrial bosses were in many cases one and the same person. Indicative of the nature of tobacco factory work, facility managers were sometimes described as "overseer[s]."[91] Robert himself acknowledged that manufacturers often employed harsh white overseers and used the whip when "the devil would set into a Negro."[92] These industrial slaves typically lived on the factory grounds or in the facility itself and thus were rarely far from their master's or overseer's eye.[93] Tobacco factory workers beset by harsh masters responded in ways similar to their fellow slaves on Piedmont plantations; they had ample opportunities to slow down their work, turn out poor-quality goods, break tools, or if conditions were intolerable, to run away. In 1861, one of Sutherlin and Ferrell's leased slaves found either factory conditions or the separation from his Madison, North Carolina, home unbearable, and he stole away from the Danville facility. A planter in Bachelor's Hall, west of the city, wrote Sutherlin that he had spotted the man traveling through the countryside, apparently homeward bound.[94] Another leased tobacco worker, Richard, fled the Sutherlin and Ferrell factory, returning to the Carrington plantation to complain to his master's son that the factory overseer whipped him without cause. Richard's negotiations pitted owner against lessee and waded the shallow waters between paternalism and capitalism in the tobacco South.[95]

Although they followed the same routines and produced similar products, not all local tobacco manufacturing enterprises were as substantial as those of Sutherlin and his fellow Danville manufacturers. As demand for bright wrappers grew, small factories sprang up across the rural portions of the three counties. In 1860, Caswell was home to 11 small and medium-sized factories, ranging in size from Zenith Page's operation, with 19 workers, to the larger firm of Graves and Vernon, which had 48 hands. In total, county operations employed 352 workers (out of a total of only 466 people working in manufacturing) and produced $345,400 worth of finished tobacco.[96] In addition to Danville's factories, rural Pittsylvania County was home to a number of smaller operations;

between the city and the county, there were 39 facilities in 1860.[97] These country manufacturers included William Finney, who worked with leaf on his Museville plantation in the northwestern district of the county, and Samuel Swanson, who produced tobacco at a Swansonville facility.[98] Halifax was home to fewer tobacco factories, but Charles Clark ran a small operation, an unknown planter opened a twist factory at Oak Level in 1847, Philip Howerton ran a lump tobacco factory at Halifax Courthouse, and Beverly Barksdale III constructed a substantial brick factory at the crossroads of Brooklyn in the southwestern corner of the county in 1855.[99] Labor arrangements at these factories were just as complex as among Danville operations. Zenith Page owned only 11 slaves ten years old or older and must have leased his remaining 8 workers, and William Finney owned 32 slaves and leased 2 other hands, while leasing out 3 of his own workers to a fellow planter.[100] On an even smaller scale, it is certain that a number of tobacco farmers produced handmade tobacco products, such as cigars, for home use and for sale to neighbors and friends.[101] These small and medium-sized tobacco factories sold products similar to those of Sutherlin and Ferrell, turning out twists and lumps destined for regional and national markets.

In her study of labor arrangements in the antebellum Virginia tobacco industry, historian Suzanne Schnittman concludes that "tobacco manufacturers were more similar to their northern counterparts than they were to southern planters"; they were "employers first, masters second."[102] Although this may have been the case for the commonwealth's largest manufacturers, small operators, such as Finney and Page, were anything but clear-cut industrialists. Among his many other avocations, Finney worked as a slave trader, operated an iron forge, and planted bright tobacco.[103] Indeed, the owners of most rural factories seem to have been agricultural opportunists; they were planters who turned to tobacco manufacturing to supplement their farm incomes, to diversify their activities, or to more fully employ an expanding slave population.[104] The history of these rural factories, as Hahn notes, has a "blurry quality"; as these operations were not always consolidated under one roof, they relied on owned and leased labor, and their planter owners were often involved in a wide variety of economic activities.[105] In these cases, tobacco manufacturing developed as an outgrowth of preexisting tobacco cultivation, rather than vice versa, and it is thus possible to conclude that these individuals were, literally, planters first and factory owners second. Indeed, planters who opened factories invariably continued cultivating tobacco, and if they failed at tobacco manufacturing, they carried on planting. The proliferation of regional tobacco manufacturing establishments during the antebellum period seems, at least in part, a response to the lucrative nature of bright tobacco sales. Planters who realized the demand for bright lumps and twists, and who could organize enough labor, were determined to capture a portion of the profit made from their raw produce. Many of the rural and urban operations were fleeting affairs as planters realized factory operations were time-consuming, they experienced difficulties with labor, or the expected profits failed to materialize. Of the twenty-nine Danville-area manufacturing facilities listed in the 1850 census, only six were in existence ten years later. The survivors were joined, however, by thirty-three new firms, as enthusiastic planters and merchants continued to chase bright tobacco wealth.[106]

Opening a small tobacco factory did not require a substantial capital investment. The essentials for manufacturing tobacco included a labor force, typically slaves from the owner's plantation or hands leased from neighbors; a structure to house workers and tobacco, an empty barn or storage shed often sufficed; a few screw presses and molds to shape the final product (although a simple hole in a tree in which to place a prizing beam might do in a pinch); flavorings, with licorice the most common; and wooden boxes to store and ship the finished tobacco.[107] Although from the postwar period, an 1869 auction inventory of the defunct Stanfield, Hancock, and Featherston chewing tobacco factory in the community of Leasburg, in Caswell, illustrates the bare bones nature of these small regional manufactories. The factory building fittings included one hydraulic press, two screw presses, a set of scales, tobacco driers, levers for opening boxes, and a licorice boiler.[108] The equipment demands were modest enough that a number of large Southside tobacco planters could afford to enter tobacco manufacturing on a small scale, transforming their plantations into sites of value-added production. A small facility located on a plantation actually provided the planter/manufacturer with advantages, as he could control at least a portion of the raw material used in manufacturing from seed to the final packing of the finished product. Thus, in modern business terminology, small manufacturers such as Finney were "vertically integrated," an arrangement that firmly linked agricultural and industrial processes.

Along with patent medicines, tobacco products were one of the few common antebellum purchases that were consistently sold under a name brand, and local manufacturers large and small branded some of their products. Some local factory production went to county merchants, who retailed lumps, cigars, smoking tobacco, and twists in their stores, but the relatively small amount consumed in the three counties barely put a dent in regional production.[109] Manufacturers shipped the vast majority of their products outside of the Piedmont; Philip Howerton sold boxes of his manufactured tobacco to Baltimore and Petersburg agents, small producer John Hatchett marketed his brands in eastern Virginia, and Sutherlin and Ferrell sold tobacco to merchants in almost every major American city, from New York to New Orleans to San Francisco. Thus branding was an important mechanism linking tobacco products and the Southside in distant consumers' minds.[110] Manufacturers had to ensure that consumers pleased with the quality of their product connected that satisfaction with their region and returned to their brands when they next purchased tobacco. The typical plug of tobacco traveled from the manufacturer to an urban agent, then on to a wholesaler or jobber, who sold the plug to a store owner.[111] Branding helped manufacturers maintain a product identity, a provenance, through these multiple exchanges. Like winemakers intent on establishing territorial reputations, the Southside's manufacturers worked to link smooth bright tobacco products to the Piedmont hills surrounding Danville. And to a great degree, these brands expressed difference in produce that literally sprang from the soil (what we might be tempted to label *terroir* today).

Manufactured tobacco brand names ran the spectrum. A number of brands bore fanciful or catchy names; local examples included "Bette Walker," "Palmira 10," "A Bell Twist," "Azalia Twist," "Rail Road," and "Andersons Ice Cream."[112] "Negro Twist," a bright plug that Sutherlin and Ferrell marketed in the Fort Worth, Texas, area, testified to its plantation origin in its name.[113] Other labels bore the names of the manufacturers themselves, making tobacconists pseudo-celebrities. Sutherlin and Ferrell marketed "William Thomas Twist" and "JMS Star" (after William's brother and partner, John M. Sutherlin), two of the company's highest-quality yellow products, and other eponymous twists and bars—from "Reid" to "A. B. Watson"—abounded in the marketplace.[114] Best-quality bright leaf brands often bore names that emphasized their use of yellow or "fine" tobacco, their Southside origins, or that suggested their smooth taste. Among these illustrative brands were "Golden Pomegranate," "Peaches and Cream Twist," "Old Virginia Fancy Twist," "Gold Leaf," "Fine Star," and "Yellow Bar." (By the late 1850s, the terms "fancy" and "fine" had become all but synonymous with yellow color.)[115]

Both products named after local manufacturers and those touting the inclusion of bright leaf tobacco branded the Southside region along with a particular tobacco product. Like California fruit and vegetable growers in the latter part of the nineteenth century, who shipped their oranges and raisins in crates painted with scenes of the state's sunny valleys and stunning vistas, Southside tobacco manufacturers attempted through branding to convince consumers that the best tobacco came only from their region.[116] Although these early tobacco brands did not include pictures of growers' fields (as far as can be determined), their wording, to borrow from historian Douglas Sackman, "turned the land-scape . . . into a brand."[117] Whether they bought and chewed "Negro Twist" or "Old Virginia Fancy Twist," branding encouraged merchants and consumers in stores from Maysville, Kentucky, to Mobile, Alabama, to think about the origins of their tobacco.[118]

While branding associated certain tobacco products with Southside manufacturers and the Piedmont landscape, bright tobacco's most powerful advertisement lay in the leaf itself. Brands adorned the crates that manufacturers shipped to the merchants and occasionally served as a reference in transactions between store owner and customer, but the wrapper—the intact leaf wrapped around a finished lump or twist—on each individual plug acted as a visual representation of quality.[119] Manufacturers from Danville and other regional factories finished their products, from top-shelf plugs to middle-grade tobacco, with yellow wrappers. These yellow wrappers survived the pressing process without losing their color or texture, and the resulting golden plugs stood out amid the traditional dark twists and lumps that crowded merchants' shelves.[120] Even if consumers chose to discard the wrapper rather than chewing it along with the rest of the plug—as many did—it had served its purpose by attracting attention. As a combination of environmental and cultural expressions, yellow wrappers made Caswell, Pittsylvania, and Halifax products unique. Like the shiny peel of an orange or apple, golden yellow tobacco wrappers promised a flavorful plug, and as was the case with fruit aesthetics, the bright leaf was perhaps more important for its appearance than its taste. Strong twist and lump sales in the late 1850s indicated that consumers found these wrappers and the bright fillers that were often inside them appealing.

Merchants made the consumer preference for bright products clear in their correspondence with manufacturers. Tobacco agent Jason Dowell in Baltimore asked Finney to send him fancy plugs because "Bright tobs. are scarce."[121] Likewise, a Charleston retailer sent out a circular requesting Southside bright tobacco, declaring that "good bright medium Pounds, and fine bright Pounds . . . are much enquired after," and that "of Fancy Pounds and Twists, we have not a pound, and they are greatly needed, as enquiries are made for them daily, without the ability to furnish them. Shipments of these grades cannot be made too early."[122] A Cincinnati mercantile made a similar observation, writing that they could "sell a thousand boxs [sic]" of "choise [sic] bright" tobacco if a Danville factory could provide them.[123] An agent of the New Orleans company Patton, Smith, and Putman made clear the importance of bright wrappers in particular. He wrote that the company was interested in Sutherlin and Ferrell's entire product range but insisted that "the wrapper in every instance [be] very bright & lively." Patton, Smith, and Putman were convinced that yellow tobacco served well as its own advertisement and suggested that consumers as far removed from the Southside as New Orleans were conscious of the reputation of bright leaf tobacco.[124] The Charleston, Cincinnati, and New Orleans markets were all located near tobacco production centers. Charleston was closer to the fire-cured tobacco belts of Georgia and eastern North Carolina than to the three counties; Cincinnati lay near the substantial Kentucky dark tobacco belt, which was rapidly expanding, as well as the new Burley district; and Louisiana growers cultivated their own distinctive perique tobacco, and New Orleans was a natural entrepôt for western tobacco, but bright leaf manufacturers were beginning to make inroads in these antebellum markets through branding and their emphasis on a mild, smooth-tasting chewing tobacco.[125]

It was not only Southside manufacturers who used bright leaf in their manufactured products; Richmond and Petersburg factories competed for the three counties' yellow tobacco as well. Halifax growers in particular shipped a large portion of their best tobacco east. Among the buyers was James Thomas, Richmond's largest manufacturer, who sought bright tobacco as wrappers for his plugs. Like his Southside counterparts, Thomas drew on bright leaf's yellow color as a particularly potent advertising device for his manufactured tobacco. Wrapped in "gold" leaves sourced along the Virginia and North Carolina line, Thomas's products dominated new western markets in the goldfields of California and Colorado during the late 1850s. Thomas's nephew also created a new plug brand, "Lucky Strike," that connected the discovery of golden leaf in Piedmont fields to prospectors' desires to strike it rich in goldfields of a different sort. The "Lucky Strike" plug line morphed into pipe tobacco and eventually the famous cigarette brand following the war, but the brand name originated in the same chewing tobacco manufacturing and marketing common among antebellum Danville and Richmond factories.[126]

As regional and national demand increased, manufacturers pressured growers in the three counties to concentrate on cultivating top-quality bright tobacco. Buyers from the larger factories roamed the sandy ridges above the Dan, Hyco, and Bannister rivers searching for bright crops and making offers to farmers to purchase their tobacco as it hung in their barns. As enterprising growers heard of growing demand, they also wrote to prominent manufacturers touting their crops, promising leaf that could "compete with Slades best," "very fine" tobacco, and "yellow cured with coal."[127] Manufacturers' demand for yellow tobacco provided growers with strong profits and also gave them some leverage in selling their traditional dark tobacco. Farmers sometimes insisted that buyers take their darker leaf along with their bright tobacco, as when Halifax's Thomas Barksdale refused to sell his yellow leaf to Sutherlin and Ferrell unless the company also bought his entire crop of dark tobacco at the extortionate rate of $10 per hundredweight, and John Garland demanded $16 per hundredweight for his entire crop, although only 4,000 pounds were yellow tobacco.[128]

Manufacturers sought bright tobacco just as avidly as planters sought high sale prices. Factory owners worked hard to build relationships with regional bright leaf growers, often visiting them on their farms to discuss business. Manufacturers such as Sutherlin made arrangements to purchase farmers' loose tobacco on site and paid for hauling the leaf to the factory, deals that eliminated the planters' need to buy or make hogsheads and hire wagoners or rollers. As one Caswell grower wrote to another, sale to local manufacturers saved "a good deal of time & expense" and thus allowed him "to make larger crops." Planters who had favorable arrangements with factory owners often recruited their bright tobacco–producing neighbors to sell to the same manufacturer, the end result being networks of bright leaf growers tied (loosely) to particular factories.[129] Manufacturers also placed instructions on raising, curing, and packing bright tobacco in local papers. An 1857 article in the *Milton Chronicle* advised growers to take care with their yellow tobacco cultivation and recommended that they pack fine wrapper quality leaf in small boxes rather than in hogsheads to preserve quality. As further enticement, the article promised potential sales for such wrappers from $40 to $100 per hundredweight.[130]

By 1860, with the stimulus of strong bright leaf sales, tobacco was more important than ever to Southside farmers and planters. Federal census takers visited the region's farms that year, and they recorded that more than 90 percent (2,651) of the region's 2,936 farms grew at least some tobacco. This tobacco dominance is even more impressive considering that the census counted a number of newly settled farms that had not yet produced a season's crops, tracts left idle while in the legal limbo of estate management, the farms of the elderly and infirm, and town smallholders who owned only a few head of livestock and a garden.[131] All things considered, it seems safe to conclude that almost every capable Southside farmer produced tobacco and, though the census did not differentiate between types of tobacco, that that crop was increasingly the more lucrative bright leaf tobacco.

The wide-scale adoption of bright leaf as the region's staple crop changed planters' conceptions of agriculture and environment. The crop's requirements led to an increasingly regimented agriculture, one in which growers labored on carefully selected plots of land, managed seed selection over multiple generations, and followed elaborate curing guides. All of these efforts worked to transform tacit knowledge into formal knowledge, with varying success. For bright leaf growers, tobacco was no longer a flexible and ubiquitous cash crop; rather, the new leaf was an intensive staple that demanded expert, professional farming. This crop culture also brought agricultural production and older ideas of land stewardship into increasing conflict. Planters understood that the sandy ridge land that made the best tobacco was ill suited to sustained row crop agriculture; it was steep, erosive, and low in nutrients. In general these ridges were the most stable in pastures, forests, and mixed farming, a land use mosaic that most regional landowners had adopted prior to the emergence of bright leaf. Bright leaf made a certain economic sense—these lands became more valuable in tobacco cultivation than they had been under older practices—but farmers also worried that their practices threatened other aspects of farming. As bright tobacco farmers increasingly brought these soils under the plow, they embraced a system that treated portions of the landscape as mediums of tobacco production rather than as integrated portions of larger farms.

This agriculture also united tobacco growing and local manufacturing. Tobacco had always been a commodity bound for distant markets, tied to transatlantic trade and global demand, but local bright leaf manufacturing further connected field and factory. Farmers and planters, who were themselves sometimes manufacturers, came to see their tobacco as an industrial resource, a plant raised and packed to suit Danville and Milton manufacturers and particular consumer tastes. As a consequence, manufacturers worked to associate bright yellow tobacco with the southern Piedmont landscape, an advertising connection that linked soil and leaf for merchants and consumers from New York to San Francisco. The success of these campaigns increased demand for bright leaf and hastened growers' exploitation of the border's sandy soil.

The region's farmers and planters were bright leaf crazy as the Civil War approached. They turned land and labor toward the crop, traded tips and techniques with neighbors and through newspapers and agricultural journals, and they increased production whenever practical. The October 21, 1859, issue of the *Virginia Echo*, a Halifax County newspaper, illustrated the prominence given yellow tobacco along the state border. In that issue, an article relating the high prices obtained by county planter James Moody for a hogshead of tobacco on the Richmond market actually appeared before a description of a paramilitary attack on the federal armory at Harpers Ferry, Virginia.[132] In Halifax, for one day at least, it seems that even John Brown and the clouds of war took a backseat to bright leaf.

THREE

Bright Leaf, Bright Prospects

MAKING PEACE WITH THE IDEA OF YELLOW TOBACCO

AMERICAN AGRICULTURE FACED a crisis of confidence in 1840, just as a few Virginia and North Carolina farmers began to experiment with making brighter tobacco. A severe economic panic initiated a lasting depression in 1837, and over the following six years farm prices plunged to historic lows while credit dried up. One New York observer of the collapse of the city's financial institutions that had relied on commodity trading and agricultural loans declared, "At no period of [New York's] history has there been as great a degree of general distress as there is at this day." Virginia and North Carolina tobacco farmers did not escape these economic miseries. The general depression that lasted through 1843 spurred interest among American farmers in programs of agricultural reform and intensification, and "improvement" movements gained traction in communities from New England to the Deep South. This reform impulse was particularly vibrant in the older districts of the Atlantic Seaboard that faced competition from fresh western lands as well as general economic troubles; from Vermont to Georgia, reformers worried about farm survival. Agricultural anxieties led to widespread but amorphous ideas of betterment, debates about the proper use of labor, and programs of renewal and modernization, self-sufficiency and hope.[1] Against this national backdrop of fear tinged with promise, bright leaf tobacco culture expanded during the antebellum period,

gradually supplanting dark tobacco on Piedmont farms and plantations.

This southern agricultural reform impulse, along with favor for traditional tobacco cultivation practices and a general rural inertia, had the
potential to hinder the transition from one form of tobacco to another.
In fact, bright leaf culture developed at a time when many regional agriculturalists sharply criticized the methods of southern planting and
farmers' overreliance on tobacco and other staples. By the 1840s, Caswell, Halifax, and Pittsylvania counties had an active contingent of
modernizing planters and farmers intent on spreading the message of
agricultural reform across the Southside. As in other sections of the
country, southern agricultural reform was a widespread yet rather amorphous movement intent on making southern agriculture more profitable
and sustainable; in Virginia, the reform movement originated among
large planters in the Tidewater region in the early nineteenth century
and slowly spread into the Piedmont, gaining steam as farm prospects
dimmed following 1837. Along the Virginia–North Carolina border, the
message of agricultural reform initially challenged the development of
bright tobacco, but in the end reform advice would encourage the new
crop's diffusion.

Contradictions between the tenets of agricultural reform and bright
tobacco culture pose an obvious question: Why would farm communities concerned with bettering their husbandry turn to a crop that so
heavily taxed both environment and farmscape? Was the effort to make
bright leaf a rejection of reform, a perversion of reformers' messages, or
was something else at play? In some rather obvious ways the growth of
bright tobacco culture flew in the face of the reform movement: the soil
and ripening demands of making bright tobacco encouraged farmers to
plant on poor, steep land, to forgo crop rotation, and to devote their time
and energy to curing leaf and clearing land rather than on fencing and
manuring. In other, sometimes subtle, ways however, the crop satisfied
modernizers' critiques of dark tobacco culture. The new crop brought
land that was formerly all but worthless into profitable cultivation; raising and curing yellow tobacco demanded that farmers understand regional soils and the elaborate art of charcoal or flue curing, a focus on
expertise and artisanal knowledge that appealed to reformers; and most
important, the premium prices bright leaf commanded made marginal

farms suddenly profitable. This profitability allowed many people to stay put on their farms rather than migrating to the Old Southwest's more fertile lands. As if by miracle, the new tobacco variety made many small, poor farms pay. Following the Civil War, bright tobacco culture would contribute to many of the conditions that antebellum reformers feared— from catastrophic soil erosion to a decline in farm diversity and profits— but the reformers' message actually stimulated the prewar growth of the crop.

Virginians' antebellum interest in intensified agriculture was part of a larger southern trend most evident in the Upper South, and though spurred on by the Panic of 1837, the movement had deep roots in regional thinking. Southern agricultural reformers drew much of their inspiration from the writings of such eminent British agriculturalists as Jethro Tull and Charles Townshend, but Virginia had more than its share of agricultural experimenters, including philosophe Thomas Jefferson and the more practical George Washington. And the voice of the Piedmont, Patrick Henry, reportedly declared, "He is the greatest patriot who stops the most gullies."[2] Indeed, an important part of the role of cultured planter-gentleman favored by so many early commonwealth statesmen was an intellectual interest in all things agrarian. Perhaps the first of a new breed of "scientific" agricultural reformers was John Taylor of Caroline County. A lawyer, Revolutionary War officer, nephew of Edmund Pendleton, member of the Virginia House of Delegates, and a United States senator by the age of thirty-eight, Taylor was also one of the first systematic American agronomists. His 1813 agrarian manual, *Arator*, called for manuring with vegetable matter, excluding animals from croplands, and producing feed specifically for penned livestock. Claiming that "our country is nearly ruined" by shoddy farming, Taylor envisioned a land made wealthy and fat if farmers would only deign to follow his reforms. Although his agricultural advice proved influential among wealthy Virginia planters, most of Taylor's assumptions regarding soil fertility were technically incorrect. Some of Taylor's methods did benefit soil fertility, though he rarely understood the science behind the results. For example, he asserted that vegetable matter alone was capable of restoring fertility if left on cropland. *Arator* called for farmers to raise more clover, but at the time no one understood that as a nitrogen-fixing legume, clover utilizes bacteria on its roots to transform atmospheric nitrogen into a usable form, in turn assisting other plants to draw more of the element from the ground.[3]

Following in Taylor's footsteps, Tidewater agronomist and planter Edmund Ruffin proved even more influential. Like Taylor, Ruffin believed the salvation of southern agriculture lay in intensified farming. He understood that much of the Atlantic Seaboard's arable land suffered from some degree of soil exhaustion brought on by careless or exploitative agricultural practices undertaken in challenging local environments, and he was particularly concerned with the sustainability of southern agriculture and the compatibility of modern farming methods with the institution of slavery. Ruffin believed that a Malthusian crisis faced the region as old farming methods and a focus on nonedible farm crops, such as cotton and tobacco, prevented food production from keeping pace with rapid population growth. This situation could only result in one of two outcomes: general poverty or out-migration to the fresher lands of Alabama, Mississippi, and Texas. As a solution to the ills of traditional agriculture, Ruffin touted a series of reforms that included crop rotation, a movement away from a general reliance on tobacco and cotton, manuring, the extensive use of cover and leguminous crops, an end to the common range, and the use of marl—the fossilized remains of marine life—to lower soil acidity.[4] The point of all of these prescriptions, according to Ruffin, was not "to draw from the land the greatest *immediate* production and profit," but rather to ensure "the greatest *continued* products and profits" from the southern landscape.[5] He espoused his views in an influential agricultural journal, *Farmers' Register,* which he published on his Prince George County plantation beginning in 1833, and in a widely read treatise, *An Essay on Calcareous Manures* (1832).[6]

Agricultural reformers tended to have conflicting feelings about tobacco as a crop. On one hand, as one of the few consistently profitable crops in the Piedmont, tobacco provided a predictable return that funded reform efforts on many farms. On the other hand, tobacco represented the staple-crop system that undermined reformers' calls for diversity. In other words, tobacco itself was not inherently a "bad" crop, but Ruffin and his cohorts believed that the mentality that invariably accompanied tobacco farming prevented agricultural growth and sustainability (though they would not have used the term). Ruffin labeled the plant part of the "exhausting culture" that depleted the South and forced the region's children west, and he called for eastern farmers to reject tobacco and cotton monocultures in favor of more diversified and sustainable systems, and reformer John Hartwell Cocke declared the crop "the Bane of Virginia Husbandry."[7] Regional modernizers, such as Halifax reformer James Bruce, echoed Ruffin's and Cocke's advice. In an 1847 address to his fellow agriculturalists, Bruce urged Southside planters and farmers to sell their tobacco and slaves and invest their profits in intensive farming and local industry.[8] Both Ruffin and Bruce understood the appeal of short-term tobacco profits—Bruce himself remained a substantial tobacco planter and slaveholder through the Civil War—but they feared that the staple crop system was unsustainable.

Like Taylor and Ruffin, Southside landowners feared population out-migration. A large agricultural population, declining soil fertility, the accretion of the best Piedmont lands in the hands of well-off planters, the enticement of a gold rush in Georgia, and the availability of new land in the Old Southwest combined to push or pull small farmers and the landless westward. White populations (the bulk of regional residents who had a choice concerning their mobility) in Caswell, Halifax, and Pittsylvania actually declined between 1830 and 1840 and increased only marginally over the next two decades.[9] Departing planters listed their lands for sale in the local papers and headed west, despite claiming that their abandoned grounds were "as fine TOBACCO LAND as any in the country."[10] An 1841 article in the Milton Chronicle, entitled "Going to Texas," encapsulated local unease over out-migration. In describing the inherent hazards of moving to the Old Southwest the piece displayed the Piedmont paranoia over population loss. The anonymous author cautioned would-be emigrants that poverty, sickness, and even death awaited those bound for Texas, where a harsh climate and Indians challenged newcomers—according to the article, the logical decision was to remain in the Southside, where poverty at least came accompanied by a modicum of safety.[11]

The letters and private diaries of regional planters expressed similar concerns in more visceral fashion. In the mid-1840s, George Jeffreys of Caswell's Hyco Creek district lamented that a neighbor, William Bethell, was planning to move, a decision made even more troubling because Bethell had recently married Jeffreys's daughter. Jeffreys wrote in his diary that he "had many painful & gloomy reflections about Mr. Bethells buying land in Louisiana—My Lord has it come to this[?]"[12] Likewise, George Clement's son Charles abandoned his Pittsylvania home for the goldfields of Sacramento in 1850, where he wrote his worried parents about the opportunities for young, single men on the West Coast.[13] And in 1859, Pittsylvania planter Peter Hairston wrote his brother George, who was thinking of relocating to the Texas cotton belt. Peter criticized George's plan, pointing out the advantages of his present location in a sarcastic counterfactual passage: "I think your desire to go to Texas perfectly natural and rational. Your large and expensive family and total want of means to provide for them under a new country desirable; where they can have plenty of elbow room—As to some 40 or $50,000 worth of Bank Stock, one hundred negroes and several thousand acres of land—you know they are nothing—There in Texas every man you meet would be your friend doing every thing they could to promote your interest and you would have every comfort provided for you without any trouble on your part, in all of which respects you know how badly off you are where you now are. And then when people go to a new country, they are always so well satisfied and never cast any longing looks back to the country they left."[14] Jeffreys's, Hairston's, and Clement's worries reflected broad concerns about the future of Southside economy and culture. These fears made the profitability and sustainability of regional agriculture a vital issue; planters hoped intensified agriculture would make Piedmont farms as appealing and valuable as fresh western lands and keep Virginia's and North Carolina's sons and daughters at home.

The campaigns of Ruffin and like-minded reformers met with some success during the antebellum era, but as the poverty and erosion of the postwar tenant South so vividly displayed, reformers' efforts ultimately failed. Even before the outbreak of the war, reform efforts were largely limited to particular southern districts, most of which were located in the Upper South.[15]

Historians of southern agriculture and the environment have bandied around a central question concerning this southern improvement movement and its demise for quite a few decades now. Why was the agricultural improvement movement, personified by John Taylor and later Edmund Ruffin, unable to rescue southern agriculture from its worst ills? Or, to phrase the question in a different manner, why were

Depression-era government officials, such as the Soil Conservation Service's Hugh Hammond Bennett, forced to ask similar questions and produce similar criticisms of southern agriculture during the 1930s as Edmund Ruffin had in the 1830s?[16]

There have been a number of explanations of the arc of southern agricultural reform, the first coming well before Bennett began his critical examination of the South. Avery Craven, in *Soil Exhaustion as a Factor in the Agricultural History of Virginia and Maryland* (1926), declared the agricultural reformers relatively successful in Virginia and Maryland and argued that the Civil War was the crisis that undermined their makeover of southern agriculture.[17] A decade later, Arthur Hall challenged Craven's claims that Virginia reform efforts were widespread, arguing that a general failure of agricultural organization prevented wholesale adoption of erosion control methods and crop rotation.[18] Long after Craven and Hall, Carville Earle definitively refuted Craven's notion that the advice of reformers was significantly more ecologically sound than the traditional long-fallow agriculture that characterized southern farming, but he avoided addressing Craven's assertion that agricultural reform was widely adopted by Upper South farmers.[19] The same year (1988), William Mathew's *Edmund Ruffin and the Crisis of Slavery in the Old South* echoed Hall, claiming that most small and middling farmers, and probably the majority of large planters as well, never encountered or put into practice the tenets of improved farming.[20] Mathew contended that very practical considerations, such as high costs, the limited availability of amendments, and shortages of labor, made Ruffin's plans mere pipe dreams. *Larding the Lean Earth* (2002), by Steven Stoll, also pointed out several of the inherent weaknesses in the programs of southern reformers. The very core crops of English sustainable husbandry—leguminous clovers and pasture grasses—grew poorly in the hot weather and acidic soils that prevailed throughout much of the South. Until the advent of cheap lime and new crop varieties in the twentieth century, Ruffin's call to cover the earth in clover could have had little impact on the southern Piedmont, much less the plains of south Georgia or the sand hills of eastern South Carolina.[21] In the most recent examination of this debate over agricultural reform, Lynn Nelson (*Pharsalia*, 2007) has argued that the programs of agricultural reformers could work in central Virginia. What these plans could not do, however, was maintain southern planters in the style to which they had grown accustomed (or the style of their neighbors, whom they envied).[22] To borrow from the agrarian Andrew Nelson Lytle, a farm geared around sustainable husbandry was "not a place to grow rich; it [was] a place to grow corn."[23] Yet most Southside farmers were not content to grow only corn . . .

So what then was the fate of the agricultural reform effort in the Southside? The reform message certainly circulated widely in the region: Southside planters and farmers who looked to print sources for advice were bombarded with the messages of agricultural reform. Agricultural journals were among the most popular sources of farm advice during the antebellum period. These journals combined editorials, how-to articles, and political guides, along with comments from subscribers, and local newspapers often reprinted journal articles alongside their own essays on modern farming. Among the journals to which the region's planters and farmers subscribed were Ruffin's *Farmers' Register,* the *Southern Planter* out of Richmond, and Baltimore's *American Farmer.* Twenty-four residents of Caswell, Halifax, and Pittsylvania were among the initial subscribers to the inaugural volume of the *Farmers' Register* in 1833; influential agriculturalists including Robert Wilson, William Hatchett, and Abisha Slade took the *Southern Planter;* and in 1849 the editor of the *American Farmer* noted that a large number of Pittsylvania farmers were among the recent subscribers.[24] These agricultural journals not only counted Piedmont residents among their subscribers, they regularly carried articles about the three counties and pieces written by Southside planters and farmers, indicating that local consumption of reform advice was not merely passive.[25] These articles touted the various reform campaigns, from the use of marl and manure to crop rotation and the need for a fence law to close the common range, and a high percentage offered either tips on tobacco cultivation or recommendations to avoid its culture altogether.

A number of local farmers, especially among the more prominent planters, also belonged to agricultural societies that emphasized modern farming methods. Although these societies ran the gamut from informal meetings in individual homes to elite organizations with prominent citizens as elected officers, most subscribed to agricultural journals for the benefit of their members and met on a regular basis to discuss modern farming techniques, alternative crops, agricultural labor, and farm economics. County and local organizations that advocated modern farming methods included the Caswell Agricultural Society and the Clover Agricultural Association, and some prominent planters belonged to larger regional or state organizations, such as the Virginia State Agricultural Society (VSAS).[26] Regional reformers not only participated in these larger agricultural societies, in several cases they headed them. James Bruce became president of the Union Agricultural Society of Virginia and North Carolina in 1854, and John Edmunds of Halifax, along with Edmund Ruffin and future governor of Virginia Henry Wise, was a founding member of the VSAS in 1851. In 1859 the society elected Edmunds as its president.[27]

Edmunds's 1853 address to the VSAS, reprinted in the *Southern Planter*, provides an illuminating window into the intellectual connections of the Southside reform movement. His speech addressed the soil resources of the commonwealth and argued that only a systematized approach to agricultural education would "let [southerners] develop those resources that the physical and material greatness of the Old Dominion shall be commensurate with her ancient renown and her historic glory."[28] Edmunds's speech drew on an astonishing body of contemporary American and European agronomic literature. He cited regional authorities, such as Ruffin (*An Essay on Calcareous Manures*) and professors from the University of Virginia, but he also drew on a wide variety of texts from Scotland, Germany, France, England, and the American Northeast, including John Norton's *Elements of Scientific Agriculture*, Mary Somerville's *Physical Geography*, Justus von Liebig's *Organic Chemistry*, Cuthbert Johnson's *Farmer's and Planter's Encyclopaedia*, J. B. Boussingault's *Rural Economy*, Henry Coleman's *European Agriculture*, David Low's *Elements of Practical Agriculture*, and various articles from the *Transactions of the Highland and Agricultural Society of Scotland* and the *Edinburgh Quarterly Journal of Agriculture*.[29] Although Edmunds's erudition was no doubt exceptional, his access to and use of such wide-ranging sources illustrates the connections of agricultural reform in the Southside to the broader movement in the South and beyond the region. Edmunds and his fellow reformers were no simple parochial squires; they were concerned with conditions in Caswell, Halifax, and Pittsylvania, but at least a few of them connected those worries to agricultural literature and movements across the state and the globe.

Regional modernizers such as Bruce and Edmunds who advocated agricultural reform in the southern Piedmont also traveled to adjoining counties to address reform issues. In 1847 Bruce gave a keynote speech to a gathering of the Mecklenberg (Virginia) and Granville (North Carolina) agricultural clubs, in which he advocated intensive farming and industrial development. Tackling a primary concern of many reformers, Bruce pointed to slavery as an institution that weakened modern farming efforts, though he placed the blame on slaves themselves rather than on the economic calculations of masters. Bruce declared, "The truth is, so far, no moral incentive has yet been found strong enough to stimulate the negro to energy or industry, and we are almost persuaded to believe that he is the genuine son of Canaan." Bruce concluded his address with a call for club members to heed the message of agricultural reformers as an avenue to escape the evils of slavery, to save slaves from themselves.[30] These concerns over the place of slavery would continue to plague the reform movement throughout the antebellum period, and they contributed to the popularity of bright tobacco, which did not challenge the institution in the same way that other agricultural alternatives did. By 1854 Bruce had risen to the office of president of the Union Agricultural Society, a club composed of prominent Virginia and North Carolina planters. In an 1854 address, he reiterated his belief in the importance of agriculture that drew on scientific studies of soils, weather, and amendments and particularly praised the growing use of guano to rehabilitate old fields. Bruce warned his audience that the particularities of the southern environment, from hot summers to high soil erodibility, would make northeastern versions of sustainable husbandry difficult, but he admonished listeners to do their best to bring diversified farming to the southern Piedmont.[31]

Some essential questions remain. Did regional farmers follow Bruce's advice? If so, to what extent did planters and farmers implement these calls for modern, intensive farming? Just what sort of farmer found reform advice appealing? And what did reform mean for tobacco fields? Firm answers to these questions are difficult if not impossible to obtain, but there is a good deal of circumstantial evidence that landowners across Caswell, Halifax, and Pittsylvania put some reform ideas into widespread practice during the antebellum period. And although most reform-minded landowners seem to have been wealthy planters, a few were middling or even small farmers. These agriculturalists implemented contour and deep plowing, ditched and terraced hillsides to slow erosion, rotated crops, worked soil amendments into their fields, experimented with new and improved breeds of animals and plants and new types of agricultural machinery, and worried over the best form of labor on a modern farm. All told, the agricultural reform impulse in the antebellum Southside seems stronger and more widespread than historians have argued was the case in the South in general. Many reform practices were expensive, and thus wealthier planters were positioned to practice improved agriculture to a greater degree than their poorer neighbors, but reform advice and practice saturated the region. The Southside farmer who had not heard reform advice and contemplated at least some of it must have been isolated indeed. That these efforts came to little over the long term was due more to the persuasive lure of making bright tobacco, and its particular demands on soil and people, than to a more general failure to attempt a reform of Southside agriculture.

Some evidence for the implementation of agricultural reformers' advice can be found in regional plantation records and recorded cultivation practices. The most basic reformer advice dealt with how to turn and move earth, touting deep and contour plowing and the use of modern tillage equipment—calling on farmers to better the most basic of agricultural tasks, putting plowshare to soil. One of the classic stereotypes of prewar southern agriculture is its reliance on shovel plows for soil preparation and cultivation.[32] If the Southside was typical, historians may want to rethink this characterization. For at least a few decades prior to the Civil War, shovel plows were only one of many groundbreaking tools employed on regional farms.[33] Shovel plows were shallow, light instruments that skimmed the surface of the ground, turning over only a few inches of topsoil. They were simple, cheap, and easy to repair. Frequent shovel-plowing pulverized topsoil and created a hardpan several inches below the surface; both conditions promoted soil erosion during hard rains as the loose topsoil liquefied and sloughed off the lower strata of packed earth. Despite this characterization of southern tillage practices, evidence from the Southside—ranging from farm equipment inventories to blacksmith records—suggests that many antebellum dark tobacco growers used a wide variety of plows. Although these records are biased toward large plantations where sources survived, they are supported by scattered accounts from smaller farms. After the early 1800s, almost every farm with surviving records contained medium-depth dagon (or Cary) plows, coulter plows designed to cut through roots and move soil to the base of established crops, and subsoil plows that broke up lower soil horizons without turning over topsoil, in addition to the classic shovel plows.[34] The large subsoil (or "iron beam") plows in particular were heavy and expensive pieces of machinery designed to minimize soil erosion. Their narrow shank cut through hard-packed ground to loosen the soil up to two feet in depth, an aerating action that promoted deep drainage and good tilth.[35] A comprehensive 1852 inventory of Ruth Hairston's numerous plantations in the Piedmont of Virginia and North Carolina reveal the abundance and diversity of plows on some properties. Between her twelve tobacco plantations, Hairston owned an astonishing 282 plows of all shapes and sizes.[36] Whether forged on the plantation by slave blacksmiths or purchased from regional manufacturers, these plows (along with the ubiquitous hoes, shovels, and mattocks) served as the primary tools of interaction between laborers and land. Shovel plows were cheap and common, but a substantial portion of landowners felt the more expensive, deeper plows were a worthwhile expenditure.

Area planters put these plows to a number of uses advocated by agricultural reformers. Such planters as William Sims, William Grasty, and Charles Coleman cut ditches across the slopes of their fields to slow erosion and promote even drainage and placed brush and stumps in burgeoning gullies, and others used their coulter plows to cut contour furrows around hillsides rather than plowing straight up and down slopes.[37] Beginning in the 1840s, landowners also built terraces and dug deep contour ditches to control erosion, though historian E. M. Rowalt has argued that these earthworks were often too shallow or too narrow to function effectively.[38] Planter James Bruce's terracing system was perhaps the most elaborate example of these early efforts. Bruce had his slaves dig substantial ditches throughout his fields along the Dan River in southern Halifax. These ditches had wide embankments along their downhill side to direct runoff around the slope and into natural ravines or depressions, and Bruce left the earthworks uncultivated so that grasses and shrubs could stabilize the soil. Although his terracing and ditching measures were extreme in their expense and labor demands, Bruce's efforts reflected a widespread ideology of soil conservation through drainage construction.[39] These efforts were all too necessary in the southern Piedmont, where rainfall rates, slope, and soil types combined to make regularly cultivated fields as much as five hundred times more erosive than land covered by original forest.[40]

This emphasis on soil conservation and erosion control was also present in period rental agreements and labor contracts. Landowners often stipulated that farm renters pay close attention to their cultivation to ensure that precious topsoil remained in tobacco and corn fields rather than washing into streams and gullies. When Wyatt Wallace agreed to rent a Pittsylvania farm from George Clement in 1845, he promised to keep up the property's buildings and fences, to cut only the timber necessary for farm upkeep, and "to cultivate the land well and in the best possible manner so as to keep it from washing."[41] William Terry's 1841 contract to serve as worker and overseer on Elizabeth Spraggins's Halifax farm contained similar language. Terry swore to supervise Spraggins's three slaves and "to Cultivate the Land to the best advantage to prevent washing it."[42] William Armistead of Caswell imposed exceptionally detailed restrictions on H. C. and Parham Moon in an 1846 rental agreement. Armistead required that the Moons keep the tract's fields in a three-year rotation (tobacco, wheat, wheat); he forbade the sale of hay (which would remove soil-building grasses from the land), wood cutting, and free-ranging stock; and he demanded that the crops "be well cultivated, the hill sides well trenched, the low grounds . . . be well drained."[43] Although we cannot be certain that these strictures were always observed by tenants, these and similar contracts indicate that concern for soil conservation was more than a passing preoccupation for the region's planters and farmers; at least in a few instances these issues were matters of such importance that landowners insisted on their codification through legal agreements.

A number of farmers also employed regular crop rotations to prolong soil fertility and reduce the soil pathogen load. Tobacco historian Joseph Robert has written that the typical dark tobacco farmer during

the first half of the nineteenth century planted tobacco in a wheat, clover, corn, wheat, clover, tobacco rotation, thus raising leaf only one year
out of every six on a particular piece of ground.[44] Though this may have
been the ideal, there is little indication that Southside farmers followed
such an elaborate rotation schedule, but most dark tobacco farmers did
employ some sort of regular field rotation, rarely keeping tobacco on the
same field more than two years in a row, except on the richest bottomland. Corn and wheat were the most common crops rotated with tobacco,
but Southside rotations also included oats, rye, flax, cotton, sorghum,
grass hay, sweet potatoes, and cowpeas. John Edmunds's crop rotation
provides one example from Halifax. In the late 1830s, Edmunds followed
a four-year cycle on his tobacco land, following tobacco with wheat and
then two years in clover.[45] Caswell's George Jeffreys derived his rotation
of oats and tobacco from Taylor's *Arator*.[46] The region's reform-minded
farmers were particularly enthusiastic about the inclusion of clover and
redtop (another nitrogen-fixer) in their rotations. These cover crops provided high-quality hay for horses, cattle, and sheep (which in turn produced more manure), and the crops' nitrogen-fixing ability increased a
field's fertility. Clover in particular was quite an investment for all but
the largest planters; a bushel of seed could run as high as nine dollars
during the 1850s. But despite the high cost, area landowners such as
Bruce, Sims, Coleman, Nathaniel Ragsdale, and Rufus Owen all regularly planted clover on their tobacco plantations.[47] Southside farmers'
enthusiasm for clover—which proved difficult to grow on the region's
acidic soils without the addition of expensive lime—illustrated the lengths
they would go to in pursuit of agricultural improvement.

Some planters looked beyond the boundaries of their property to
maintain soil fertility. One of the most popular purchased amendments
was guano—the dried droppings of seabirds from the arid archipela-
goes of the Pacific Ocean, where the extremely dry climate preserved the
waste almost indefinitely. The Chincha Islands off the Peruvian coast
had guano deposits 200 feet deep, veritable mountains of the phospho-
rus- and nitrogen-rich droppings, dried to the consistency of chalk. American farmers experimented with using guano on their gardens and
fields as early as 1824, but it was not until the 1840s that the amendment
gained widespread popularity. Among the first regions to adopt guano
was the old tobacco belt of the Virginia and Maryland Chesapeake, where

farmers used it to rehabilitate their abandoned tobacco fields, and the fertilizer quickly spread west into the Piedmont. Guano was expensive, but it gave worn out croplands an immediate boost of fertility, a benefit for which some wealthier planters were willing to pay.[48] By the 1850s the Richmond and Danville Railroad had standardized shipping rates for "artificial manure" among its list of twenty-six fixed-rate goods, and regional newspapers ran regular guano advertisements from local merchants.[49] Sales receipts indicate that planters applied guano only to limited portions of their land, as purchases were usually modest; regional planters and farmers often bought a ton or less of the amendment at a time.[50] These small quantities suggest that planters either applied guano to a single field or portion of a field, or that they combined the purchased amendment with other forms of soil improvement. Nevertheless, the widespread use of at least limited quantities of guano reflected farmers' knowledge of technical "fixes" for declining fertility touted by some reformers.

A more common soil amendment in the three counties was calcium carbonate in one of two forms: lime (calcium oxide) or plaster (calcium hydroxide). Whether derived from pulverized limestone or from the plentiful marine shell resources of the Chesapeake Bay, lime and plaster lowered soil acidity and permitted farmers to successfully grow crops such as clover, which demanded relatively neutral ground. Plaster and lime were cheaper than guano, but on the highly acidic lands that characterized the southern Piedmont, farmers had to apply heavy doses to achieve satisfactory results. Mathew claims that there was very little use of lime, marl, or plaster in the antebellum region and points to a supposed lack of these substances as evidence that agricultural reform was not practiced to any significant extent. At least in the Southside, Mathew's claims appear incorrect; the amendments seem to have been fairly common in Caswell, Halifax, and Pittsylvania in the decades leading up to the Civil War.[51] James Bruce operated a plaster manufactory in Halifax and also imported plaster from the east for resale. In a two-month stretch of 1840 alone, Bruce purchased twenty-six tons of plaster from eastern Virginia.[52] Agriculturalists considered plaster, along with guano and clover seed, so essential to the success of Southside agriculture that it passed duty-free through the toll sluices and canals of the Staunton and Dan rivers during the 1850s.[53] Farmers intent on implementing agricultural reform needed these amendments to combat their acidic soils, and they were willing to pay for them.

Farmers without the funds to purchase outside amendments, or those who wanted to supplement their applications of guano and plaster, turned to materials present on their farms and in the surrounding forests to enrich their tobacco and grain fields. Most prominent among these fertilizers was barnyard manure. Livestock waste from winter pens, barnyards, and stalls had long been the lifeblood of English agriculture and the convertible husbandry of small northeastern farms, in which farmers penned and fed their stock throughout the winter to hoard waste, supervised their animals in the grazing of fallow ground following harvest, and kept livestock in confined "folds" overnight, which concentrated manure on specific portions of the landscape. But as Steven Stoll has pointed out, this intensive recycling of manure was largely absent from southern farms large and small. In the South the practice of free-ranging stock, the poor performance due to climactic conditions of the high-quality forage grasses needed to sustain penned stock, a general reluctance among southern planters to invest in convertible husbandry, and the relatively low cost of land and high cost of labor all served to hamper the accumulation and systematic application of manure.[54] In addition, heavy doses of barnyard manure produced poor-quality tobacco—the leaves grew dark and heavy, were slow to ripen, and tasted harsh.[55] Despite these limitations, Southside farmers regularly used manure in some form or fashion, spreading it in spot applications if not across their entire arable ground. In a few examples, William Sims ordered his slaves to haul wagonloads of manure from his stables to his fields during the late winter; George Jeffreys placed his new plant beds on sheep pens rich with waste; and the small farmer Daniel Merritt, who owned no slaves, spread manure composted with leaves on his cornfields.[56] Stoll is likely correct in his assertion that the South in general lacked systematic, intensive manuring, but the practice probably existed throughout the broader region as it did in the Southside as a supplement to other fertilizing and amending measures. Although these farmers failed to record their thoughts concerning the application of manure, it is likely that they viewed manuring as an activity that both facilitated soil conservation and made economic sense in certain situations—but not as an activity central to raising tobacco.

Manure was not the only source of fertility found within the boundaries of the farm. Almost any organic matter worked into the soil improved its texture, moisture-holding ability, and fertility. Some farmers practiced green manuring: the plowing under of tobacco stalks, cover crops such as grass, or weeds to slowly rot in the topsoil.[57] Most tobacco producers also relied on the fertilizing power of wood byproducts. They used the calcium carbonate, phosphorus, and potassium released by burning wood to give new ground a boost of fertility, plowed ashes into acidic portions of cropland, scattered sawdust and wood chips on their fields after cutting firewood and fence posts in the winter, and coated their plant beds and even their tobacco seeds in ashes.[58] Other materials served as impromptu amendments as well. Landowners hauled straw from their grain crops back to their fields after threshing out the grain or mixed the refuse with manure to create a rich compost. Other planters supervised their slaves in the labor-intensive task of transporting wagonloads of leaf mulch and pine needle from the woods to cleared land where they could be worked into the soil.[59] Although taken alone these spotty applications could not maintain soil fertility indefinitely, they did work with other amending practices to improve poor ground and increase or sustain crop yield.

Planters intent on boosting the productivity of their lands also kept a diversity of animals, sought out new breeds of livestock and strains of crops, and experimented with alternative cash crops and animals. Following the northern craze for merino sheep in the 1840s and early 1850s, the *Milton Chronicle* touted the breed as a potential alternative to hogs, cattle, and tobacco. The article declared that merinos were growing in popularity in the Southside, and as an example of the breed's profit potential cited a regional farmer who made between $750 and $1,350 each year from the sale of wool and lambs from his flock of 170 sheep.[60] Although sheep largely disappeared from the Piedmont's post–Civil War landscape, in 1850 more than half of regional landowners raised some sheep, from small flocks of 10 or fewer animals to John Sims's 300 head.[61] Area farmers also produced Oregon peas, goats, flax, and turkeys, and Pittsylvania and Caswell farmers experimented with that old southern agricultural alternative, silk, during the antebellum period, though their sericultural efforts came to little.[62] Farm families raised a wide variety of fowl as well; the Clarks of Halifax kept chickens, geese, turkeys, and ducks, and farm-raised birds of some sort graced their table more than once a week.[63] The most common avenue to stock improvement was through on-farm selective breeding and the introduction of a few "improved" breeds of traditional livestock. George Jeffreys's efforts were typical: in 1845 he made notes on the importance of carefully selecting sheep for fleece and meat traits, and he ordered several varieties of roosters to introduce new bloodlines into his plantation flock.[64]

James Bruce extended his livestock improvement efforts to the most ubiquitous and least managed southern domestic animal: the hog. Southern farmers and planters had long allowed their swine to roam the communal woods and roads of their neighborhoods. The animals provided for themselves throughout the year, rooting and rummaging for mast, roots, and garbage and breaking through the occasional fence to pillage corn and wheat fields. Stockowners rounded up their animals in the late fall or early winter, selected a few hogs to slaughter for the year's pork supply, and cropped the ears of the remaining animals in proprietary patterns before turning them back on the commons.[65] Bruce implemented a hog-raising system that combined these traditional uses of the common range with feeding and penning. He turned his hogs into the woods in the fall when white oak mast was heaviest but penned his animals in the evenings and continued to give them "as much corn as they will eat."[66] His efforts paid off in heavier hogs at killing time. Bruce's surviving records from the antebellum period document the slaughter weights of 82 hogs; the animals averaged 161 pounds each.[67] Although far lighter than modern market hogs—which are raised in confinement, administered growth hormones and antibiotics, and have undergone decades of selective breeding—these weights compared favorably to those of other local hog producers. A sample of 399 hogs killed by farmers in Caswell, Halifax, and Pittsylvania revealed an average weight of only 139 pounds each, making Bruce's hogs almost 16 percent heavier than average.[68] His hog weights also bettered concurrent averages across the South. Geographer Sam Hilliard's examination of antebellum slaughter weights throughout the region found an average of 146 pounds for 11,212 animals.[69] Bruce's system demonstrated the value of more intensive stock management, but the low carcass weights evident in the records of other regional landowners suggest that his efforts were atypical and probably not worth the effort.

Enterprising planters and farmers also turned to machinery to aid in their agricultural labors. Tobacco remained a crop reliant on hand labor and draft animal work, but farmers built or purchased wood and iron presses to pack their hogsheads as tightly as possible and sought improved plows to prepare their fields. The region's grain culture proved slightly more amenable to mechanization. Local merchants offered guano and lime spreaders for amending fields, mechanical reapers for harvesting ripe wheat and oats, and threshing machines for separating the seed from the chaff.[70] Three Halifax planters—Charles Cabaniss and neighbors Philip Howerton and Thomas Easley—built their own horse-operated mechanical threshing machines. The men marketed their inventions throughout the region but seem to have been largely unsuccessful because their machines proved both expensive and unreliable.[71] These pieces of equipment were costly and rare enough that even wealthy planters shared their use, as when Charles Coleman's neighbor, a Mr. Harris, loaned him a threshing machine to process his wheat crop in 1843.[72] These mechanical farming aids remained uncommon in the Southside during the antebellum era, but their limited uses fit into the broader message of agricultural intensification advocated by reformers.

Nowhere was the impulse to experiment and improve more evident than in the orchards and vineyards of planters and farmers across the region. Landowners large and small planted multitudinous varieties of apples, peaches, grapes, pears, cherries, apricots, and plums in stands that at once filled farm larders and served as displays of horticultural knowledge and husbandry.[73] Almost every plantation and farmhouse had its orchard located behind the farmyard, along the drive, or tucked into an oddly shaped field corner, and landowners used the fruit fresh, made large quantities of cider, and fed surpluses to livestock. Farmers grafted these trees and vines themselves, selected likely looking volunteers from a neighbor's stock, or if as wealthy as William Sutherlin, ordered saplings and cuttings from commercial nurseries. In 1856 Sutherlin ordered four varieties of apples, four of peaches, one nectarine, two plums, and one apricot to supplement his plantation orchard.[74] The substantial planter Elijah Hundley had a similar interest in apple varieties; in one lot he bought fifty grafted trees of sixteen varieties, including "English pippin," "Pryons red," and "green cheese."[75] Smaller farmers also planted as extensive orchards as possible. Daniel Merritt, a tanner and small farmer, owned no slaves during the 1840s and 1850s, but he planted a number of apples, peaches, pears, plums, and cherries on his small Halifax farm, and he repeated his horticultural efforts after he moved to a farm on the outskirts of Milton in 1845.[76] Whether a large planter such as Sutherlin or a small farmer such as Merritt, landowners who invested their energy and money in fruit trees and vines made long-term commitments to improving the land that they tended.

Prominent agricultural reformers made demanding calls for diversified farming, and surviving records suggest that they attempted to implement their advice on their own farms. Both James Bruce and John Edmunds raised a number of cash and subsistence crops on their home plantations. In 1850 the two planters owned horses, dozens of milk cows, oxen, beeves, swine, large flocks of sheep, and they both grew wheat, corn, oats, sweet potatoes, Irish potatoes, and cowpeas. Vincent Witcher, an advocate of reform in Pittsylvania County, produced an even wider variety of farm products. In addition to the above crops and livestock, over the preceding year Witcher's slaves grew hops and flax, sheared wool, made butter and cheese, and gathered eighty pounds of beeswax and honey. Despite the diversity of all three plantations, Bruce, Edmunds, and Witcher still depended heavily on tobacco as a cash staple. In 1850 Witcher produced 12,000 pounds, Edmunds 38,500 pounds, and Bruce— who agitated against the crop's cultivation on a number of occasions— raised an astonishing 71,925 pounds of leaf, perhaps the product of more than one hundred acres of tobacco.[77] Even reformers intent on escaping the clutches of tobacco could not give up the crop.

Perhaps no planter epitomized the region's engagement with agricultural reform better than Halifax's Charles Coleman.[78] In the mid-1840s, Coleman owned a substantial tobacco plantation, Woodlawn, near the hamlet of Clover (its very name evoking agricultural improvement) in the county's northeastern corner. From his white frame house on the bluffs above the Staunton River, Coleman guided a miniature agricultural empire where he diligently labored to put reform advice into practice (fig. 7). Coleman was the master of both land and people; at his death in 1849 he owned ninety-four slaves and more than 2,100 acres. His family had farmed portions of Halifax for several generations, and his brother, Ethelbert, was a substantial planter in the community as

FIGURE 7: Charles Coleman, owner of Woodlawn, pictured here, was a dedicated antebellum agricultural reformer. By the 1930s, when this picture was taken, Wood-lawn estate had fallen into disrepair, symbolic of the inability of Southside agricultural reform to break tobacco's grip on the region, as well as environmental and economic troubles. Courtesy Library of Virginia.

well. Most important for our purposes, Coleman kept an agricultural diary describing the successes and trials of intensive farming.

Although he was not a regional reform spokesman, Coleman was connected to the ideas of the reformers and to local groups of "modern" farmers, and he put those ideas into practice on his own land. He sub-scribed to regional newspapers and read agronomic texts, such as "Chap-tal's agricultural chemistry," which he judged applicable to "any country & . . . excellent."[79] He was also an active member of Clover's small agri-cultural club, which met in its members' houses. But these public and intellectual commitments to agricultural reform paled beside his practi-cal application of reformers' advice on his own plantation; for Coleman, reform was first and foremost an action that took place on the ground. Woodlawn served as a place of production and a vast experiment ground, where Coleman tried various conservation and efficiency measures, worked to mold fields and forests into more productive spaces, and sought self-sufficiency, profit, and peace of mind.

One of Coleman's principal occupations was increasing farm diversity. Under his management, Woodlawn contained an astonishing variety of crops and livestock. Plantation slaves managed domestic animals of all sorts, including mules, hogs, horses, sheep, cattle, ducks, and chickens. Coleman's fields grew the traditional Southside staples—tobacco, corn, and wheat—but he also produced a number of other crops on a smaller scale for both sale and domestic use. These lesser crops included sweet potatoes, rye, oats, flax, cotton, cowpeas, clover hay, watermelons, and turnips. The plantation garden was also critical to plans for a diversified and independent estate. Among its vegetables any given season were pumpkins, melons, English peas, peanuts, cucumbers, cymblings (squashes), Irish potatoes, sweet corn, beets, asparagus, strawberries, salsify, sugar beans, red peppers, snap beans, cabbage, radishes, lettuce, collards, celery, and parsnips. As on so many reformers' farms, Coleman's orchard was a center of horticultural diversity and experimentation. He planted grape vines, rows of raspberry canes, figs, gooseberry bushes, apricots, plums, and a peach orchard, but his real obsession was with apples. In the spring of 1843, Coleman's neighbor, Dr. Norton, taught him how to graft trees, and over the next five years he grafted and planted more than 200 apple trees of at least ten varieties. Coleman carefully recorded the numbers and location of each winesap and pippin, and he clearly enjoyed experimenting with new varieties of fruit and grafting techniques. The interest Coleman exhibited in listing and describing crops and livestock reflected his belief in the importance of a diversified plantation. His comments on these efforts displayed pleasure in the intellectual exercises of intensive farming, but they also stressed the value of personal and plantation independence.

Coleman's efforts to increase Woodlawn's animal and plant diversity were mirrored in his campaign to improve the estate's soil and prevent erosion. He reserved the rich bottomland along the river for his principal crops, tobacco and corn, and rotated his fields on a regular schedule, alternating clover with crops that depleted nitrogen. Woodlawn's slaves spent a portion of each winter digging ditches to drain the low ground and cutting other trenches across the slopes of the higher ground to turn water off fields. Coleman also used fire to improve his estate. Regular burnings kept the plantation's pastures in good condition by killing weeds, lowering soil acidity, and promoting palatable new growth, and slaves cut brush out of orchards and woodlots and burned the trimmings over plant beds or other sites that benefited from ashes. Coleman purchased plaster, spread barnyard manure, hauled leaves and straw onto his fields, collected ashes to spread on his plant beds, and even saved the chaff winnowed from his wheat crop to plow into the earth. Obtaining an adequate supply of amendments was a constant source of concern. In the winter of 1843 he worried that "the manure turns out nothing like as well as we expected in the stable lot, nor will it, I fear, in the farm pen," and in response he built a manure composting pen, complete with a roof to block the rain.[80] Like many of his fellow planters in Caswell, Halifax, and Pittsylvania, Coleman was not content to simply plant a field until it no longer produced and then abandon the old ground for new land.

Unlike James Bruce, Coleman seemed little concerned about the place of slavery in reformed southern agriculture. His diaries are remarkable for their dearth of direct references to the slave labor force that provided the muscle and grist that kept Woodlawn functioning. Coleman viewed his slaves as extensions of his will, simple beasts of burden who carried out his directives. Reflecting another reformer stance on slavery, Coleman seemed to envision the institution as just another tool to better landscapes rather than as a relationship that degraded people and place. The terms he used reflected this attitude. He invariably referred to the plantation's slaves as his "hands," and when describing such black labor as ditching or stripping tobacco, he invariably used the pronoun "we." Typical entries began "we commenced cutting wood" or "we killed hogs."[81] His diary showed little worry over his slaves' well-being or sense of community. When an unnamed epidemic struck Woodlawn in the summer of 1844, killing two slaves and reducing ten others to their sickbeds, Coleman made but a brief comment and failed to list the names of the dead or ill, despite the calamity of an illness that killed or incapacitated more than an eighth of his workforce. The same autumn, an escaped brush fire burned up the year's corn supply and much of the plantation's slave housing, and Coleman calmly noted that he planned to rent out a number of his slaves for a short time while he placed the estate back in order. Whether his slaves were laboring in the fields or dying of illness, Coleman reduced men and women to another variable in the calculus of farming.

Like Ruffin, who regularly calculated the benefits of reform within the southern slave system, Coleman saw no contradiction between plantation slavery and agricultural reform, perhaps because of ingrained racial attitudes but also no doubt because his activities at Woodlawn were economically successful.[82] Of course improvement measures at Woodlawn, from diversified crops to efforts to sustain soil fertility, were not without setbacks. Cutworms attacked Coleman's corn; cockles invaded his wheat fields; hornworms infested the tobacco crop every year; powerful storms caused the Staunton River to periodically inundate bottomland fields, carrying away corn and tobacco; and on the occasion mentioned above, a fire set to clear shrubby ground escaped and burned up Woodlawn's stable, a full corn crib, and several slave houses. However, these difficulties and the cost of reform measures did not hinder the overall economic success of Coleman's improvement efforts. Although his diary does not include detailed accounting records, Coleman economic prospects seemed to improve each year. By 1848, Woodlawn's herds and crops were more numerous and varied than ever, and these diversified activities did little to affect the plantation's tobacco production. Coupling reformer's advice with the traditional staple seemed to pose little difficulty for Coleman; his 1848 crop of leaf totaled 32,900 pounds, the heaviest of the decade. Coleman saw little incongruous in slavery coupled with agricultural reform: in his mind the two worked in Woodlawn's political economy.

Although Coleman's diary is concerned with the daily business of farming and not with the attitudes and ideologies of the diarist, it does exhibit his general satisfaction with efforts to improve the land he owned. As a prominent planter, Coleman sought profit, but he also desired permanence and even beauty. A sense of satisfaction permeated entries recording good harvests, orchard expansions, and projects to improve the landscape; Coleman's diaries exude a sense of moral worth connected to careful and intensive farming. Reflecting Ruffin's calls for sustainable farming and planting, this "agricultural morality," exhibited in Coleman's diaries, and in the records and journals of a number of other regional farmers and planters, complicates the notion of southern planting as a simple calculation of profit.[83] Certainly some landowners depleted the natural resources of their lands and moved west, valuing the maximization of profit over the importance of place and permanence, but not all. Coleman is an example of an equally powerful conservatism, intent on "bettering" the agricultural landscapes of the South.

In this chapter I have gone to great lengths to explore the impact of regional reform, if only to highlight just how transformative bright leaf would be on this nascent agriculture. If the messages and methods of agricultural reform had made inroads in the antebellum Southside, shaping the practice of Charles Coleman as well as smaller farmers, what then were the effects of bright tobacco culture on these farm intensification efforts? In a word, cataclysmic. The new crop seems to have significantly undermined reform efforts along the Virginia and North Carolina line. And it did so by both competing with and appealing to certain tenets of the reform movement. Bright tobacco became the enemy within agricultural reform, undermining the message of improvement while assisting farmers in achieving some of its goals. Yellow tobacco's cultivation methods directly conflicted with the soil and timber conservation efforts involved in agricultural intensification, but the crop's culture also entailed following selected reform advice and countered farmers' concerns by producing healthy profits on marginal lands. In a cruel twist of irony, the crop that eventually contributed so much to regional erosion, stream siltation, deforestation, and declining agricultural prospects owed part of its popularity to an agricultural movement advocating farm sustainability.

Despite the crop's eventual success, elements of bright leaf culture undermined the soil conservation efforts of agricultural reformers in undeniable ways. For example, farmers interested in making bright leaf cultivated their tobacco fields almost exclusively with shovel plows, with erosive consequences. Bright tobacco experts, such as the Love brothers of Pittsylvania County, argued that the key to successfully producing light-colored tobacco lay in the composition of the region's topsoil: deep plowing that intermixed the sandy surface with the red clay below led to lower-quality leaf because it altered soil stratification and slowed drainage. These modified soil traits were valuable in some cropland but produced poor bright tobacco.[84] In the Southside the turn to shovel plows was not primarily the product of a lazy or backward agriculture; rather, the return to this older technology stemmed from efforts to create a particular form of tobacco. George Jeffreys's experiments plowing bright

leaf land during the 1840s reflected the tensions between the conservation advice of reformers and the practices of bright leaf farmers. Jeffreys
looked to his Caswell neighbors, such as the Slades, and tried to repli-
cate their successful crops. He advocated shovel-plowing over methods
of deeper cultivation in his diary, writing, "It [the shovel plow] gives as
much loose dirt as the one horse dagons with this decided advantage
that they do not *turn over* the soil but leave it where nature has placed
it—on the *surface*."[85] This shallow cultivation produced fields suitable
for raising very light tobacco, but it also posed serious risks. Frequent
passes with a shovel plow left pulverized topsoil overlaying a clay hardpan. Heavy rains saturated the loose sand in the upper soil horizon and
then flowed horizontally along the clay below, forming gullies or galled
spots on hillside fields. As an educated agronomist who had published
advice on the value of deep plowing, Jeffreys was not blind to this reality.
In *A Series of Essays on Agricultural and Rural Affairs* he had written
in no uncertain terms that the "evil" of erosion "has principally arisen
from the practice of shallow ploughing. If ever a material reformation
takes place in our system of agriculture, it will chiefly be founded on
deep and horizontal ploughing."[86] He noted the gullying and sheet erosion that accompanied his new shallow-plowing campaigns but believed
that other elements of reform advice would help him overcome the dangers of ignoring his earlier critique. He wrote of his plans to plow out
the gullies that formed, sow oats and other cover crops on the damaged
areas, and keep livestock off the problem slopes until the land could once
again return to production.[87]

The adoption of bright leaf also undermined crop rotation on Piedmont farms. Virginia and North Carolina farmers had long avoided
heavy manuring or rotating their dark tobacco too frequently with such
nitrogen-fixing crops as clover and cowpeas, as excess nitrogen made for
a harsh, heavy cured leaf. If dark tobacco farmers used manure and
legumes cautiously, bright leaf farmers completely shunned the dung
cart and these valuable crops in their rotations because yellow tobacco
was even more sensitive to excess nitrogen than was dark leaf. Bright
leaf planted after clover refused to ripen properly, remaining green well
into the fall, and tobacco that did not begin to yellow in the field was almost impossible to cure bright. Other traditional farm crops, such as
oats and wheat, fell out of favor in tobacco rotations as well. Although

these crops did not fix nitrogen, they took up valuable field space and changed soil structures in ways that complicated making yellow tobacco. Bright leaf demonstrated reliable productivity year after year on the same ground (at least at first), and farmers who switched a ridge top or hillside field from bright leaf to wheat for a few years lost a significant amount of profit. In addition, decomposing organic matter left in the soil from the roots and stalks of these rotation crops increased the land's moisture retaining capabilities. This was a beneficial trait for most crops but a detriment to making bright leaf, in which the plants demanded rapid soil percolation to grow and ripen properly.[88] Crop rotation still occurred on Southside farms—farmers needed oats for their horses and mules, clover for their cattle, and wheat or corn for their families' bread—but tobacco increasingly became a crop apart, divorced from farm subsistence and sustenance cycles.

Regional reformers had long been concerned about deforestation and timber shortages, and curing yellow leaf used more wood than producing dark tobacco.[89] As with dark tobacco, bright tobacco culture required farmers to clear woodland for fields, build split-rail fences to keep hogs and cattle from trampling the crop, erect log barns for curing the harvested leaf, and cut staves for hogsheads. Construction of a typical tobacco barn alone used more than 150 logs, with each log sixteen to twenty feet long by six inches in diameter.[90] The new curing techniques involved in bright leaf placed an additional burden on the regional timber supplies. Bright leaf curing lasted significantly longer than traditional fire curing, adding days to the time farmers burned wood in their barns, and experts advised curers to build hotter than normal fires to create a bright yellow product. Estimates of the wood needed to cure a barn of tobacco vary widely based on the type of wood used, its seasoning, weather conditions, how tightly the barn was filled, and the ripeness of a given crop; historic farmers and modern tobacco historians record from 1 to 8 cords of wood per barn as the amount of fuel needed to cure bright leaf. To give an idea of just how much wood went into curing tobacco, one geographer has estimated that the average mid-nineteenth-century family consumed 17.5 cords of wood per year for their combined fuel needs. At the upper estimate of 8 cords of wood per barn, curing just two barns of tobacco taxed the resources of local forests almost as much as the annual cooking, heating, and building needs of an entire family. Given the 20,204,052 pounds of cured tobacco reported produced in Caswell, Halifax, and Pittsylvania counties in 1860, if we take a contemporary calculation of 1,200 pounds of tobacco per barn and a moderate estimate of 4 cords of wood to cure each barn, then over the course of the year the region's planters burned 67,348 cords of wood heating their tobacco barns. This is a gross estimate to be sure—curing demands fluctuated due to weather, wood type, and so on, only a portion of the tobacco produced in 1860 was bright leaf, some growers used flues while others used charcoal—but it suggests the tremendous quantity of wood burned in the process of curing tobacco. Although there are no similar estimates for dark tobacco curing in the same region, commentators were uniform in their agreement that making yellow tobacco used significantly more fuel than more traditional curing methods.[91] The rising use of charcoal to heat bright leaf barns also consumed more wood, as a given unit of charcoal produced less thermal energy than the wood that braziers burned to produce it.[92] Curing a barn of bright tobacco with this fuel could consume eighteen bushels of charcoal or more.[93] Whether growers used curing fuel in the form of charcoal or cord wood, bright leaf cultivation almost certainly increased the demand for wood and put greater pressure on the counties' forests and woodlots.

In these and other ways, efforts to make bright tobacco seemed anathema to the messages of reform, yet the crop offered palliatives. If the culture of bright leaf ignored many reformer precepts, it embraced other elements of the agricultural improvement agenda. The new tobacco easily fit into the existing agricultural infrastructure, drew on available regional resources, appealed to reformers' ideals about farmer expertise, and was profitable on even the poorest of soils. Bright leaf posed obvious problems: it disrupted crop rotations, continued tobacco's dominance of local farm economies, and made erosion an even greater threat. But it countered these difficulties in the most important regard imaginable: it made more regional tobacco farms profitable. For these reasons, bright leaf became popular even with reform-minded Southside farmers and planters. Bright tobacco illustrated a commonly overlooked aspect of agricultural reform: portions of the reform agenda that proved profitable were invariably the ones adopted, for reform's underlying raison d'être was the creation of profitable farms. Reformers preached a sustainability rooted in economics as much as in conservation, and in

the Southside early bright leaf culture was (temporarily) more successful than any other revitalization effort.

Perhaps bright leaf's most appealing characteristic was the manner
in which it shifted physical and mental farm geographies, transforming
"waste" land into rich land. Since Euro-American settlement, dark tobacco had flourished in the region's limited stretches of rich bottomlands, and farmers and planters largely relegated their corn and wheat
crops to the marginal, but more abundant, uplands. This landscape arrangement was far from universal of course, but farmers with access to
both bottomland and upland naturally preferred to plant tobacco—their
highest-value crop—on the richer ground. Faced with bright leaf's nutrient and drainage demands, in the late 1840s and 1850s farmers moved
their new tobacco fields onto the sandy ridges and slopes and planted
corn (and less frequently wheat) on the heavier soils of the low grounds.
For bright leaf farmers this shift seemed almost too good to be true:
bright leaf, which brought a much higher price than dark tobacco, grew
on land formerly deemed all but useless, and by moving their grain
fields to more fertile ground, they were able to produce more of these
crops with no increase in labor. Indeed, George Jeffreys argued that farmers who turned their bottomlands along the Hyco River in Caswell to
corn production could produce a crop every year for half a century before
yields decreased.[94] The region's typical sandy upland soils produced as
little as ten bushels of corn or seven bushels of wheat per acre, compared
to yields as high as sixty bushels and seventeen and a half bushels per
acre, respectively, on the rich brown Congaree loam of bottomlands.[95]
The shifting arable landscape that accompanied the expansion of bright
leaf thus promised not only more profitable tobacco but better corn and
wheat crops as well. But this increase in productivity came with a clear
price; farmers cleared and cultivated more of the most erodible portions
of the Piedmont landscape, and the results were soon obvious.

A shift to bright leaf tobacco was also in some ways a reassuring sort
of "reform." Although bright leaf cultivation entailed important modifications to the traditional ways to grow tobacco, the new variety was familiar enough that many dark tobacco growers felt comfortable undertaking
its production. Farmers had to make some changes: they learned the
characteristics of their upland soils, how to read their new crop's growth
and ripening, and elaborate curing formulas. As important as what

changed was what remained the same: farmers continued to use the same tobacco barns, transportation networks, and many of the same labor methods as they had with dark leaf. The switch to bright tobacco did not involve the purchase of new equipment (with the exception of flues for some farms), new labor arrangements, or substantial capital investment. For the most part, slave owners worked their laborers in yellow tobacco as they had in dark, and small farmers without slaves could cultivate bright leaf as easily as they could grow the older variety. These similarities made the new crop appealing and comforting, offering all of the rewards but few of the uncertainties of more exotic agricultural alternatives, such as silk, grapes, or merino sheep.

These moderate transitions were especially palatable because they came from within the farm. The initial cultivation of bright leaf did not require expensive fertilizers, new machinery, or even rich bottomland; it demanded only stretches of relatively weak soil, an abundant commodity in the region. These minimal requirements in turn promised a basic sort of sustainability. While bright tobacco culture countered many tenets of reform wisdom focused on sustained production—from wood conservation to crop rotation—it offered in their stead the promise that farmers could profitably cultivate a cash crop on a single field almost indefinitely. This benefit was immediately apparent and universally understood.

Bright tobacco culture's emphasis on expertise and the artisanal nature of yellow tobacco production also appealed to modernizing farmers. Unlike some southern staple producers, who Ruffin complained were "land killers" out of general *"ignorance,"* bright tobacco growers emphasized careful attention to land, crop, and technique.[96] Good bright leaf was the product of a constant attention to detail, or so growers told themselves: farmers had to select appropriate seed varieties, determine proper soil types, and follow elaborate curing regimes. Curing in particular demanded a farmer's experience and constant vigilance. The curer had to take into account the conditions of each barn of tobacco and the surrounding environment: high temperatures and humidity called for a cooler fire, windy days required adjustment to doors and flues, rain forced the farmer to add more wood to the furnace, and so on. The curer also had to account for the minute differences in each cutting of tobacco. As an assemblage of organic once-living matter, each barn of tobacco was slightly different; some cuttings were greener, some contained more tar, some housed heavy crenellated leaves while in other barns they were thin and light. The farmer had to adjust temperature and draft to best cure each crop.[97] A postwar tract captured the demands of early flue curing. The pamphlet, written by the Love brothers of Pittsylvania, warned would-be bright leaf cultivators that expertise had to be coupled with a devotion to producing a quality finished product: "Many sleepless night have we passed in the barn, sometimes never leaving a curing from the time it was housed until the coals were in their last slumbering embers."[98]

This appeal to the artisan nature of experience coupled with an emphasis on scientific technology—via flues and careful temperature regulation—was especially resonant with agricultural reformers. As historian Benjamin Cohen has argued, reformers were particularly enamored with a "Georgic" ideal of farm improvement, the notion that agricultural modernization must come from experience laboring on the land.[99] Technical advice concerning selective breeding, soil chemistry, and curing mechanisms were important to making bright tobacco, but these "book farming" innovations were made all the more appealing by bright leaf advocates' insistence that a farmer's experience remained the most important element of cultivation. It was a touch of practical magic that was necessary to make the very best tobacco. Virginian Ellen Glasgow captured this belief in her novel *The Deliverance,* when she wrote of a Pittsylvania character, farmer Christopher Blake, who "kin pick up a leaf blindfold an' tell you the quality of it at his first touch."[100]

Bright leaf boosters also followed the agricultural reformers' models of disseminating knowledge. Experts in the new crop culture, such as Abisha Slade and his slave Stephen, traveled around the region during the 1850s sharing their techniques with interested farmers and planters and providing demonstrations of the curing process. Much like authors in period agricultural journals and speakers at agricultural club meetings, these authorities relayed practical advice throughout communities of interested farmers.[101] It was no coincidence that Jeffreys recorded his desire to publish a pamphlet once he had mastered the techniques of producing bright leaf; as an agricultural reform writer enamored with the new crop he sought to shape the emerging culture through his traditional medium.[102] Robert Ragland remembered seeing Slade exhibit his

bright tobacco at an 1856 Halifax County farmers' meeting, and another
spectator recalled Slade exhorting growers, "Boys, if you have barns to
cure, use charcoal! Use charcoal!"[103] Growers who adopted these methods engaged in a process with which reformers, from Ruffin to Edmunds,
were familiar: these enterprising farmers applied outside expertise
in an attempt to get greater profits from their land. Heeding reformers'
calls for agricultural education and scientific production, bright leaf cultivators used their newfound expertise to add value to their product.

Reformers and tobacco farmers alike also appreciated the impact of
bright leaf's expansion on the regional tobacco manufacturing industry.
From small to large, the factories that emerged in Danville, Milton,
Yanceyville, and at rural crossroads and plantations scattered across the
countryside strengthened the local economy and the region's ties to tobacco culture. As these factories turned bright leaf into lumps and twists,
they transformed a local raw commodity into a value-added product.
Despite reformers' emphasis on agricultural innovation, they were not
adverse to industrial growth that drew on local resources, especially
when the industry did little to threaten farm profits or labor forces. Indeed, local tobacco manufacturers worked hand in hand with local producers, purchasing local bright leaf directly from farmers, leasing or
buying a planter's surplus slaves, and supporting efforts to increase local production and leaf quality.[104] Improvement-minded individuals like
William Sutherlin combined tobacco farming with their manufacturing
interests, often participated in agricultural clubs and local politics, and
supported fellow planters by leasing or purchasing surplus slave labor.[105]
Small planter-owned factories located on plantations were the epitome
of several reform ideals: they transformed raw farm leaf into value-
added twists and plugs through the use of investment and expertise.
These operations brought the profits of urban industry to the countryside without challenging the plantation system. Like the northeastern
improvers described by Stoll, who believed the combination of merino
sheep production and local woolen mills would connect farm and factory to the benefit of both, Southside improvers, such as James Bruce,
argued that tobacco manufacturing and farm improvement were mutually reinforcing activities.[106] Early bright leaf manufacturing was thus a
manifestation of an agrarian industrialism in which the factory managers and the plantation owners were often one and the same, and they

sought a rural countryside that evoked both Thomas Jefferson and Alexander Hamilton.

This expanding manufacturing base and the increasing value of the region's tobacco crop stimulated internal improvements long sought by agricultural modernizers. One twentieth-century agricultural economist declared of Piedmont tobacco: "The marketing of this staple required no elaborate transportation system. It remained for cotton to arouse in the people of the Southern upland a paramount interest in securing easy connections with the ocean highway."[107] But like so many rural American in the first half of the nineteenth century, Southside boosters had pushed for better roads, clear river channels, and railroad connections, and the lucrative nature of bright leaf, in manufactured and raw form, encouraged an even greater push to improve transportation networks. Some infrastructural developments had occurred by mid-century: locks and dredging in the Roanoke River and improvements to the Dismal Swamp Canal in the 1820s had made water transport to Norfolk easier; the Franklin Turnpike, completed in 1840, brought tobacco from the western reaches of the state to Danville; and in 1842 a new stage road connected Danville and Lynchburg, speeding travel between the two important tobacco towns. Despite these improvements, Southside planters and farmers demanded a more direct overland route to Virginia's primary tobacco export cities: Richmond and Petersburg.[108] At the behest of planters, Pittsylvania County officials were instrumental in securing a rail connection between Danville and the state capital. Work began on the Richmond and Danville Railroad, which passed through the rich tobacco districts of Halifax on its way into Pittsylvania, in 1850, and the railroad was completed in 1858. The new line "excit[ed] much interest" among regional growers, who began using the railroad before its completion.[109] In 1855, the partially completed line carried 41,588 boxes of tobacco plugs, and by 1857 planters and manufacturers had made the Richmond and Danville the busiest tobacco transportation route in the commonwealth, surpassing much older railroads and canals.[110] Completion of the Richmond and Danville Railroad did not satisfy regional transportation boosters. Danville interests built feeder roads connecting tobacco-producing districts in Pittsylvania and Caswell to the town's manufacturers; North Carolina growers sought an extension of the railroad through Caswell into their state (a spur would connect Danville to Greensboro during the Civil War); and northern Pittsylvania growers wanted a rail connection between their region and Lynchburg (the Virginia Midland Railroad would eventually link Danville and Lynchburg by 1873).[111] In addition to transporting raw tobacco and manufactured plugs out of the region, the Richmond and Danville Railroad and improvements to local roads and rivers also brought finished goods, guano, and plaster into the Southside, facilitating exchanges that appealed to many reformers.[112]

The most appealing characteristic of bright leaf was the crop's profitability; unlike most agricultural alternatives, yellow tobacco produced profits from the weak Southside soils. Bright leaf cultivation "redeemed" the poorest uplands, transforming them from stretches of waste into the region's most important arable farmlands. This transformation did nothing to improve soil fertility, but it did "improve" the land in another way: the bright tobacco boom increased the economic value of poor land and the return on labor while allowing for the richer land to be used for diversified purposes. As bright leaf cultivation expanded in certain portions of Caswell County, land prices increased as much as sixty-fold, rising from fifty cents per acre to more than thirty dollars per acre over the course of just a few years.[113] The case of H. P. Womack demonstrated the potential effect of the booming tobacco economy on farmer thinking. In 1847, Womack left his Caswell home for land in Lincoln County, Tennessee, where he wrote that he was pleased with both the country and his new opportunities. Six years later, as bright tobacco prices gained steam, he wrote to his relative, Pleasant Womack, of his desire to return to Caswell to make his fortune. Asking Pleasant to send him a description of available farms and land prices, H.P. seemed intent on returning to his native county.[114] If the agricultural reform movement's most important goal was to preserve the viability of agriculture and rural populations in the old Southeast, then the profit potential of bright leaf held tremendous appeal. Womack's migration and return symbolized the reformer fantasy: the tide could turn and bring wandering Virginians and North Carolinians home again.

The local press played an important role in touting the profitability of bright leaf. An 1857 *Milton Chronicle* article on Abisha Slade's amazing profits emphasized the rewards of remaining in the Piedmont and adopting the new crop. The paper informed its readers, "It comes to us from unquestioned authority, that Capt. A. Slade of Caswell N.C., and his two brothers, have sold their entire crops of Tobacco, lugs included, to a Lynchburg manufacturer, for the extraordinary price of $35 per hundred lbs." The anonymous author then calculated the return of this bright crop. "Capt. Slade, we are informed, estimates his crop at 18,000 or 20,000 lbs. It is the product of the labor of some ten hands. If it should turn out to be 30,000 lbs., he will realize from each laborer the unprecedented sum of $700." At a time when prime slaves sold in the neighborhood of $1,000, a profit even half as large as the *Chronicle*'s speculative figure would have been a tremendous windfall. For readers who failed to connect this extraordinary return to the increased appeal of remaining in the Piedmont, the article elaborated on the diminishing need to emigrate in order to find profitable agricultural lands: "Can the cotton fields of Louisiana, the sugar plantations of Cuba, the rice fields or the turpentine Districts of the Carolinas, boast of larger profits?"[115] Similar price announcements appeared in other issues of the *Milton Chronicle*, despite editor Charles Evans's advocacy of diversified farming. In 1855, Evans had agitated for the local production of "Meat! Meat!! More Meat!!! and less Tobacco," yet the appeal of extraordinary bright leaf prices demanded the attention of Evans and his fellow regional editors.[116] In his letter attempting to convince his brother George to refrain from moving to Texas, Peter Hairston pointed to these robust returns as an argument against emigration, writing that high tobacco prices meant a bright future for the Southside.[117] For reform-minded planters and farmers, the steep prices commanded by yellow tobacco provided valuable support for their claims that there was a future in regional agriculture.

These high prices, along with bright tobacco's fit within the existing farm landscape and the adaptability of the crop to marginal lands, attracted dark tobacco growers to the new staple. Across the three counties planters like William Bailey, Elisha Barksdale, Joel Hubbard, and William Sims continued to cultivate traditional tobacco varieties using old techniques, but they also began to experiment with raising and curing yellow leaf. Likewise, tobacco manufacturer William Sutherlin and his fellow agricultural industrialists produced dark twists and chews, but they also sought local bright leaf, and many began to grow their own fine tobacco.[118] This bright leaf revolution attracted regional planters and farmers of all sizes, with neighbors of varying economic statuses experimenting with making yellow tobacco. In northern Caswell the large planter Jeffreys learned how to cure his crop from Abisha Slade, who, with only the ten hands he shared with his two brothers, was more middling farmer than elite planter, and both Charles Coleman—who owned almost a hundred slaves—and the yeoman Vincent Shelton, Jr., experimented with coal fires in their curing barns.[119] As the antebellum period drew to a close, bright tobacco and dark tobacco shared the Southside landscape, but a fundamental agricultural transition was underway.

Of course the development of yellow tobacco culture did not force farmers to abandon strict agricultural reform in favor of the new staple, but it did undermine the advice of Taylor, Ruffin, Bruce, and Edmunds by obliquely addressing some of the reform movement's most pressing concerns. Bright tobacco challenged intensification efforts that, if not impractical, were certainly quite difficult to implement on Piedmont farms. Such farm inputs as guano and clover seed were expensive: during the 1850s, one bag of guano (200 pounds) cost between four and five dollars, and clover seed brought as much as nine dollars per bushel.[120] In addition, crops key to the convertible husbandry practiced on many northern farms grew poorly in the Southside, from cool season pasture grasses to legumes. Perhaps most significant, local farmers and planters had a strong historic attachment to tobacco. For more than a century they had staked their fortunes on producing cured leaf along the region's watercourses and hollows. Caswell, Halifax, and Pittsylvania transportation networks, towns, plantations, and the rural countryside had been shaped by tobacco, and it was to be expected that regional farmers would be reluctant to abandon or lessen their cultivation of the crop. Bright tobacco worked within the natural and built environments of the Southside, subtly modifying farm routines, field placement, and crop rotation. In the end the new variety would prove transformative, altering regional economics and ecosystems, but for all its fanfare in newspapers and at market, during the antebellum period the transition to bright leaf on local farms and plantations was a quiet revolution.

If the new tobacco type made a certain sense based on the rural built environment, many growers also found its culture fundamentally comforting in a way that other cash crops were not. Bright tobacco was a less ideologically threatening form of agricultural alternative than grape vines, grain, or sheep. At a point in time when southern defenses of slavery grew more strident and unyielding, with such apologists as Virginia's George Fitzhugh declaring the institution a "positive good" rather than a "necessary evil," bright tobacco did little to challenge existing labor relationships.[121] The new tobacco culture did not ask planters to reduce their slave workforces, nor did it force white landowners who owned no slaves to purchase bonded labor. Unlike agricultural reform campaigns to refashion the southern plantation, bright tobacco promised to use black hands in traditional ways.

And some reformers, such as Ruffin, believed slavery was a tool that could actually "better" the land. For these masters, slavery was an institution ordained by nature, in which lesser people labored for their betters to make a more perfect landscape. As Mark Fiege has put it, these improvers believed a hierarchy of people to be only natural; in their eyes "inequality was inherent in nature and was a source of improvement." Even reformers (James Bruce in particular) who paid lip service to the notion that modern agriculture would eventually eliminate southern slavery were almost universally slave owners themselves, and their holdings in human beings tended to increase rather than decline.[122]

Bright leaf ultimately relieved farmers of the need to ponder the viability of slaves as shepherds, viniculturists, stewards of manure, or silk weavers, and the crop eliminated questions of what to do with slave labor during the months when wheat and barley fields all but cared for themselves. Bruce's worries about the future viability of slavery and Coleman's obvious efforts to ignore the issue were rendered less pressing by profitable tobacco culture within the bounds of existing farms, barns, and markets. Yellow tobacco thus refashioned Southside agriculture in a distinctly "southern" fashion, and local cultivators seemed comfortable with this new wine in old bottles. Although the spread of bright tobacco quelled a burgeoning agricultural revolution in the three counties, in part by co-opting agricultural reform goals, the Civil War would challenge the existing agricultural paradigm by restricting tobacco markets and, eventually, by deciding the slavery question that reformers found so perplexing.

FOUR

Tobacco Goes to War

WHEN GENERAL ROBERT E. LEE SURRENDERED the Army of
Northern Virginia at Appomattox on April 9, 1865, Confederate president Jefferson Davis was sitting in a house tobacco built: Danville manufacturer William Sutherlin's mansion on the southern bluffs of the
Dan River. Following the Confederate government's retreat from Richmond to Danville on April 3, Sutherlin had offered his house to southern officials, and for seven days the structure became the seat of the last
capital of the Confederacy. Sutherlin, a major figure in both the city's
wartime tobacco industry and regional military activities, demonstrated
his wealth by spreading a gracious table before Confederate leaders and
extending them every hospitality.[1] At the outbreak of the war, the small
tobacco town seemed unlikely to one day find itself the seat of a fleeting
(and fleeing) nation—even if only for a week—yet the same geographical and environmental conditions that contributed to the growth of
bright tobacco culture made the region an important source of supplies
and men for the Confederate war effort and one of the most vital links
between Virginia's eastern front and the Deep South. The Richmond
and Danville Railroad, which remained operational until the very last
days of the war, hauled food, tobacco, and troops to the front lines and
connected the Confederate capital to more southern locales. The conflict also challenged the primacy of tobacco culture in Caswell, Halifax,

and Pittsylvania, but it eventually cemented the place of bright leaf in the local agricultural and commercial landscape.

The Civil War presented both problems and opportunities for to-bacco agriculture and manufacturing along the border. Throughout much of the South the war proved a punctuated environmental distur-bance "akin to a natural disaster" that lasted for four years.[2] Combat and its attendant societal reordering destroyed forests, gouged holes in the earth, consumed vast herds of livestock, and spread disease, in addition to killing people. The conflict was less destructive in the Southside, behind the lines for most of the war, but it did threaten to change re-gional agriculture. Among the most serious difficulties were the closure of northern and foreign markets to southern tobacco, the conversion of several of Danville's warehouses and tobacco factories into military pris-ons, pressure on farmers by Confederate and state authorities to pro-duce foodstuffs in the place of tobacco as the war progressed, and ever-increasing diversions of white and black agricultural labor into the war effort. Despite these restraining factors, the war ultimately proved stimulating to bright leaf production. Regional growers and manufac-turers remained well behind the front lines for the duration of the war, whereas fighting forced many of their competitors in cities like Rich-mond and Petersburg to temporarily close or curtail production. A smaller market for regional tobacco products—essentially limited to portions of the Southeast—put pressure on all tobacco manufacturers and brought increased attention to quality, and as a result bright leaf's reputation as the finest tobacco type led to high prices and increased market share. In many ways bright tobacco agriculture and the slavery (at least until Emancipation) that so shaped it continued pretty much as if there was no conflict raging just a few counties to the east.[3]

Historians have described Danville and the surrounding country-side during the Civil War in two quite disparate ways. Frederick Siegel portrayed Danville as "a major economic beneficiary of the war," and he attributed the boom largely to the expansion of the tobacco manufactur-ing industry.[4] Nannie Tilley, in contrast, pointed to "disrupted manu-facturing interests and lack of transportation facilities," turmoil that she believed "caused serious disturbances in the tobacco industry."[5] The truth is probably somewhere between these two viewpoints. The constricted wartime market reduced demand for dark tobacco and pushed some traditional growers out of tobacco and into grain and livestock production. At the same time, bright leaf continued to sell well, with prices for quality leaf keeping pace with the war's dramatic rate of inflation. More than anything, the conflict spread the reputation of the region's bright tobacco and its manufacturers' brands throughout the South as Danville area "yellow leaf" appeared in markets from Atlanta to Mobile to New Orleans, continuing regional manufacturers' antebellum marketing efforts.

For bright tobacco the trial of wartime was in many ways similar to that of the debates over agricultural reform. The conflict had the potential to change how local residents thought about local environments and agriculture; they might have turned away from tobacco and toward grain or other crops for the reasons that follow. In short, they could have reconceptualized the "best" use of the land. But preexisting ideas about the value of tobacco, the limitations of local soils, and the cultural weight of existing farming practices ultimately resisted the turmoil of wartime. National pressures failed to trump regional environmental ideas. The Civil War neither doomed bright tobacco nor ensured that it would become the dominant form of regional agriculture, but the conflict did accentuate antebellum cultivation and marketing practices.

The war's outcome created new questions. Emancipation threatened the human foundation on which tobacco growers had built regional agriculture; small farmers and planters alike may have planted tobacco, but large plantations utilizing slave labor were lucrative and expanding during the antebellum era, bucking the general trend in the rest of Virginia and North Carolina. At the same time, emancipation offered former slaves the potential opportunity to become tobacco farmers in their own right, working for their own futures. Although the war altered the antebellum pattern of growth, eventually creating a landscape worked by wage labor, sharecroppers, and renters, it helped build the reputation of bright tobacco and expanded the market for bright leaf products by creating severe competition among the South's tobacco producers, a competition that favored tobacco of the best quality. It was these wartime trends and the associated tobacco manufacturing that propelled the growth of bright leaf during Reconstruction. The issue of agricultural labor would loom large after Appomattox, but the Civil War reinforced the importance of tobacco in the regional economy, thus assuring that landowners would vigorously pursue answers to the labor question.

If the majority of white residents in Caswell, Halifax, and Pittsylvania had had their way, there would have been no Civil War. The region's numerous slave owners valued the institution and resented northern interference in questions of southern labor and race, yet the general prosperity of the 1850s bright tobacco boom made tobacco planters and manufacturers, the bulk of political leadership in the three counties, reluctant to risk division of the Union under any but the direst of circumstances. Local leaders like Sutherlin, the majority of whom were Whigs and moderate Unionists, feared abolition, but they also worried about the stability of a hastily assembled southern government, the loss of tobacco markets, potential changes to tax rates, and the general unrest of war.[6] As in other reaches of the greater South reluctant to commit to secession, the April 12–13 siege of Fort Sumter and Lincoln's April 15 call for 75,000 troops to quell the brewing rebellion resulted in a sea change in public opinion.[7] Southside leaders threw their hats in with the Confederacy, and the region remained staunchly partisan throughout the course of the war. Pittsylvania County alone committed twenty-two companies to the southern cause, as communities sent their farmers, overseers, merchants, sons, and husbands off to the front, in units with such colorful names as the "Turkey Cock Greys" and "Pigg River Invincibles."[8] In all, "nearly four-fifths of the county's [white] men of military age served in front-line Confederate units" at some point during the conflict.[9] Halifax and Caswell men volunteered in similar numbers.[10] Danville resident Robert Withers commented on the martial fever in the city as the troops boarded trains for the eastern front in 1861, and William Tredway, captain of Company I in the Fifty-Third Regiment of Virginia Infantry, recorded crowds of well-wishers sending soldiers off in a festive atmosphere. Soldiers paraded the streets, and "each man had at least one large trunk well packed as if taking a trip to the Greenbrier White" (an elite resort hotel in the mountains). The Richmond and Danville Railroad, the same line that carried guano and plaster into the Southside and tobacco out of the region, was now carrying its farmers and planters to the eastern front.[11]

Once the troops departed, the border settled back into tobacco cultivation shaped by a new set of political and economic constraints. The
coming of war severely constricted tobacco markets; the Union naval
blockade of southern ports hampered foreign trade by the summer of
1861 and tightened over the following years.[12] Some tobacco was smug-
gled across enemy lines, perhaps most effectively in southeastern Virginia and eastern North Carolina where the front fluctuated with
regularity, but this contraband trade was a tiny trickle compared to the
antebellum torrent.[13] This contracted market threatened reduced prices
and a temporary halt to tobacco cultivation and industry. Pittsylvania's
William Sours noted the lack of attention to local business as the threat
of war loomed, writing, "All Business are at a stand and only warlike
preparations are being made instead of the busy hum of machinery."[14]
Many planters and their agents worried that there would be little demand for tobacco, and market conditions soon after the onset of hostilities seemed to support their pessimism. James Bruce managed to find
buyers for his hogsheads in the spring of 1861, but at modest prices; his
tobacco (probably dark leaf) sold for a little less than three dollars per
hundredweight.[15] Around the same time, merchants Williams and Carrington of Richmond informed Halifax planter William Bailey not to
expect strong sales of his leaf. They cautioned that a flooded market at
the onset of the conflict, as planters dumped their 1860 crops in hopes
of making a sale before trade ground to a halt, meant his tobacco might
bring as little as a dollar per hundredweight if he insisted on an immediate sale.[16] Despite the general excitement and unease engendered by
the war and fears of depressed markets, there were few immediate
changes in the Southside's agricultural routines. Tobacco farmers continued their annual tasks; they plowed and prepared the land and planted
their fields when the "season" wet the soil in the spring.

Within months the gloomy economic forecasts of the war's initial
days turned out to be largely incorrect. Although the conflict limited the
outlets for southern tobacco, it also created shortages in production, as
tobacco cultivation in Virginia's Tidewater, northern Piedmont, and much
of Maryland was disrupted by war, and many farmers in more marginal
areas shifted from tobacco to foodstuff production in anticipation of
feeding the Confederate armies. In addition, the substantial Kentucky
tobacco trade was cut off from southern consumers. Tighter wartime

markets also placed a greater emphasis on quality, strictures that the Southside's bright leaf growers and manufacturers were ready to meet. Sutherlin's correspondence and business papers highlight the opportunities the war created for bright tobacco. His agents across the South noted that consumers demanded good bright leaf, and merchants from New Orleans to Columbus to Fredericksburg wrote the manufacturer requesting products that were as "bright as possible."[17] A fellow manufacturer from Leatherwood along the Henry-Pittsylvania line wrote Sutherlin that it was the nature of a crowded marketplace to select for the best product, and he advised Sutherlin to follow his lead and "get the fine wrappers if you can."[18] A Lynchburg correspondent offered similar advice, stating of dark tobacco plugs, "Nobody seems to want these grades . . . bright tobacco . . . is the sort that is *wanted*."[19] In consequence of this growing preference for yellow tobacco products, Sutherlin's sales expanded throughout the Deep South as the war progressed. In addition to New Orleans and Columbus, he soon had agents in Atlanta, Memphis, Augusta, and Mobile, as well as smaller towns like Eufala and Huntsville, Alabama, and Albany and Americus, Georgia. Sutherlin was not alone in his efforts; even in rural reaches of the lower South his agents encountered other Danville salesmen hawking manufactured bright leaf.[20] Danville warehouse owners T. D. Neal and T. J. Talcott were among Sutherlin's Southside competition marketing bright leaf in the Deep South, though unlike Sutherlin, their company specialized in selling unprocessed yellow tobacco to southern manufacturers. In spring 1864, Neal and Talcott forged a partnership with an Augusta merchant to sell their tobacco across the Georgia Piedmont.[21] Wartime marketing also increased an antebellum trend: Danville manufacturers sent wagon trains of their twists and lumps across the region, touting the fine color and taste of tobacco produced in the Southside. These wagon trains allowed manufacturers to cut out the middleman fees of antebellum tobacco agents, all while emphasizing quality and provenance.[22]

By late 1861, Sutherlin and his fellow bright leaf manufacturers faced an unanticipated problem: demand for regional tobacco was so strong that they were having difficulty filling orders. Sutherlin's brand names—from "Yellow Bar" to "Old Virginia Fancy Twist"—touted his company's reliance on bright leaf, and his buyers scoured southern Pittsylvania, southwestern Halifax, and northern Caswell for quality crops to meet these advertising promises.[23] As early as August 1861, agent Johnson Owen warned Sutherlin that the business might have to cut corners to meet demand. He advised his employer that Atlanta consumers were less discerning than in other reaches of the South; the company might be able to use low-quality tobacco for plug and twist fillers "if the Wrapper is bright," but he cautioned that such as trick would not work everywhere, as buyers in cities like Memphis demanded only "good stock."[24] Sutherlin seemed to take Owen's advice; he bought two- and three-year-old crops and hail-damaged leaf, and he even tried to work with tobacco that had been cured poorly or packed in too high order (or too moist, a condition that often led to mold). Even this poor tobacco brought higher prices than bright leaf had at the war's onset.[25] As the war progressed, area manufacturers faced additional problems as the Confederacy's overburdened and undermaintained rail system became increasingly tied up with military traffic and licorice and other common tobacco flavorings became more difficult to procure.[26]

A public disagreement between Sutherlin and one of his business partners in 1863 provides a window into this intra-South trade and illustrates the tensions of the tight wartime market. James Millner, a small tobacco reseller from Pittsylvania County, entered an agreement with Sutherlin to market Danville-manufactured twists and plugs in Georgia. With Millner acting as the partnership's Georgia agent and Sutherlin providing most of the capital, the two men purchased either 745 or 1,300 (the number varies in the two accounts) boxes of low-grade finished tobacco, which they attempted to sell at a 100 percent profit on the Augusta market, relying on the reputation of Danville twists to make the sale. By all accounts, Sutherlin and Millner planned to parlay war shortages into speculative profits. The boxes failed to sell at the anticipated price, perhaps because the partners ignored the growing emphasis on quality products as the war progressed, and each partner accused the other of dishonest dealing. Millner claimed that he was unable to sell some of the leaf, and Sutherlin asserted that Millner had sold the entire lot, lied about the quantity and price, and pocketed the difference. In March, after efforts at arbitration failed, Millner published a pamphlet in which he accused Sutherlin of besmirching his good name by blaming Millner for the poor sales. Millner claimed that Sutherlin used the deal to dump poor-quality tobacco out of state, while trying to gouge his partner in the process, thus threatening the name of both Millner and Southside bright leaf. (If Owen's and Sutherlin's plan to adulterate Atlanta-area manufactured tobacco was any indication, Millner had cause to question Sutherlin's attention to quality.) He degenerated to name-calling, referring to Sutherlin as a "knight of tobacco trash" and an "ass in a lion's skin" and in a last bit of bluster predicted that "buzzards will roost upon his tomb."[27] Millner's attack brought into question Sutherlin's personal character and also challenged the manufacturer's dedication to making first-rate manufactured tobacco.

In a business where quality of product and quality of character went hand in hand, Sutherlin felt the need to respond in kind to Millner's accusations. In his own pamphlet, he described market conditions in Georgia, claiming that inflation and scarcity had pushed the prices of manufactured tobacco to between seventy-five and eighty cents per pound; with an honest effort Millner should have been able to get equivalent prices. According to Sutherlin, Millner displayed an "utter disregard of truth." An experienced local businessman and politician, Sutherlin used calm insinuations rather than brash diatribe throughout his pamphlet, though near the end of the publication he did resort to base race-baiting, accusing Millner of using his stolen profits to purchase a "*fancy negro girl.*"[28] (Sutherlin's use of the word "fancy" was a play on words that would have been quite apparent to white tobacco planters, who used the term to denote both slaves and tobacco light in color, or "high yellow.") Although Millner and Sutherlin's argument was in some ways a petty squabble, their case illustrates both the ability of regional manufactures to find markets for their products during the war and the difficulties inherent in the process. This public disagreement also demonstrates the assumptions of connections between personal character and the quality of an agricultural product that underlay tobacco cultivation and manufacturing. Both Millner and Sutherlin believed that the proper type of land made good bright tobacco and that only quality merchants were suited to make and sell that tobacco.

Responding to demand from manufacturers such as Sutherlin, many farmers and planters in the three counties still believed tobacco their most profitable option, and they continued to plant and cultivate the leaf, especially the more valuable bright variety, despite Confederate appeals for increased food production.[29] Although total tobacco production trended down in 1861, the year's crop of bright leaf seems to have been exceptionally large. Anderson Willis of Caswell wrote Sutherlin that war rumblings had done little to limit planting in his neighborhood, where his farm alone cured around 30,000 pounds of bright leaf. He boasted "we cured more fine Yellow than I ever did."[30] Farmers with a long history of viewing tobacco as the only certain cash crop in the region clung to the crop as security in unsettling times, and Confederate officials complained that farmers spent too much time and energy on bright leaf and paid too little attention to producing food.[31] In 1863, farmer George Jones was serving in the army but directing operations on his Pittsylvania farm through letters to his wife. He instructed her to continue selling tobacco despite requests from the Confederate government that farmers switch to foodstuffs, stating, "I had rather had it [tobacco] than Confederate bonds," as good leaf was appreciating faster than bonds at the time.[32] Caswell's William Hatchett gave his brother similar advice, declaring that the region's farmers should "make all the tobacco we can" as "the article is bound to sell high."[33]

As the war progressed, both growers and manufacturers placed increasing emphasis on the importance of producing bright tobacco, relying on the quality of area leaf to outsell the cheaper produce of other southern regions. The Danville warehouse of Neal and Lucas assured farmers that, despite the blockade, demand for bright leaf remained high among the seventeen tobacco factories operating in the town and promised that farmers who brought them "bright wrappers" could command "good prices," but their advertisement said little about lesser grades of leaf.[34] The account books of Robert Wilson bear out these claims. The Pittsylvania planter received premium prices for his entire crop in 1863, selling seventeen hogsheads of bright filler at eight dollars per hundredweight and six loose lots of moderately bright leaf for ten dollars per hundredweight.[35] Other planters did much better. Caswell resident A. Wellis received seventy-five dollars per hundredweight for his 1862 crop of exceptionally yellow tobacco, the sort of profit that made farmers wealthy overnight.[36] A resident of Danville noted that the speculation in bright tobacco in its raw and manufactured forms created a boom in the town, claiming that, "the war-prices paid for tobacco soon drew to [Danville] a crowd of tobacco speculators from both Virginia and

North Carolina."[37] Although Danville was the center of this boom, some of the small rural manufactories, such as the one located on William Finney's Museville plantation, also remained active throughout the war.[38] The emphasis on the importance of quality production further acceler- ated the turn to bright leaf over the traditional Virginia varieties. Dis- playing this bright leaf fever, merchant John Booker of Richmond wrote his client William Sims of Halifax that good tobacco was selling well despite the war and rhapsodized about a future of open access to north- ern and foreign markets: "Could the Blockade be Raised; what times, we should have. Glorious, Glorious, beyond conception."[39] For manufactur- ers like Sutherlin, contemporary profits were sufficiently "glorious." By 1863, Sutherlin's business was going so well that he pondered diversi- fication, contemplating spending $110,800 on a cotton plantation, com- plete with sixty-five slaves and a full complement of farm equipment, near Montgomery, Alabama.[40]

It was not only regional tobacco farmers who refused to let the war break their reliance on tobacco. Consumer demand for tobacco in the southern states during the war reflected a continuation of the nation's mid-nineteenth-century obsession with tobacco consumption. Whether as chewing plugs, snuff, pipe tobacco, or cigars (cigarettes would not become popular before the 1890s), Americans consumed the plant in enormous quantities.[41] By 1859, United States manufacturers produced roughly two pounds of finished tobacco products for every man, woman, and child (including slaves) in the nation. Almost all of this manu- factured tobacco remained in the country, where it was supplemented by foreign cigars and an undetermined quantity of home-manufactured chew and cigars.[42] Foreign travelers often commented on the ubiquity of chewing, smoking, and spitting in the antebellum United States, from the rural South to eastern cities. Charles Dickens, during an 1842 visit to the nation's capital, remarked with distaste on the omnipresence of tobacco in even the most formal settings. "In all the public places of America, this filthy custom is recognized. In the courts of law, the judge has his spittoon, the crier his, the witness his, and the prisoner his; while the jurymen and the spectators are provided for, as so many men who in the course of nature must desire to spit incessantly. In the hospi- tals, the students of medicine are requested by notices upon the wall, to eject their tobacco juice into the boxes provided for that purpose, and not to discolour the stairs."[43] This widespread usage continued during the Civil War. Union soldiers in particular commented on the ubiquity of tobacco in the South. As one Massachusetts soldier given to hyperbole wrote, "The little girls in these parts about seven or eight years old chew tobacco like veterans and babies smoke before they are weaned."[44] To-bacco was everywhere, used by almost everyone.

The outbreak of hostilities may have increased per capita consumption among the nation's men. Hundreds of thousands of soldiers from across the South (and the North as well) spent much of the war stationed in northern and eastern Virginia. These troops lived and fought near the Virginia and North Carolina tobacco belt, received many of their supply trains from Danville, and sought distraction from the alternating terror and boredom of the front in activities as varied as cards, sports, and religious meetings. Perhaps the most common diversion was the consumption of tobacco. As Bell Irvin Wiley, the famed historian of Civil War camp life observed, "It is doubtful if any single item except food, water, and letters from home was so highly cherished by Johnny Reb as 'the delightful weed.'" He goes on to suggest that southern generals feared "an anticipated cut in tobacco" nearly as much as they did the enemy. Southside families sent care packages that included tobacco to relatives in service or in northern prisons, which they gratefully consumed, shared, or sold. The war thus provided the Southside with expanded marketing opportunities across the South due to the loss of competition from tobacco districts behind Union lines and with a large and handy body of consumers in the form of troops serving in the eastern theater. Tobacco was a ubiquitous part of American life, and the Civil War only encouraged the habit.[45]

Although tobacco consumption was all but universal, not all regional residents supported the continued cultivation of a nonessential farm product during the trials of the war. Confederate officials expressed frustration at the prevalence of tobacco fields as the war progressed. Opponents of wartime tobacco production appealed to the patriotism of farmers and condemned growing the crop until after the Confederacy achieved independence. Charles N. B. Evans, editor of the *Milton Chronicle*, was among the most outspoken tobacco opponents. In the fall of 1863, an editorial in the paper declared that continued tobacco cultivation kept food prices higher than would be the case if every farm turned to growing

foodstuffs. The *Chronicle* laid the blame for high food prices on the tobacco manufacturer as well as the grower, noting that factory owners with
wallets fattened by wartime profiteering were willing to pay inflated prices
in order to feed their workers. According to Evans, "It would be a glorious
deed for this Southern Confederacy if every Tobacco Factory in it were
burnt to the ground and their very ashes scattered to the four winds of
heaven. . . . Our idea is that the people can do better without tobacco than
meat and bread."[46] The article reflected the realities of a conflict quickly
turning in the Union's favor, but it also emphasized the continued appeal
of high-quality tobacco, even on a dramatically constrained market.

Although tobacco remained an appealing cash crop throughout the
course of the Civil War, grain production provided a profitable wartime
alternative to tobacco cultivation, and there is substantial evidence that
a number of farmers and planters in the three counties turned to grain
cultivation out of either patriotism or practicality, while reducing their
tobacco acreage, especially that of less-profitable dark tobacco.[47] Grains
(corn included) were particularly attractive alternatives for several reasons: most farmers had antebellum experience raising wheat, corn, oats,
and to a lesser extent rye; these staples kept well; the crops could be used
for both human and animal consumption; and as the war progressed,
all types of grain found ready markets in regional towns and with Confederate quartermasters. Even districts devoted first to bright leaf often
produced surpluses of grain. William Ayers of South Boston wrote the
Danville quartermaster in 1862 that the Confederate government could
procure "as much Corn in this neighborhood as you want"; in South
Boston and neighboring districts, commercial tobacco and grain cultivation were not mutually exclusive.[48] Unlike with tobacco, a failure to market grain was not a complete loss as farmers could feed the surplus to
their livestock. In addition, increased grain production required little in
the way of changes to farm layout and management; farmers simply
planted their tobacco fields in wheat or oats, and if storage proved necessary, they could temporarily stockpile grain in their idle tobacco barns.
The three counties also contained an established infrastructure of grist-
mills, which could convert wheat and corn into value-added products:
flour and meal. Five mills were located along one fifteen-mile stretch of
Countryline Creek between Milton and Yanceyville alone.[49]

Before the war, almost every tobacco farmer had raised substantial amounts of corn, oats, and often wheat (wheat was especially popular in Halifax County), in addition to lesser quantities of rye. In 1860, for example, farmers in the three counties reported the production of 1,455,674 bushels of corn, 605,731 bushels of oats, 531,857 bushels of wheat, and 6,042 bushels of rye (table 4.1).[50] Increased production of these crops required little additional investment in farm equipment. Most farmers already owned scythes and cradles (for harvesting wheat, oats, and rye), and the same shovel plows, coulter cultivators, hoes, and drags used in tobacco culture worked for raising grain as well. Farm wagons, rail cars, and bateaux hauled barrels of flour and corn just as easily as they had hogsheads of tobacco. In short, the region had a prewar grain infrastructure, though a small-scale one, and thus Confederate officials' hopes that the southern Piedmont might become a Confederate breadbasket had some material foundation.

Environmental realities, however, posed more concrete challenges for expanded grain production. The region's poor hill soils produced low per-acre yields of all three major grains. Just as particular soils and climates encouraged continued tobacco cultivation, they worked to discourage larger grain crops. While the South's richest lands could produce yields of 60 bushels of corn or 20 bushels of wheat per acre, the Piedmont's sandy ridge lands were much less productive. In the districts surrounding Danville, farmers often harvested only 10 bushels of corn or 7 bushels of wheat per acre. And though the richer bottomlands opened up by bright tobacco's movement to the hills producing healthy grain crops, these landscapes were of limited expanse. To resume plant-

Table 4.1.

Grain Production in Caswell, Halifax, and Pittsylvania Counties, 1860

COUNTY	CORN (BU.)	OATS (BU.)	WHEAT (BU.)	RYE (BU.)
Caswell	403,288	116,888	110,227	1,846
Halifax	533,012	229,790	237,518	731
Pittsylvania	519,374	259,053	184,112	3,465

Source: Compiled from Kennedy, *Agriculture of the United States,* 104–106, 154–156, 158–160.

ing grain on hillside fields would reverse the farm rearrangement that farmers had found so appealing. Much of the region's grain production had been consumed on the farm, either eaten by planters and slaves or fed to livestock, and these trends continued during the war. Farmers understood these soil limitations—they had after all dealt with them for years in many cases—and continued to have their doubts about the "suitability" of widespread grain production in the southern Piedmont environment. In order to overcome these perceptions, the price for grain would have to skyrocket while bright tobacco markets struggled (which, as we have seen, was not the case).[51]

Demand for livestock also increased during the war as the conflict cut southeastern connections to the surplus pork and beef of Kentucky and portions of central Tennessee.[52] In particular, the market for meat swelled in the Southside, with its influx of northern Virginia refugees and its position as a supply base for Confederate forces fighting in the eastern theater. As early as fall 1861, there were localized shortages of pork despite the presence of swine on virtually every farm and plantation.[53] Some farmers increased their stock-raising efforts, selling pigs, cattle, sheep, and poultry on the Danville and Milton markets, or shipping their stock to Richmond on the railroad. Planter William Sims sold dozens of turkeys in Richmond for $1.30 a head in early 1861, a price that justified the freight charges on the fowl; another Halifax farmer recorded cattle selling for as much as $200 a head on local markets in 1864; and yet another planter sold nearly 4,000 pounds of beef on the hoof to the Danville Sutler's Department for thirty cents per pound the same year.[54] As the war drew to a close, livestock products also occasionally served as cash; Pittsylvania County officials accepted bacon as payment for taxes due in 1865.[55] Throughout the war there was good money to be made in meat.

Large plantations had long kept substantial herds of hogs and cattle, but the war encouraged small and middling farmers to raise as many animals as possible too, both to sell and for their own subsistence as inflation made provision dearer. Three examples serve to illustrate the diversity of animals on local wartime plantations and farms. An inventory of William Sims's Halifax plantation in 1863 included 38 horses, 63 head of cattle, a flock of 191 sheep, and 165 hogs. (Sims owned turkeys, and probably chickens as well, but they were not included in the inventory.)[56]

Pittsylvania's Berry Hill plantation, owned by Ruth Hairston, contained 13 horses, 37 cows, 54 hogs, and 103 sheep in 1865.[57] Sims and Hairston were large planters, but some smaller farmers also possessed a number of animals. When John Muse of northern Pittsylvania County passed away in 1864, he held but four adults and six children in slavery, yet his estate included 5 horses, 15 cows, 3 oxen, 66 hogs, 30 sheep, 30 geese, and 2 beehives.[58]

Grain, hay, and livestock may have been profitable alternatives to tobacco during the war, but this produce faced the danger of additional wartime taxes and confiscation, threats from which tobacco was exempt. In June 1863, the Confederate government circulated an order to the region's quartermasters that imposed a 10 percent "in kind" tax on certain agricultural commodities. The order authorized government commissaries to inspect local farms and take one-tenth of the surplus grain, livestock, fodder, and fiber available on each property.[59] On Samuel Wilson's plantation in Pittsylvania, the quartermaster visited in 1863 and 1864 and confiscated bacon, corn, sorghum, oats, wheat, hay, and fodder, providing Wilson with a small quantity of salt in return.[60] A North Carolina agent's visit to Pleasant Womack's Caswell farm in 1864 garnered a similar variety of agricultural produce. The official took a portion of Womack's corn, fodder, oats, peas, wheat, cotton, and bacon.[61]

By the second half of the war, this tax-in-kind system failed to produce the quantity of foodstuffs necessary for the continued supply of the military, and the Confederate government resorted to more draconian measures. Quartermasters in the three counties were authorized to seize supplies from farms at below-market prices, or for IOUs, and in all cases to pay with Confederate paper money, which was depreciating at a rate that made the bills all but worthless. If farmers refused to yield the demanded crops or animals, officials were authorized to impress them without payment.[62] In a typical broadside advertising the new rules from November 13, 1863, Quartermaster Jason Paxton, in charge of securing supplies from the southwestern Virginia Piedmont, laid claim to all surplus corn, rye, oats, hay, fodder, and straw in the district. Paxton instructed local farmers to deliver these goods to regional military stables, where they would receive government-established rates, ranging from $1.30 for one hundred pounds of wheat straw to $4.00 for a bushel of shelled corn.[63] Officials visited Samuel Wilson's plantation again under the new rules and seized wheat, oats, rye, hay, fodder, and wool.[64] Army officials also impressed mules and horses for cavalry and artillery units, further undermining local farmers' ability to cultivate their fields.[65] Quartermasters on occasion commandeered local fields more directly, as when they pastured horses and mules on plantations in northeastern Halifax County during the months before and after the Battle of the Staunton River Bridge in 1864. During the campaign, William Sims complained of the depredations of soldiers stationed on his property to guard the bridge. He lamented, "They take every thing on the plantation in the way of fruit & vegetables they can lay their hands on. I dont think by the fall I will have a potatoe."[66] Danville resident Robert Withers remembered the thoroughness of the quartermasters' agents in his memoirs forty years after Appomattox. He described the "inspections of corn cribs and smokehouses all through the country by officers of the Quartermaster and Commissary Departments delegated for the purpose, and all surplus food [was] carried off."[67] These confiscations and taxes in kind certainly blunted the profitability of raising foodstuffs during the war, and added to tobacco's appeal.

The turmoil of the war also affected the most important "natural" resource on Piedmont tobacco farms: human bodies. Slaves had long recognized the power of having ultimate control over their bodies and labor, using various forms of resistance to carve out their own space within a slave system that covered the spectrum between brutal compulsion and paternalism.[68] A vital part of slaves' agency lay in their ability to exploit the legal definition of their muscular energy as property. The most defiant act of agency, "stealing" oneself, whether to the woods for a week to escape an unbearable situation or making the long trek toward freedom, challenged masters' claims on bodies as commodities. There is anecdotal evidence that runaways increased in number in the Southside as the war wore on, especially once Union forces moved closer to the region.[69] Slaves who stole themselves engaged in a unspoken but devastating critique of slavery's environmental foundations: they rejected masters' assumptions about the naturalness of bondage, the nature of the relationship between black bodies and the Southside environment, and the hierarchy of man. And they turned their understandings of the Piedmont environment into tools to create some sort of personal liberty; intimate knowledge of the surrounding forests, fields, rivers, and wild

food sources facilitated escape, and familiarity with tobacco and its culture led to an understanding of when the absence of labor was the most
devastating (and hence powerful). As the war progressed, changing
southern political geography and emphasizing the importance of tobacco, it also changed the nature of running away.

The majority of slaves did not run away, of course, but the demands of
the Confederate army also withdrew black bodies from the fields and contributed to a shortage of white overseers needed to direct slave labor. Like
plantation masters, state and Confederate governments viewed the nature
of slave bodies in terms of muscular energy. Slave impressment was widespread in the three counties, as in the rest of the state, as southern officials demanded that planters loan their laborers to the cause. An October
3, 1862, act permitted state officials to impress slaves for military work—
though it exempted slaves on plantations producing only grain—and
promised masters of impressed slaves sixteen dollars per month in exchange for their service. Southside slaveholders faced a total of seven impressment orders during the remainder of the war; the commonwealth
ordered the first three, and the final four were direct orders of the Confederate government (and affected Caswell as well).[70] Among other tasks, impressed slaves built defensive works, laid and repaired railroad lines,
worked in the Danville hospital and arsenal, and assembled rifles in Belhartz Hall's Pittsylvania firearms shop. Some of these work projects were
quite large. The first commonwealth impressment in October 1862 drew a
total of 873 slaves from Halifax and Pittsylvania to labor on the earthworks surrounding Richmond. From 1862 to 1864, construction of an
extension of the Richmond and Danville Railroad from the latter city to
Greensboro, North Carolina, relied on as many as 2,500 slave laborers, at
least 300 of whom were impressed from Pittsylvania County.[71] Before the
Battle of the Staunton River Bridge in 1864, rebel forces had reported 300
slaves digging earthworks along the river, and Confederate officials also
impressed slaves from the surrounding countryside to work on the fortifications surrounding Danville in the last year of the conflict.[72] These impressed slaves worked in the regional environment in familiar ways; they
toiled in the Piedmont sun and rain to move earth and build infrastructure for military projects much as they had cultivated regional landscapes.

Local slaves' opinions of impressment went unrecorded, and likely
ranged from relief at temporarily escaping the plantation to fear of the

uncertainties of military work, which often entailed family separation, but the thoughts of white slaveholders were quite clear. The profitability of wartime bright leaf production caused many planters to balk at slave impressment. Pittsylvania and Halifax planters complained that the war's demand on white labor, from soldiering to raising food, made the work of every slave necessary, and tobacco planters seemed particularly reluctant to part with their workers. Under each impressment order, Virginia's individual counties were given target numbers of slaves, and the southern Piedmont proved the worst region of the state at meeting the Confederacy's call for military workers.[73] The head surgeon of the Danville hospital, frustrated with his inability to hire or impress slaves from the surrounding countryside, sent the following complaint to the commonwealth's chief surgeon: "Having failed after diligent efforts to procure colored men and women in sufficient number to meet the demands of the hospital, I respectfully suggest the expediency of authorizing the Quartermaster of this Post to impress the hands of Planters engaged in cultivating tobacco. It would I presume be inexpedient to cripple the agricultural force of the farmers who are raising breadstuffs and other subsistence supplies, but there is a very considerable number of persons in this region . . . who have turned a deaf ear to every appeal to the patriotism and have appropriated their best hands to the production of tobacco during the present season. It would be fit and proper to make these men bear a share of the necessary burdens of the war."[74] Many farmers were reluctant to yield their slaves to the war effort, but tobacco producers were the most militant. This reluctance to lease their hands to the government was due in large part to the continued profitability of bright leaf cultivation.

Equally troubling to many planters was the difficulty in finding overseers during the war. As adult males of fighting age who usually lacked the resources to purchase a substitute, the region's overseers were particularly susceptible to both martial fever and conscription. Planter William Sims, who had hired a replacement for himself at the start of the war, complained that two of his overseers were mustering for the army. Noting similar instances in the Black Walnut community, he worried that the impending war "has broken up almost entirely this neighborhood."[75] The shortage of qualified overseers worsened over the course of the war, causing white planters and their wives anxiety over the state of agricultural production and their own safety when surrounded by so many slaves. Sims's aunt, Phoebe Bailey, wrote him in spring 1864, requesting his assistance in locating a new overseer to help manage her plantation, which over the course of three years had lost four overseers to the draft.[76]

By fall 1863, Union army threats on Richmond and the relative security of the Southside caused the Confederate government to order the majority of Union prisoners then in the capital transferred west to Danville. Danville's wartime prisons posed two challenges to local agriculture. A large body of prisoners consumed a good deal of food that might otherwise have fed local citizens or Confederate soldiers, and—perhaps more important for tobacco production—southern officials housed these prisoners in buildings used for tobacco handling and manufacturing. In November, roughly 4,000 prisoners of war traveled the Richmond and Danville Railroad in boxcars, to be housed in six tobacco warehouses or factories that town officials had cleared for their arrival. The Hollands, longtime tobacco industrialists, owned three of the buildings.[77] It is unlikely that the structures were available because of a decrease in the city's tobacco auctioning and manufacturing during the war; authorities probably confiscated the structures out of necessity. Tobacco-related buildings were the largest and sturdiest in town, in addition to being centrally located close to the rail line. A number of Richmond tobacconists and warehousemen—among them James Thomas, Jr., the capital's largest antebellum tobacco industrialist—moved their operations to Danville during the war to avoid the risk of the front lines. With tobacco storage and manufacturing space at a premium in the crowded city, it seems likely that government seizure of the warehouses interrupted the active business of the Hollands and other owners, perhaps forcing the city's manufacturers to buy more leaf directly from farmers or to partner with other manufacturers.[78]

Whatever the prisons' effects on tobacco storage and manufacturing, the facilities certainly increased the danger of epidemic disease and exacerbated local food shortages. By early 1865, the prisoner population had grown from 4,000 to almost 7,000 men, this in a town that held only 6,000 residents at the outbreak of the war.[79] This dramatic population increase, augmented by war refugees from northern and eastern

Virginia attracted to the town's safe location, led to a number of serious sanitation issues.[80] Confined 200 or more to a floor in the warehouses, with poor clothing, slim rations, and almost no access to clean water, Union prisoners rapidly succumbed to an epidemic of smallpox or a more gradual death by scurvy or dysentery. In his examination of the town's prisons, historian James Robertson calculated that out of the 7,000 men imprisoned during Danville's fifteen months as a detention center, only 3,000 survived.[81] Conditions were exceptionally difficult for black prisoners of war. Confederate authorities confined all black prisoners to one floor of prison number three, in especially cramped quarters. Prison officials forced them to labor on local defensive works, where they dug trenches around the city in the hot sun beside impressed slaves, work from which white prisoners seemed exempt. Prisoner Alfred Roe claimed that black troops faced the additional threat of being pulled into the local tobacco economy, as planters who claimed prisoners were runaway slaves could take the men as laborers for their plantations, an action that some unscrupulous planters may have used to recoup their loss of tobacco labor through impressment.[82]

Poor sanitation and overwork promoted the spread of such diseases as smallpox and dysentery, and malnutrition was also a chronic problem in the prisons. Food supplies moved through Danville regularly, traveling from the region's fields and pastures to the front lines at Richmond and Petersburg, but only subsistence amounts remained in the region by the last year of the war. George Putnam, a prisoner in 1864 and 1865, wrote of the Danville and Richmond Railroad, "The one-track road was very fully employed with the trains from the South bearing to Lee's army such supplies as were still to be secured in the almost exhausted Confederacy."[83] Faced with a general shortage of supplies for town residents and the unceasing demands of Confederate troops in the eastern theater, food for Union prisoners received the shortest shrift. Survivors recorded appalling fare during their detention. Prisoners ate rough cornmeal made from kernels and cobs ground together, the occasional ration of trashy rice infested with weevils and maggots, a few peas, slaughterhouse trimmings such as beef eyes and lungs, and on rare occasion the men were served a soup of "musty rice and spoiled cabbage." The men augmented their slim rations with lice and the occasional trapped rat.[84] Supplies were so scarce that the guards supervised prisoner foraging expeditions in the countryside surrounding the town, such as an excursion to gather persimmons that W. H. Newlin recorded during his stay in Danville.[85]

The demands of the Confederate army, the needs of the Danville prisons, and continued competition from tobacco production all contributed to food shortages and drastic inflation in the Southside as the war progressed. As early as fall 1862, prices for livestock and foodstuffs were on the increase. Halifax farmer and tanner Daniel Merritt complained of stolen hogs and corn priced as high as $12 a bushel in his community; by the following autumn, he recorded sales of sheep at $40 each and cattle at $200 per head. Likewise, William Sims bought sugarcane seed in hopes of producing his own molasses in 1863, when faced with prices of $11 per gallon on local markets.[86] By 1864, a barrel of flour commanded $150 on the Danville market.[87] High prices, poor quality, and small quantities of foodstuffs provoked some complaints. Even when faced with accounts of the conditions in Danville's war prisons, some county residents argued that Union internees were no worse off than the average Southside farmer. William Sours, a native of Pennsylvania who had moved to Pittsylvania County in the late 1850s, wrote his northern kin that he had little sympathy for the prisoners: "So far as rations was concerned they got the same the soldiers got but corn bread and beef did not suit those who never lived on corn."[88]

Inflation worsened as the war dragged on, and food prices climbed beyond the reach of many residents; Confederate greenbacks became so devalued that one silver dollar bought seventy paper bills in Danville by the war's conclusion.[89] A Richmond merchant sent the following price list to a Halifax County planter in January 1865, illustrating the extent of the food shortage and the severe lack of faith in Confederate currency:

flour	$750–$800 per barrel
corn	$80 per bushel
cornmeal	$90 per bushel
peas	$65–$90 per bushel
dried apples	$65 per bushel
dried peaches	$120 per bushel
beans	$100–$110 per bushel
onions	$120 per bushel

turnips	$25 per bushel
Irish potatoes	$40–$50 per bushel
sweet potatoes	$50–$60 per bushel
beef	$4–$5 per pound
poultry	$5–$6 per pound
molasses	$50 per gallon[90]

Other shortages also troubled regional agriculture. A limited amount of guano was still available, but by 1863 it was lower-quality Mexican produce rather than the Peruvian article and cost a prohibitive $130 per ton. A dearth of civilian powder and shot also limited locals' ability to turn to hunting to supplement their diets, even though by the war's end a farmer complained that "the woods and fields of this vicinity are literally filled with game."[91] These inflated prices and a general lack of foodstuffs at any price created a good deal of hunger along the border. Danville's warehouses held stockpiles of rations, but that food was for Confederate troops rather than civilians. Additional food supplies, meager though they were, went to the thousands of Union soldiers incarcerated in Danville's prisons. According to Jeffrey McClurken, hunger and growing impoverishment in the countryside caused serious discontentment among area troops concerned about their families at home, though this situation did not lead to widespread desertion.[92]

By the beginning of 1865, the outcome of the war seemed all but certain. That many landowners continued to produce bright leaf despite the general despair, food shortages, and rampant inflation in the southern Piedmont, suggests just how durable notions of the best use of local landscapes were. Sherman's Georgia campaign and Union successes in eastern Tennessee cut Piedmont tobacconists off from much of the Deep South market, but bright lumps and twists still sold well in places that remained connected to Danville by rail, such as the Carolina and eastern Georgia Piedmont. In January 1865 agent Thomas Patrick could still write that "tobacco is selling very well" in Augusta, despite the threatening presence of Sherman's army in Savannah, an observation confirmed by Augusta retailers requesting additional boxes of manufactured tobacco from Sutherlin the same month.[93] The Deep South addiction to Southside bright leaf remained strong even in the face of the Union army and defeat. Perhaps the most adamant declaration of local farmers' commitment to tobacco culture lay in their decision to plant a crop in the spring of 1865. Although transplanting did not take place until after the surrender, planters and farmers had prepared their plant beds and fields throughout Caswell, Halifax, and Pittsylvania during the last winter of the war. Although tobacco was firmly rooted in the region's past, Southside farmers also believed the crop to be the key to their future.[94] The Confederacy may have been coming to an end, but tobacco endured.

The fall of Richmond finally brought the leading edge of the war home to the Southside in the first week of April. As Lee began abandoning his defensive lines near Petersburg, the Confederate government made plans to retreat to Danville along the Richmond and Danville Railroad. Along with the Confederate archives and perhaps as much as half a million dollars of gold and silver bullion, President Davis and his cabinet reached Danville on April 3, where they met in Sutherlin's mansion.[95] Davis's arrival sparked wild rumors in the Southside, with rampant speculation on the future of the Confederacy and the proximity of Union troops. Rural people, black and white, flooded the town on whispers of stockpiles of soldiers' rations in its warehouses and were barely restrained by Confederate officials. Lee's surrender at Appomattox on the ninth spurred Davis to flee Danville, which he and the vestiges of the government did the next day. Upon departure, Davis ordered the main bridge across the Dan burned, along with the town's warehouses and stores, in order to slow Union forces and deny them supplies. Garrison commander Colonel Robert Withers refused Davis's orders, in part due to requests from tobacconists Sutherlin and T. D. Neal, but general looting broke out as crowds of refugees forced their way into the warehouses and the homes of the wealthy in search of food. A number of looters were killed in an accidental explosion while searching the arsenal, but Danville's commercial infrastructure, including its numerous tobacco factories and warehouses, survived the turmoil largely intact.[96] By April 27, the Sixth Corps of the Army of the Potomac finished its march through Halifax and Pittsylvania along the Richmond and Danville line and occupied the town without opposition. The corps printed a small newspaper on a confiscated press and informed the town of the rules of occupation, forbidding any resistance and in turn promising that Union troops would cease foraging in the surrounding countryside. Soldiers camped in the town itself and on the surrounding hills of Pittsylvania and Caswell.[97]

Several Union soldiers stationed in Danville blamed local planters and agricultural practice rather than the war for food shortages and were less than complimentary of regional farming methods, especially those associated with raising tobacco. Their observations present an interesting "outsider" perspective on local conceptions of land use and environmental change. W. H. Newlin, who escaped from a Danville prison late in the war, described the countryside for several miles along the Dan River west of town, prime bright leaf land, as barren and desolate old fields grown over with a thick growth of secondary shrubs.[98] These barren stretches could have been either fields abandoned as part of regular tobacco cultivation practices, land awaiting bright leaf, or stretches of ground temporarily vacated as the war in the Southside heated up. Where locals saw enriching ground, Newlin saw degraded land.

Alanson Haines, of the Fifteenth Regiment of New Jersey Volunteers, was more directly critical. Haines's unit was part of the Union force that occupied Danville after Appomattox, and he and his fellow soldiers camped a few miles south of town in northern Caswell County. Haines was glad to see the end of fighting, but he was less enthused about his environs, revealing particular disgust with the agricultural practices of the rural Piedmont. He described the region's farmland as "miserably cultivated, and only with the view to get the most from it for the present crop, regardless of the future." Haines went on to connect locals' abuse of the landscape to their abuse of former slaves. He noted that desperate farmers and Confederate foragers had stripped clean all edible crops from an already slovenly farmscape, and as a result, "colored men and boys crowded our camps, asking for employment, and saying nothing of wages if they were only fed."[99] Lewis Foster of the Ninth New York Artillery, stationed in the town, made a similar observation, noting: "There are lots of Darkies with us in camp. They do all of the work for the boys; bring wood and water, and wash for us for their bread, which we can afford to give as we have plenty. They say the Rebs told them that the Yanks had horns and tails like cattle."[100] Whereas local planters viewed intense use of land and mastery over enslaved people as two sides of one paternalist coin, these northern observers reversed the equation. Haines and Lewis equated poor stewardship of the land with the past dominion of masters over their slaves, a critique leveled at patterns of land use central to the production of bright leaf. But Union critiques of the relation between Southside landowners and the environment would do nothing to change the devotion of planters and farmers to bright tobacco.

These critical observations of the Southside landscape continued during Reconstruction. John Dennett, a correspondent for the *Nation*, passed through Pittsylvania and Caswell counties in 1865 as part of a trip chronicling the condition of the conquered South. Dennett was not impressed by regional agriculture; his account limned a rural landscape worn out by a long war and feeling the effects of intensive bright tobacco cultivation, implying that military conflict and agriculture had waged simultaneous war upon the landscape. As soon as he crossed the Staunton River and rode south into Pittsylvania, Dennett encountered signs of severe erosion. He complained that the road was often impassable "where the rain had cut deep channels and gulleys in the earth, or had washed bare the rough ledges of rocks."[101] Across this eroded landscape, patches of old field had grown up in tangled forest. Crossing into Caswell County, he found the prospect gloomier still: "The country is the poorest I have yet seen, with crops that seem less abundant and healthy than those further north, and with less timber."[102] Despite the poverty of the soil, throughout this desolate landscape Dennett found the presence of tobacco. He commented on the fine planters' mansions surrounding Danville, noted Chatham's stately courthouse where farmers gathered, and described meeting black wagoners driving their oxen home from the Danville warehouses.[103]

Dennett's observations on the Southside were far from unique in the years following the war. Edward Pollard, another regional traveler during Reconstruction, also had harsh things to say about an unnamed corner of the Virginia Piedmont. He lamented the region's "galled hills and old fields, worn to exhaustion by the plough and hoe in the culture of tobacco and corn. It is a level and barren picture. The old field pines, the broom sedge and the persimmons are the memorials of 'improvement' under the past system of slavery."[104] John Trowbridge, roaming the same region in 1867, expressed a similar sentiment. Of the rural countryside he noted, "A striking feature of the country is its 'old field.'

The more recent of these are usually found covered with briers, weeds, and broom-sedge,—often with a thick growth of infant pines coming up like grass." Trowbridge believed he knew the cause of the land's impoverishment; he concluded, "Tobacco has been the devouring enemy of the country."[105] George Bagby, riding the Southside's back roads in the 1870s, connected the ragged regional landscape to bright tobacco culture in particular. He declared of southern Pittsylvania, "In the bright tobacco belt . . . there has been a great advance in lands, and an astonishing prosperity among individuals once very poor." This economic success, however, came at the cost of land abuse. Bagby mocked the tobacco region as the realm of the "sweet Virginia gulley, the rich Old Dominion gall, and the lovely scrubby blackjack [oak]." He noted that the border was wealthier than the surrounding dark tobacco districts, but his critique of the regional environment suggested that bright tobacco threatened the foundations of its own success.[106]

Northern commentators in particular brought assumptions that the southern defeat in war stemmed at least in part from an inferior way of life; in agricultural regions this implied a critique of farm practices. These assumptions in large part drew on antebellum travelers' characterizations of southern slave agriculture, with Frederick Law Olmsted's *The Cotton Kingdom* serving as the classic example.[107] Haines, Dennett, and their fellow commentators held an incomplete understanding of the environmental and agricultural realities of bright tobacco cultivation. Their observations at once combined northern farmers' general aversion to the extensive methods of plantation agriculture, tied to a firm belief in free-soil ideology, with a sort of grim admiration of the destructive power wrought by the Union war effort. Danville's gullied and overgrown fields simultaneously symbolized the "right" and the "might" of the northern cause. But the writing of these commentators suggests that they knew nothing of the soil preferences of the region's dominant crop, its value in comparison to agricultural alternatives, or its grip on the collective Southside imagination. These contextual understandings made all the difference in interpreting the landscape. Land that Union soldiers and travel writers viewed as hillside thickets alternately appeared as prime tobacco land in the eyes of Southside farmers. That is not to say that these critical observations did not represent a basic reality of the region's political economy. This neglect, whether intentional or brought on by exigencies of war, had if anything accelerated during the conflict. Erosion, deforestation, and soil depletion did make all types of Southside farming less productive over time, even if bright leaf tobacco seemed to do well on ground far too weak for grain crops. Outside observers did misunderstand much about farming in the bright tobacco belt, but they did not completely misrepresent the agricultural realities along the border.

The end of the Civil War brought both threat and promise to the Southside. Slavery was dead, but northern and European markets were open to bright leaf once again, and the region's manufacturers had been hardened by the fierce competitiveness of marketing during the conflict. In the fall following the war's end, William Sours expressed both the promise of the bright tobacco economy amidst the rubble of the war and the growing discord of Reconstruction in a letter to his brother in Pennsylvania. Sours wrote: "Imagine the country devastated by two ravenous and beligerant armies of greater numbers than in former wars every thing devoured from sucking Pigs to lean milch cows, all grain and vegitables destroyed, and trodden under foot and all avenues of trade shut out with thousands within to feed and then you will only have a faint idea of real Hard times. but we think the nigger song of hard times come again no more will in cours of a year or two will be realised. excepting with the freed Nigger if they are not colonized will remain a lasting monument of root Hog or die Their condition today is a thousand times worse of than when they were servants and thousands of them already say so."[108] Sours's confidence that better times were coming for whites was matched by Sutherlin's belief in the bright future of yellow tobacco. The industrialist greeted occupying federal forces with the same hospitality he had shown the fleeing Confederate officials. Sixth Corps commander General Horatio Wright dined and visited with the Sutherlins following the city's capitulation, and the manufacturer entertained General George Meade, of Gettysburg fame, and his entourage as they passed through the city en route to North Carolina.[109] Sutherlin's actions reflected his desire to return to the business of making and selling tobacco across the nation and the globe. Sutherlin survived the war with the bulk of his wealth intact, despite losing sixty-three slaves, and by mid-June he had already reestablished his contacts in New

York and London in preparation for sending manufactured bright leaf to the Northeast and overseas.[110] For Sutherlin, his fellow manufacturers, and the region's bright tobacco growers, the end of the war meant a return to business.

The most significant result of the Civil War for Southside agriculture was that the conflict ultimately changed so little. The war that swept the South held the potential to remake the landscape. It could have turned farmers away from tobacco in favor of grain; cut off access to all markets; and destroyed the barns, warehouses, and factories that made up the tobacco infrastructure. Most important, a war over the very meaning of American identity held the potential to change landowners' minds about the best use of Southside environments, to make planters rethink their reliance on a single staple crop. But the Civil War did none of these things. Bright tobacco survived the war, just as it had co-opted agricultural improvement, and in its survival convinced many landowners that it was destined to drive the local economy, that its ascendency was only natural.

The war did, however, free almost 40,000 slaves in the three counties.[111] After Appomattox, the African American half of the Southside population would work to determine just how much freedom they actually possessed and their potential place within tobacco culture, and the white half would labor to resume agricultural life on antebellum terms. Both black and white Southsiders took for granted that bright tobacco lay at the center of their collective future. Despite this general confidence in the potential of yellow tobacco, a key question remained. What form would Southside tobacco labor take now that the slaves were free? And how would that new labor structure affect the landscape?

FIVE

Fire in the Fields

RECONSTRUCTING LABOR AND LAND
FOLLOWING THE CIVIL WAR

IN HER 1929 HISTORY of Pittsylvania County, Maud Carter Clement recorded an incident that she believed summarized the struggles of Southside Reconstruction. According to Clement, the following encounter took place in Chatham sometime in the late 1860s:

> On another court day when the town was full of men, armed as before, one of the negroes had some trouble with a white man and the blacks grew very ugly and sullen and gathered in a mass at the lower end of Main Street. The white men gathered at the upper end of the street and the divergent crowds began approaching one another. The white men were led by a man named Whit Bradshaw who carried a long pistol, and as he came opposite the courthouse he waved the pistol around his head, crying, "Clear the way," and brought his gun down to aim on the approaching negroes. The latter began running in every direction through alleys, behind stores, anywhere to get out of reach of those guns. But not a gun had been fired. The white people had no desire to hurt their former slaves. When they saw danger threatening they met it coolly and thus averted it.[1]

Despite Clement's assertion that freedpeople were both the source of danger and inherently safe from the potential of white violence, Pittsylvania's emancipated slaves had good reason to fear Bradshaw's pistol. Black struggles for economic and social independence met harsh white opposition during the years following the war. Emancipation launched a long and tangled negotiation over exactly what form labor would take in the three counties, a struggle fought in the courts, at the polls, in the tobacco fields, and in the offices of the Bureau of Refugees, Freedmen, and Abandoned Lands (BRFAL, more commonly known as the Freedmen's Bureau).[2] And it was a battle fraught with undertones of racial violence that no amount of historical whitewashing could erase.

Other episodes of violence more directly knitted together nature and labor. In 1867 in Caswell County, for example, the pressures of the tobacco harvest provoked bloodshed. A white farmer named John Blackwell demanded that freedman Griffin Cobb work on a Sunday, perhaps because Blackwell felt his tobacco was ripening too quickly or feared that an early frost loomed. Cobb refused, asserting his right to the traditional day of rest, and an enraged Blackwell attacked him "and broke a wavy tobacco stick [used to hang plants in the barn] over" the freedman's head. Refusing to meekly accept a beating, Cobb fought back. Blackwell ran to his house and retrieved a gun, returning to fire at Cobb who fled to the surrounding woods. Cobb took his case to Freedmen's Bureau officials, claiming that he feared Blackwell would kill him. The freedman had reason to fear his employer: Cobb mentioned in his statement that Blackwell had a history of violence against his agricultural workers, "as he killed Littleton Gwyn after the surrender."[3]

As Clement's court day vignette and the argument between Blackwell and Cobb illustrate, the specter of violence undergirded all regional labor struggles, and these struggles in turn were firmly rooted in the Southside environment. The details and demands of bright tobacco cultivation exacerbated violent propensities as high tobacco profits encouraged landowners to secure labor. White landowners too poor to hire laborers raised their own small crops of tobacco rather than resort to wage work, and thus larger landowners turned to freedpeople as the only available source of tobacco labor, a decision encouraged by the historic regional association of tobacco and slavery. The seasonal rhythms of tobacco placed further tensions on labor relations, as in all forms of agriculture the biology of plants and the nature of people's lives were intimately connected. Growing tobacco required year-round labor, but at certain crucial periods timely work was essential or a year's crop might be lost. The tasks of transplanting, harvesting, and curing bright leaf all called for punctual and strenuous effort, and the uncertainties of free labor threatened white landowners' control over production. It was during these crucial periods—when weeds grew quickly, storms drenched the earth, tobacco ripened, or frost loomed—that freedpeople had the greatest power to flex their social and political muscles. It was also when white landowners were most likely to turn to violence in order to retain their control over labor, tobacco, and land.[4] Reflecting these rhythms, night-riding and other confrontations became seasonal activities. Even the riot at the courthouse might have had environmental connections hidden in the narrative. Although Clement did not record the season in which the violent episode took place, similar instances occurred when crowds of black and white residents gathered for winter contract-signing.

If the demands of the tobacco season provoked violence, the region's physical landscape made systematic aggression possible. The entire region—with the exception of the growing town of Danville—was rural, a sparsely settled countryside that hindered the effective organization of black wage workers and tenants, who were soon more scattered across the three counties than they had been as slaves. In the postwar period workers followed tobacco, and as a crop that experienced few economies of scale, tobacco fields spread across the landscape in small patches. This same rural geography that hindered black defense and organization aided the efforts of night riders, who attacked isolated cabins before melting back into the woods and fields.

Historian Mark Fiege, in his excellent reconceptualization of crucial moments in American history as intrinsically environmental, has pointed to Reconstruction as a period in need of more attention from environmental historians. Fiege notes, "Efforts to remake people's relationships to one another often have involved efforts to remake their relationships to nature, and Reconstruction was no exception. The environmental history of Reconstruction centered on the just landscape that the freed people desired and the racially repressive and exploitative landscape that reactionaries tried to impose."[5] The Reconstruction experience in the Southside certainly gives truth to this observation, with the landscapes figuring most prominently being sites of tobacco production. By following the role of tobacco in shaping one region's experience following Emancipation, we begin to understand the importance of local environments on understandings of bodies, labor, and freedom. And we can tease out some answers to important questions about the ways in which people responded to a very limited sort of liberty.[6]

Planters who sought to build a landscape around the needs of bright leaf turned their postwar energies to reordering the labor needed to work the land. This chapter probes the dramatic ways in which labor relations, so central to the interactions of people and landscape, changed in Caswell, Halifax, and Pittsylvania in the first few years following Emancipation. Most local African Americans first moved from slavery to some form of wage labor and then on to sharecropping, shifts that reflected both struggles over racial control and the demands of bright tobacco, which remained the region's economic engine. This ordering of labor reflected conceptions of land and work that had begun to coalesce during the antebellum period. After Appomattox the power of large planters appeared broken, but this change proved more illusory than real in the three counties. White landowners largely retained control of the landscape, and they would spearhead bright tobacco's expansion and the growth of the local tobacco industry in the decades to come. But first they had to master the natures of both people and place made un-certain by the war; it was a process they would undertake, in many cases, by any means necessary.

At the conclusion of the Civil War, Pittsylvania, Halifax, and Caswell, like the rest of the South, faced a bleak prospect. Broad swaths of south-ern infrastructure lay in ruins; Virginia's and North Carolina's port cit-ies had faced fire and federal occupation; agricultural pursuits had suffered from four years of general neglect while some farmers were away at war; and slavery, the basis of the region's agricultural economy, had been eliminated.[7] Livestock, another pillar of regional agriculture, suffered decimation from contagious diseases, such as glanders, spread by the massed horses and food herds of the Union and Confederate armies. Large planters retained their land in most cases, but the loss of their slaves erased a large percentage of their personal estates. Wide-spread southern bank closures restricted the credit available for re-building plantations or hiring labor. To compound these difficulties, an unusually high percentage of the region's soldiers had either died in the war or returned home as invalids (a Pittsylvania regiment was one of the hardest hit in the Confederacy's bloodiest debacle, losing 40 percent of its soldiers as casualties in a single afternoon as part of Pickett's charge at Gettysburg).[8] Many a family faced the difficult task of the Merritts, who brought the body of their son James home to Halifax from its temporary burial place at the eastern front. His father Daniel wrote, "We brought James home and bur[ied] him . . . in back of the garden under a walnut tree[.] he [was] nearly decayed[,] head & feet & hands off[.] sad sight."[9] The destruction of war was as harmful to property as to human life; Benjamin Simpson, of Danville, described his father's situation at the end of the war as a typical one for local planters: "His property which mainly consisted in slaves, was swept away and all that he had left was about one hundred and fifty acres of land."[10]

Although the future seemed dim at first glance, the region's immediate economic prospects were actually brighter than those of much of the South. The three counties were behind the battle lines for the vast majority of the war, Union troops quickly repaired damage to the Richmond and Danville Railroad, and towns like Danville, Milton, and Halifax Courthouse largely escaped the burning and looting from both sides that befell so many southern urban areas. The culture of bright leaf that had continued through the course of the war promised a palatable transition into Reconstruction, as the market for regional tobacco products expanded with renewed access to northern and foreign markets. Tobacco also remained a crop that farmers could produce with little in the way of capital or equipment. The economic hardship caused by the wartime destruction of expensive equipment, such as the sugar mills, cotton gins, and rice mills necessary for the production of other southern staples, had few parallels in tobacco cultivation.[11] If farmers had a standing tobacco barn and could organize enough labor to tend their crop, they could resume raising tobacco. Even a burned barn was not an inordinate hardship; a few workers with axes and access to timber could build one in a couple of weeks. The main resources necessary for quality bright tobacco production resided in the region's soils, its farmers' (and former slaves') environmental knowledge, and a supply of strong backs, commodities that the war could not easily destroy.

A resumption of tobacco—and especially bright leaf—production was enticing following the end of the war because demand remained extremely high. Just as farmers who were able to continue production throughout the war profited, growers who raised a crop during the first years of Reconstruction saw tremendous gains at market. During periods of shortage bright tobacco prices shot above even wartime levels. In just a few representative examples from the first two Reconstruction years, Joel Hubbard of Halifax sold twelve lots of bright tobacco for as much as $61 per hundredweight, W. C. Tate sold his crop as high as $35, Philip Howerton's tobacco sold for $14.50, and a neighbor of William Hatchett hoped to command $50 per hundredweight for his entire crop.[12] These prices compared favorably with the best antebellum sales and provided powerful incentive for farmers to continue (or begin) bright tobacco production in the cash-strapped region.

Emancipation, while certainly not unexpected by the war's end, dramatically upset antebellum labor relationships in the three counties. Some farmers tried to keep former slaves on their plantations, others quickly sought alternative forms of labor, and yet others remained indecisive, unsure of how to proceed in a tobacco economy without enslaved workers. They remained certain tobacco was the best use of many local environments, but what was the best way to produce it? Daniel Merritt, a small farmer and tanner in Halifax who had never owned slaves, expressed a general apathy in his diary, writing of his neighborhood in 1865, "Negros are all free, some stay & some go."[13] Bird Ferrell, of Pittsylvania, was more worried. He wrote to a relative that "my freed Negros have all left me," and he blamed his neighbors for luring away his former slaves with promises of wages.[14] William Sims, of northeastern Halifax, put on a brave front; he boldly proclaimed, "I am glad the institution of slavery is broken up and I think we will be more prosperous and happy than we have ever been." Sims, who had worked 163 slaves on his Black Walnut Plantation two year earlier, laid out his 1865 labor plan as well as his vision for a future of all-white agriculture. "I employ most of my negro men to work in the crops, none of the women except those about the house—five of the men I have driven off, as they were worthless, and I did not want them. I furnish all with food but charge the food to their husbands to be taken out of their wages. I expect next year to rent out all my land to good tenants and git rid of all the negroes. I want white servants mostly about the house." He prefaced his account with the complaint that too many of his former slaves remained on the plantation despite his "wish all would go, but I cannot induce them to leave."[15] Other substantial planters expressed similar desires to be rid of former slaves altogether, yet they obviously coveted their labor.[16]

Although terms of labor were of vital interest to both former masters and freedpeople following Emancipation, the question of land ownership was of at least equal importance. What use was a practical expertise with tobacco lands without land on which to practice it? Freedpeople in the three counties understood that access to land was instrumental to their ability to make a new life, and they pushed for a fair share of the countryside, but in this pursuit of land they met staunch opposition. Local whites vigorously resisted sharing their property; as one Pittsylvania planter declared to a Union officer during the occupation, "I certainly do love a nigger as a nigger, but when they set up for white folks I've no use for them at all."[17] And for most Southside whites during Reconstruction, black desires to own land were attempts to "set up for white folks." Even Freedmen's Bureau officials and northern travelers in the region seemed to have little sympathy for black desires for land, or where there was sympathy, little to no ability to transfer property. Robert Withers, a Pittsylvania man who worked briefly for the Bureau, recorded the following story concerning a freedman who felt the federal government was obligated to provide former slaves with farms.[18] Withers overheard the freedman's conversation with a Bureau official:

> "I came sir," was the reply, "to ask you to lot me my land. They tell me the niggers is all free now and has to look out for their theirselves. I've got a wife and seven children, it's now late in April and time corn was in the ground, and I want you to lot me my land so I can get to planting."
>
> The Colonel [Fletcher] looked surprised, but at once replied, "Why, I have no land to give you, your former master's land still belongs to him, and so with all the other land. It still belongs to its owners, and the Government does not own an acre of it." The negro looked at him in evident surprise, but said, "Didn't you all set Master's niggers free?"
>
> "Yes," said the Colonel, "but we had no right to take his land." I never saw disgust and contempt more plainly expressed on the human face than they were on that of

this negro. He said, "If you had the right to take Master's
niggers you had the right to take Master's land too. And what
good will freedom do the niggers if they get no land to work to
make their bread?" This was evidently a poser and the Colonel
could only reiterate his assurance that the Government owned
no land that could be distributed to the negroes just set free.[19]

Although he was unable to counter the freedman's logic, Fletcher was
quite correct in his statement that the federal government had little to
give in the way of commonwealth land. At the end of 1865, the Freedmen's Bureau laid claim to just 75,653 acres in the entire state of
Virginia, only 3,366 acres of which were located in the Piedmont. Even
this meager amount would never become the permanent property of
freedpeople. By the end of 1868 the federal government had returned all
Virginia acreage to its antebellum owners or their families, and the situation was similar in North Carolina.[20]

Regional Bureau and army representatives were overwhelmed by
the large number of freedpeople who moved into the district's towns
looking for work and aid. Poor sanitation plagued the refugee camps
and typhoid became a major killer in Danville. Although these difficulties were quite severe, many northern officials seemed convinced, like
southern landowners, that freedpeople's place was on the land. Captain
J. F. Wilcox and other officials urged freedpeople to sign work contracts
with whites, often their former masters, and return to the countryside.
In this the Bureau simply echoed the Union troops who occupied Danville at war's end; one of the first orders of the federal Sixth Corps when
it took Danville was that freedpeople remain on their farms conducting
"their work as usual."[21] Wilcox had little inclination (or ability) to requi-
sition land and seemed most concerned with keeping his branch of the
Bureau operating smoothly and within budget. He also made a con-
certed effort to work with a local black organization, the "True Friends
of Charity," to place freedpeople with white landowners.[22] Part of
Wilcox's inclination to encourage contracting centered around the Bureau's limited means and the extensive poverty of a large percentage of
freedpeople. Lacking both the resources to provide significant aid and,
seemingly, a belief that former slaves truly deserved land, Bureau officials distributed rations, gave out firewood, and arranged for medical

care for the sick. Wilcox complained that his office could barely keep up with the demands of people "in quite a deplorable condition."[23]

Of a similar opinion as Wilcox, Caswell County farmer and sheriff John Flintoff believed freedpeople needed white supervision rather than their own land. He recorded that all his former slaves stayed on with him as laborers but worried that "most of others are running about from home to home believing they are free—many of them are killed and die-ing for want of money and protection—poor creatures—I have to ride often after them and arrest them for trial, for their fighting, stealing and other meanness they are very troublesome to the white people."[24] Al-though they exhibited a great deal of concern with freedpeople's behavior, both Flintoff and Wilcox seemed to have little sympathy for their aspira-tions to land ownership. As a Bureau official, Wilcox at least had the po-tential to work for black land ownership, but his reluctance to do so was far from unique. As Jeffrey Kerr-Ritchie noted in his study of free labor in the Virginia Piedmont, during the first few years of emancipation "the BRFAL was effectively transformed from an agency that had the potential to redistribute land into an agency that primarily supervised free labor relations."[25] Although they were to remain largely unfulfilled, black desires for land ownership were not naive. As Steven Hahn has observed, mirroring Wither's anonymous freedman's point, rumors of federal distribution of lands "derived powerful credibility from federal actions, [and] spread widely among African Americans in the southern countryside."[26] Government actions, such as General William T. Sher-man's Special Field Order 15, which temporarily divided a strip along the coasts of South Carolina, Georgia, and Florida among freedpeople, had created durable rumors that every black family would receive "forty acres and a mule." Federal actions raised the persistent question: If the government could free the slaves, why could it not liberate the landscape as well?[27]

Of course a lack of land ownership did not prevent freedpeople from using the landscape. Like almost all of the mid-nineteenth-century South, unfenced portions of the three counties were commons available for transient public use. Throughout the South, African Americans con-tinued antebellum practices, hunting deer and opossums in the wood-lands, trapping rabbits along field edges, and fishing and collecting shellfish in streams and rivers. Freedpeople engaged in tobacco wage work or sharecropping could also range stock on the commons, where their animals foraged unimpeded over all unfenced lands (though laws later in the century would circumscribe this liberty). Tenants often planted vegetable gardens, and even the landless might find an untended piece of ground on an abandoned plantation for a small food plot. Postwar contracts alluded to the ubiquity of gardening; agreements such as the one negotiated between Robert Wilson and six freedmen defined the size and contents of laborers' gardens and house yards and codified stock ranging practices.[28] These gardens probably resembled their antebellum counterparts: interplanted masses of beans, melons, okra, peas, and corn, ranged by chickens, plots that efficiently used small spaces to supplement diets of salted meat and cornmeal.[29]

Freedpeople also continued to use wild plants as remedies for illness. Gabe White, who had been enslaved in southern Pittsylvania, remembered using herbaceous plants, such as "master weed," mayapple, "Peter's root," and sweet William, to treat sickness, and William Williams, born in Caswell, recalled that sick slaves and freedpeople relied on a local woman "who used roots and herbs."[30] And "Doctor" remained a relatively popular African American name in the region, perhaps drawing on the Igbo custom of labeling individuals ordained to practice herbal medicine.[31] Certain portions of the landscape were probably more available to freedpeople than others, as swampy low grounds, large tracts of forest, and the mountains of the western edge of Pittsylvania were less intensively farmed than plantation districts. While these traditional land uses continued during Reconstruction, white landowners worked hard to restrict African Americans from the most profitable use of the Southside environment—bright tobacco farming on freehold land. Getting by was one thing, getting ahead quite another.

Despite the twentieth-century stereotype of tobacco as the bastion of the small white farmer, post–Civil War tobacco farms, like their antebellum counterparts, relied disproportionately on black labor, though this particular labor organization was not grounded in any particular economic logic. As far as southern staple crops went, tobacco benefited relatively little from economies of scale. Unlike sugar, cotton, rice, or even wheat, few bright or dark tobacco tasks could be made dramatically more efficient through massed labor.[32] Although this reality meant small white or black farmers who owned land could produce tobacco as easily as large landowners, the strenuous nature of tobacco work meant white farmers looked to induce or coerce black labor whenever possible. Tobacco cultivation was never "black work" in the same way that tobacco manufacturing was, but it did carry at least some connotations of slave days. As one Chatham area small farmer told a traveler passing through immediately after the war, "D——n farming; it's enough to kill a horse; it's just fit for a nigger."[33] William Sutherlin expressed a similar belief in the necessity of black tobacco labor, whether enslaved or free, in an 1866 address to the Virginia State Agricultural Society, though he predicted that African Americans were ill suited to freedom and would eventually disappear from the commonwealth.[34] Although white planters and farmers of all sizes did not need coerced labor to make tobacco, they had become habituated to the idea of black tobacco hands. In much the same way that ideas about the best ways to use local environments came to be seen as inevitable, white landowners eventually believed black tobacco labor was natural.

Faced with the reality that land was not coming from the federal government, and encouraged or even forced by Bureau agents, most freedpeople in the three counties signed labor contracts with white land-owners during the first winter following the war. Virginia Bureau head-quarters in Richmond allowed its agents to help freedpeople negotiate contracts if necessary but encouraged officials to "leave them free to mak[e] their own bargains" if possible.[35] During the first few years, these local labor contracts took a variety of forms; there was no single contract type, and even within broad categories individual contracts often varied significantly in specific details. With this disclaimer in mind, there were three general models that encompassed most of the surviving contracts from Pittsylvania County (where surviving records are most extensive). The first model was a wage labor one. In these contracts landowners promised to pay a specified wage to workers over the course of a year, and in return they expected the laborers to work under their direct supervision. Directed labor under these terms often resulted in large gangs of freedpeople working under a white overseer, much as they had as slaves before the war. The second model paid an agreed-upon share of the crops at the conclusion of the contract, in return for either directed or independent labor. The final and least frequent model involved landowners who promised only provisions and housing in

Table 5.1.

Surviving Contracts Signed at the Freedmen's Bureau Field Office, Danville, for the 1866 Season

TOTAL CONTRACTS	CONTRACTS FOR WAGES	CONTRACTS FOR SHARES	CONTRACTS FOR PROVISIONS	AVERAGE ANNUAL WAGE
108	73	28	7	$93.31

Source: Compiled from Contracts, Indentures, and Papers Regarding Cases, Danville Field Office, BRFAL, RG 105, M1913, roll 72.

exchange for work. These contracts seem to have been reserved primarily for elderly or ill individuals, and they often came out of antebellum relationships. In the first two models, landowners usually agreed to provide housing and clothes, and in the case of share contracts the landowner typically promised to furnish seed, tools, and often draft animals. Most contracts also stipulated a stiff cash penalty if either side defaulted on any of the terms.[36]

Although all three models were present throughout the South in the years immediately following the war, Pittsylvania (and by all indications Halifax and Caswell as well) was unusual in its high percentage of wage contracts. For the 1866 crop season, just over two-thirds of the surviving Bureau contracts were for wages (table 5.1).[37] Additional contracts in various planters' papers support this percentage. In the months following the surrender, Sutherlin's Halifax farm manager hired freedpeople for wages, though he found it "quite a troublesome business"; Ruth Hairston hired at least seventy-four former slaves for monthly wages; and Robert Wilson agreed to pay wages or provide provisions for seventeen freedpeople on his farms in Pittsylvania and Caswell counties. None of these planters made share contracts.[38]

With several notable exceptions, historians dealing with labor contracts in different parts of the South have generally found that some form of share contracting dominated labor negotiations from the very beginnings of Reconstruction, and they have concluded that this was the case primarily because of a shortage of hard currency needed to pay wage labor, coupled with black resistance to the tight supervision im-

plicit in wage work and a desire to possess a stake in the process of production.[39] Indeed, at odds with the notion that tobacco districts were
bastions of small farms and relative yeoman independence, the prepon-
derance of wage contracts was most similar to contemporary labor agreements in Deep South districts dominated by rice and sugar plantations
or extremely large cotton operations.[40] This emphasis on wage work
came in part from white conceptions of the nature of bright tobacco culture and African American agricultural abilities. Bright tobacco planters pushed for wage labor out of a belief that tobacco work was exacting
skilled labor; they did not trust freedpeople to produce top-quality
tobacco without white supervision. These beliefs refuted historic black
relationships with the tobacco landscape but served to justify retention
of the land. White landowners privileged their own understandings of
agriculture and the environment over those of former slaves.[41]

Compensation under most contracts was modest, highly variable,
and gendered. Annual cash wages ranged from a low of $30 for a female
domestic worker to as high as $180 for a skilled male field hand. In the
Bureau contracts the average annual wage was approximately $93, or
just under $8 a month, a figure that compared poorly to the average an-
nual slave rental during the 1850s.[42] Of course wage figures in contracts
can be a bit misleading. In some cases landowners expected the signer
and family members to all work in the fields, in others only the head of
the household was expected to work, and in yet others each family member signed his or her own contract. Some wage contracts promised
monthly payment, but most stipulated payment via quarterly installments
or in full at the end of the year. Share contracts were similarly variable.
The most frequent arrangement was for half the crop, but in some contracts the freedman received as much as three-fourths or as little as little
as a tenth of the crop. Share agreements were also typically made with
the head of household but often involved the labor of multiple family
members. In a typical example, William Sims of Halifax contracted
with thirty sharecroppers in the Clover district in one year, almost all of
whom represented family units. In some cases share agreements stipulated that the laborer raise tobacco, but in many there were no written
crop strictures. This relative lack of crop restrictions suggests that white
landowners preferred to raise tobacco under wage labor, while allowing
sharecroppers to raise less demanding staples, such as corn or wheat.[43]

Halifax planter Samuel Wilson's arrangements with freedpeople in 1866 illustrate the variety and complexity of Reconstruction labor contracts. In total, Wilson employed fifteen workers for the duration or a portion of the year. He paid ten workers cash wages ranging from $20 to $150, and he reserved payment until the end of the year (in fact, not settling with the employees until May 1867). Not all wage contracts were with individuals. Wilson paid Ruth Wilson and her daughter Nancy together, and he made similar arrangements with Billy Wilson and his wife. Further complicating his accounts, Wilson arranged a share agreement with Stephen Millner, with the planter selling Millner's tobacco crop and keeping a portion of the proceeds. Samuel Wilson continued these labor contracts throughout Reconstruction, always favoring wage labor but never resorting to a standardized contract.[44]

An agreement between N. C. Miller of Pittsylvania and thirty-three freedpeople in 1865 demonstrated the harsh terms under which some laborers worked. Eighteen adults and their fifteen children committed to "do any kind of work [Miller] may require of us" in exchange for "one tenth part of the corn crop made on the place after fattening the pork." Miller agreed to furnish some clothing, but he retained the right to dismiss any employee at his discretion and without penalty.[45] Robert Wilson, a notoriously harsh master before Emancipation, drove a similarly hard bargain in 1869. Six freedmen agreed to work for Wilson for the 1869 season in exchange for just one-third of the crop, excluding all the fodder. Wilson was not liable for feeding or housing the freedmen, explicitly writing that he had "nothing to do with it," forbade them any visitors without his knowledge, prohibited black-owned livestock, and would permit his workers to plant a watermelon patch only if they cultivated one of equal size for him.[46]

Although these surviving contracts provide an intriguing window into early negotiations between freedpeople and white landowners, they by no means yield a complete picture of regional labor relationships. The majority of contracts may have been verbal agreements, and many of the ones that were written down have been lost over the intervening years.[47] The surviving contracts do, however, suggest a few things. First, they indicate that white landowners were interested in resuming relationships with freedpeople under terms that mirrored antebellum work as closely as possible. Wage labor tended to place black workers under white supervision to a greater extent than work for shares, an arrangement particularly appealing for white landowners convinced that bright tobacco demanded special attention and expertise. These contracts also illustrated the unfortunate reality that in most cases freedpeople had to come to terms with former slaveholders (often their own previous masters). The vast majority of freedpeople had no money, little experience beyond tobacco and corn work, and received little encouragement to find employment outside of agriculture even if they were so inclined. Aside from Danville's tobacco factories, industrial jobs were all but nonexistent in the region.

Even individuals willing to brave all these obstacles in search of work elsewhere would have had to leave their friends, family, and the land they knew behind, to divorce themselves from places of social and environmental expertise. Attachment to place has been an underappreciated force in freedpeople's tendency to remain on or near the land of their enslavement. As Jeffrey Kerr-Ritchie has noted, former slaves were not simply "homo economicus," that is to say, rational economic actors. Freedpeople often had an affinity for familiar landscapes that overpowered or conflicted with their pecuniary interests.[48] For many, a lifetime of labor on a particular piece of ground fostered a sense of attachment to the soil. William McFeely, in his study of Sapelo Island, Georgia, concluded that for many ex-slaves "the memory of the scene was as strong as that of the crime." Individuals that returned to the location of their enslavement "saw the place as separable from the oppression that had taken place there."[49] For most freedpeople the attraction of the land, pressures from local landowners and the Freedmen's Bureau to return to agricultural labor, and a lack of realistic alternatives were too much to ignore or overcome during the first post-Emancipation years; they signed contracts in an attempt to survive and to remain in a familiar and freighted landscape.

This process of negotiation and mixed contracting took place across the South in 1865 and 1866, with share contracts eventually becoming a more common form of labor agreement. A number of Reconstruction scholars have argued that share contracts were appealing to both landowners and freedpeople for several reasons. For landowners, sharecropping required little in the way of ready cash, a boon in the credit-strapped region; tenants who borrowed against their shares over the course of the year often ended up indebted to the landowner and thus were forced to sign a contract for the ensuing year; and sharecropping generally removed the onus and cost of direct management from landowners. Landowners also hoped that sharecropping would provide freedpeople with a greater incentive to care for the land since, unlike in a wage system, their income had the potential to increase under good stewardship. For freedpeople, share contracts were appealing because they typically provided greater independence than wage labor. To a certain extent black families could govern their own labor division, work at their own rhythms, and direct their own agricultural activities.[50]

But landowner opposition to sharecropping in the three counties remained staunch following the war and persisted throughout Reconstruction. At the core of white preference for wage labor was a strong belief that blacks could not raise quality bright leaf tobacco without direct (white) supervision. John Ott, a bright tobacco booster for the Southern Fertilizer Company, summarized this viewpoint in an 1875 pamphlet on the state of tobacco culture. Ott argued that freedpeople were the best possible tobacco hands—but only if carefully supervised by white men as "the negro, as a general thing, is destitute of that judgement and capacity essential to efficient management." He recommended that tobacco farmers hire blacks as wage workers, set regimented work schedules, and pay wages in a mixture of cash and food. With a supporting letter from Halifax bright tobacco farmer Robert Ragland, Ott assured readers that black labor had a promising future under the wage system. He declared, "The negro, as a field hand or domestic, will do work that no native or foreign white man will touch, and esteem it a privilege, after his week's work is done, to black your boots and drive the carriage to church."[51] With these beliefs in fine tobacco's combination of drudgery and skilled labor, area tobacco farmers worked to impose a wage labor system that coupled tight control over daily work regimens with long-term assurances of continued black labor.

A careful analysis of the Pittsylvania wage contracts reveals a system of labor that combined the elements of wage and share contracts that most benefited white tobacco growers: direct supervision of labor and indebtedness. The typical wage contract stipulated cash payment at the end of the quarter or year rather than on a monthly basis. Freedpeople worked under direct supervision of the landlord and were allowed to borrow against their wages; as with the share system, they often ended the employment period in debt. This form of wage contract allowed white landowners to tightly control black labor, and at the same time it removed the necessity for large amounts of hard currency, all while tying black workers tightly to white landowners through a form of debt peonage.[52]

Evidence that this method of delayed-wage labor served to keep freedpeople bound to particular planters can be found in the Freedmen's Bureau records. In the autumn of 1868, twenty-five freedpeople filed complaints with the Halifax Courthouse Bureau agent, charging their employer W. T. Dickerson with cheating them out of their wages. Dickerson apparently advanced his workers money and supplies against their wages over the course of the season, and at settling time the planter informed them that they were either due only a small cash payment or actually owed him money. Due to illiteracy or an understandable distrust of Dickerson's calculations, the workers sought review by a Bureau court. Following an examination of the landlord's books, Bureau officials concluded that the landowner's calculations were correct; thirteen of the twenty-five freedpeople owed Dickerson money at the end of their contracts, in one case as much as sixty-two dollars. Although Dickerson did owe ten laborers some portion of their annual wage, on balance the twenty-five laborers owed their employer almost a hundred dollars above their wages. Without a source of ready cash, the workers in debt to Dickerson were forced to sign on with him for another season.[53]

Complaints such as those made by Dickerson's workers were so numerous in the district that they clogged the dockets of the Bureau courts established at the Danville and Halifax Courthouse field offices. These courts heard all sorts of complaints from freedpeople, but the majority pertained to contract violations.[54] Out of 324 cases heard in 1868, 192 (59 percent) involved freedpeople accusing whites of violating written or oral labor contracts (table 5.2). An additional 7 cases (2 percent) involved landowners claiming that freedpeople broke the terms of the contract.[55] The disparity in these figure support Wilcox's assertion that landowners were much more likely than freedpeople to break contracts, as the latter considered the agreements "sacred."[56] As in Dickerson's case, the courts typically heard testimony from both sides, reviewed the books and written contracts if they existed, and issued a summary judgment. Bureau agents passed only cases that involved violence on to local courts.[57]

Table 5.2.

Surviving Complaints Brought before the Freedmen's Bureau Courts in
Halifax and Pittsylvania Counties, 1868

COUNTY	TOTAL SURVIVING COMPLAINTS	TOTAL COMPLAINTS INVOLVING CONTRACTS	WHITES ACCUSED OF CONTRACT VIOLATION	FREEDPEOPLE ACCUSED OF CONTRACT VIOLATION
Halifax	274	153	146	7
Pittsylvania	50	46	46	0

Source: Compiled from Halifax Courthouse Complaint Book, 1868, BRFAL, RG 105, M1913, roll 97;
Danville Letterbook, Cases Reported June 1868, BRFAL, RG 105, M1913, roll 72. A few of these contracts
involved landowners and freedpeople from Caswell County.
Note: The Pittsylvania records are incomplete.

These complaints indicate that white landowners regularly and systematically abused the contract system. In the case of wage contracts,
landowners, such as Elisha Barksdale and Richard Cunningham, often
simply refused to pay freedpeople at the conclusion of the year.[58] Other
landowners who had made share arrangements kept all of the crop or a
greater percentage than agreed upon. This sort of deception was particularly common when it came to division of the tobacco crop because of the
nature of leaf sales. White landowners generally transported the crop to
market, sold the leaf, and were then expected to divide the money with
their tenants.[59] Because the sharecropper was rarely witness to the sale,
tobacco division was particularly ripe for abuse. In one example, a Mrs.
Moon of Halifax divided the provision crops with two of her tenants but
refused to split the $1,513.94 from the sale of nine hogsheads of tobacco
before the court ordered her to do so. In another, black sharecropper
Allen Chappill used the Bureau to challenge landowner James King's
monopoly of sales knowledge, asking the court to verify that King had
actually sold their joint crop of tobacco for the amount he claimed. The
court inspected King's sales receipts and ruled that he had divided fairly
with Chappill.[60] A number of whites refused to honor contracts because
they claimed freedpeople did not work hard enough, left the farm too
often, produced poor-quality crops, or attended "political" meetings.[61]

Although contract enforcement was difficult in Pittsylvania and Halifax, the situation was even worse in Caswell (which had no Bureau office) by 1868. According to Assistant Commissioner F. W. Liedtke, stationed in adjoining Alamance County, Caswell contained no "magistrate who dares to take up any Case against any of the Enemies of the present State Government." He went on to describe the case of freedman Peter Summers, denied his promised share of tobacco by landowner Bentley Page. Summers had no luck finding a justice of the peace or Bureau agent who could help him obtain his portion of the crop. Citing Summers's case, Liedtke asserted that in Caswell "a poor man has no chance of obtaining justice."[62]

What the surviving Bureau complaint records make plain is that freedpeople and landowners pursued their interests in decidedly different ways. Freedpeople who felt wronged took the matter to the Bureau courts, hoping agency officials would prove a sympathetic audience. White landowners who became disgruntled with wage laborers or sharecroppers usually took matters into their own hands; they drove workers off their farms for sloppy work or acting "impudent" or, in a number of cases, simply assaulted them.[63] In essence, many landowners still viewed labor relations in antebellum terms, a worldview that refused to accept the legitimacy of black complaints or biracial contracts. The Bureau courts could provide some relief, but even if freedpeople had a fair chance of receiving settlement justice, their odds of receiving social justice were still all but nonexistent.

Although many labor disputes went through the Freedmen's Bureau, elements of the region's antebellum notions of planter paternalism permeated many day-to-day racial interactions. Prominent men, such as Halifax's Elisha Barksdale, seemed to feel a conflicting responsibility to control and manage black labor for their own economic interests and simultaneously to look out for the well-being of freedpeople, whom they believed too innocent or "ignorant" to do well under the conditions of freedom. During Reconstruction Barksdale served as an outside witness or legal counsel in several cases involving former slaves, often arguing on their behalf. In the 1868 case of Polly Jennings, convicted of killing her newborn child and sentenced to death, Barksdale successfully organized seventeen Clover-area planters (including Jennings's former master) who testified to her good character in a commutation petition to

Commissioner Orlando Brown of the Virginia Freedmen's Bureau.[64] Barksdale also intervened in cases more directly related to struggles over field labor. When a Halifax farmer shot and crippled a sharecropper in a drunken dispute, Barksdale lobbied to have the white farmer's fine ($150) transferred to the injured freedman rather than remanded to the county's coffers. In typically convoluted fashion, before working to provide for the black plaintiff, Barksdale had served as defense counsel for the white farmer, managing to get his client off with no jail time.[65] In a similar case, he was able to get the $200 a Halifax criminal court fined farmer Stephen Tucker for assaulting and shooting Charles Womack funneled to the freedman.[66]

While these three cases demonstrated Barksdale's efforts to look after the well-being of certain freedpeople, his was a paternalism firmly grounded in a belief in the superiority of white landowners and their preeminent claims to the productivity of local landscapes. In both shooting cases, Barksdale seemed untroubled that the white defendants received no jail time and only moderate fines for attempted murder. He also took direct action with his own black workers. In the same year that he took a legal interest in the cases of Jennings and the two injured freedmen, Barksdale was brought before the Bureau courts for violating a labor contract with freedman William Barksdale, perhaps one of Elisha's former slaves based on the surname. The court found for William, releasing him from his contract with Elisha because the white planter "did not pay him regularly."[67] Elisha Barksdale thus simultaneously worked to care for freedpeople's interests and to violate their rights, conflicting actions that reflected an ideology of planter paternalism and privilege that seems to have survived the war and emancipation intact.

Delayed-wage agreements, contract violations, and paternalistic gestures were common ways for landowners to obtain and control cheap tobacco labor, but whites in Caswell, Halifax, and Pittsylvania had other ways of coercing black work. For the first several years of Reconstruction, Bureau agents bound a number of destitute black children and orphans to white farmers. In theory these arrangements were to be made only with the consent of the parents or for children who would otherwise become wards of the state. Masters promised to clothe, feed, and shelter the child and to give instruction in a useful trade. Ideally the useful trade was a skill, such as blacksmithing or bricklaying, but most apprenticeships listed farming or housekeeping in the appropriate blank. In return, the master obtained the labor of the child (who was often as young as six or seven) until the age of eighteen or twenty-one. The master also typically promised to pay the child a sum of money, from fifty to one hundred dollars, at the conclusion of the indenture.[68] Despite contracts clothed in the language of artisanal apprenticeships, bound black children served under terms more akin to colonial indentured servants, relationships in which "servitude was a labor system, not an educational institution."[69]

Although the indentures spelled out the ideal purpose of these contracts, the surviving records from the Danville field office paint a much different picture of the reality. From September of 1865 into the early summer of 1866, Danville officials bound out at least 201 black children to white masters. Many of these children were forcibly removed from families that vowed they were able to provide for them. When Captain Wilcox took over the Danville office from Bureau agent P. J. Hawk, he was confronted with petitions from parents desperately seeking the return of their children, pleas that eventually added up to a hundred separate complaints. He wrote to his superior, R. S. Lacey in Lynchburg, that Hawk had "bound out several children without the consent of their parents," but he reassured Lacey, "I have no doubt that in most of the cases, the children are better off where they are, than with their parents."[70] Despite Wilcox's blithe assertion, he pursued the claims and began to suspect that Hawk had been systematically binding out children despite their families' ability to care for them. Two weeks later he sent Lacey a list of twelve children Hawk had bound out under suspicious circumstances and declared the he believed there to be numerous similar cases in the county.[71] He may have seen the indentures as being in the children's best interests, but Hawk most likely bound them out to receive the five-dollar fee the Bureau charged to draw up indentures. It is unclear if Wilcox ever nullified any of Hawk's indentures, and the cases demonstrate the gross potential for abuse that accompanied the child apprenticeship system.

In essence, the apprenticeship system as practiced in the three counties proved little different than slavery. Masters held absolute sway over their indentured servants, directing their work with little or no supervision from the Freedmen's Bureau; they were even authorized to whip children "moderately" until the age of eighteen.[72] Unlike true free labor, neither indentured children nor their parents had any power to end the contracts or negotiate the terms of servitude. Even in Caswell, covered by a statewide circular issued by North Carolina Bureau commissioner Eliphalet Whittlesey ordering agents to bind out only orphans or children of consenting parents, the rules were often ignored.[73] Based on the details provided in the surviving indentures, black adolescents performed many of the same tasks that they had worked at before Emancipation, laboring in the tobacco fields and as household servants, and received no real compensation outside their master's whims. An 1867 advertisement in the *Danville Register* starkly emphasized these connections between slavery and indentures. In the paper a local planter offered a reward to readers who might help him locate a thirteen-year-old runaway boy, "bound to me by the Freedmen's Bureau." Illustrating the advertisement was the same image of a fleeing black man used in the newspaper's runaway slave advertisements just a few years earlier.[74]

Behind these postwar contract negotiations, indentures, and judicial actions, the threat of white violence always lurked. Throughout Reconstruction a certain percentage of the region's landowners were willing to use any means necessary to ensure an uninterrupted supply of black labor. While there was certainly a political element to this racial violence, it often revolved directly around landowners' agricultural demands and the rhythms of the tobacco season. Cycles of fallow, seeding, growth, and harvest spawned parallel cycles of social tension, heated exchanges, and even bloodshed. Farmers' success in maintaining ownership of the landscape fueled their desires for tractable and inexpensive labor.

Like John Blackwell, who attacked Griffin Cobb, some white landowners seemed unable to accept freedpeople's ability to say no, and farmers like James Norman sometimes resorted to beating blacks who angered or offended them.[75] This violence escalated to murder on more than one occasion; Bureau agents reported a number of unresolved cases of whites killing former slaves during the first months following the war's end.[76] These outrages were especially prevalent around contract signing, harvest, and settling time, when landowners were intent on obtaining labor for the upcoming season, under pressure to get tobacco in before frost, or determined to retain as much of the crop as possible. Throughout the region, racial violence ebbed and flowed with the rhythm of the tobacco season, and with a "thirteen month season," tobacco permitted no peaceful interval between crops.

The following are a few typical encounters. During an argument at season's end, a drunken Richard Cunningham ordered his sharecropper John Brandon off his Pittsylvania farm without pay. When Brandon refused, Cunningham shot him in the stomach, permanently crippling the freedman. In Halifax, James Spencer shot at William Suttle "with intention to kill" around settling time, perhaps over a disagreement concerning the division of the crop. When Pittsylvania freedman Billy Tanner refused to sign a contract with Robert Terry for the 1869 season, Terry struck him repeatedly with a tobacco stick, cutting open Tanner's arm and head. And in October 1868, anonymous night riders left a letter for the freedpeople living at Oak Hill, one of Samuel Hairston's Pittsylvania estates, that threatened violence if they did not leave the plantation, as they were a "Den of Thieves." It seems unlikely to be a coincidence that the mob determined to drive Hairston's workers away only after the tobacco crop had been safely harvested and cured.[77] Just as these seasonal peaks in tension provoked white violence, they presented black workers with leverage to use in their fight for greater rights and a larger share of regional prosperity.

Some violence came out of a disbelief that a freedman would dare stand up to a white man, verbal reminders of the reality of emancipation triggering outrage and even murderous action. Emboldened by emancipation and the presence of the Freedmen's Bureau, and determined to obtain some of their promised rights, quite a few freedpeople resisted landowner oppression. This open resistance often triggered harsh reprisals. A typical example was the case of farm manager Edward Self of Caswell. Charged with killing a worker named Bouldin, Self blamed the incident on the freedman's "impertinence." Bouldin apparently questioned Self's farming practices and honesty in front of the farm owner, a "roguery" Self "would not stand from a white man nor a negro." After the disagreement Self lay in waiting for Bouldin and shot and killed him from ambush.[78] Self used the issue of his honor as a white man to defend his lying in wait, a behavior defined as inherently dishonorable.

Freedpeople who believed the Bureau would protect them from these acts of violence often met with a rude surprise. Overextended and tasked with ensuring that blacks remained faithful agricultural workers, the agency was frequently torn between guaranteeing freedpeople's rights and supporting the landowners who, the Bureau believed, drove local economic growth. When cases involving murder or other violence came before local Bureau magistrates they without fail referred the incidents to local grand juries, which almost always dismissed the charges. Freedmen's Bureau representatives occasionally moved from ignoring instances of racial violence to open complicity. Indeed, as one concerned visitor to Danville reported to Virginia superintendent Orlando Brown in 1868, Bureau agents and the federal troops under their command sometimes employed violence to ensure that freedpeople worked diligently. Informing Brown of the disturbing occurrences in town, C. Thurston Chase wrote: "The whites declare that the 'negroes are idle' and 'won't work.' They come in and report them to the provost-marshal—a Col. [indecipherable, perhaps Captain Wilcox]—for idleness and insolence—The officers, at their discretion, send out men to punish the negroes. 'Bucking & gaging'— 'whipping'—'tieing up by the thumbs'—&c were spoken of by the privates in 8th Pa. cav. as common modes of punishment. By all accounts the people through the country are determined to make their freedom more intolerable than slavery."[79] That Bureau agents sometimes served as physical enforcers for white landowners must have further complicated former slaves' responses to their social and economic struggles.

Pittsylvania and Halifax witnessed their share of postwar violence, but Caswell's tobacco fields and town squares were particularly bloody. With the greatest percentage of freedpeople of the three counties, several well-organized black political groups, an active contingent of resistant landowners, and perhaps the densest concentration of prime bright leaf land, Caswell rapidly devolved into a countryside of armed camps. When the new North Carolina constitution of 1868 replaced the old Caswell Court of Pleas with a board of commissioners (which held the possibility of appointed black officials), the court authorities predicted violence and unrest in their last official statement: "The Constitution of our fathers, the constitution of our happiness, we drop a Tear of remembrance for thy many blessings and now bid thee a long farewell. We turn with fearful forebodings to the future. We see general lawlessness, a most fearful disregard of public and private obligations, great demoralization, the marriage not so sacredly regarded and observe a general feeling of uneasiness as to the future and a dread when the vilest of men are exalted the nation will mourn."[80] Within five years of Appomattox, the violent struggles over the meaning of liberty and labor in the fields in Caswell spawned the assassination of a state senator, led to the impeachment of North Carolina's governor, and echoed in the halls of the United States Congress.

Although some Caswell violence took place in the context of personal relationships, such as the Self and Blackwell cases, other incidents centered around the formation of political and social organizations, such as black Union Leagues and white Ku Klux Klan "dens." And the aims of these organizations were as often about control of the landscape as about the ballot box. As early as 1865, Caswell freedpeople organized a Union League with the assistance of Albion Tourgée, a prominent Republican "carpetbagger" judge from Greensboro. The league lobbied for land redistribution and black suffrage, and most threatening to local whites, it formed and drilled militias for the defense of the county's black population.[81] Union Leagues formed from a combination of fraternal, religious, and defensive motivations, and, like most area freedpeople, the leagues sought black landownership. League branches often had kinship, common employment, or church membership uniting adherents, as did Caswell's Leasburg League, which had twenty-nine members in 1868, most of whom had belonged to the Yancey family or had worked with the Yanceys as slaves on local plantations and in Leasburg's tobacco factory prior to emancipation.[82]

Many Caswell whites resented any form of black organization, but they found martial activities especially disconcerting. In the summer of 1867, white farmer S. M. Lotten wrote to Bureau official William North at Greensboro, complaining of the prevalence and intensity of league activities. Lotton worried that Caswell's freedpeople "have been mustering and drilling their societies for what purposes I do know not." He went on to claim that league members committed nightly depredations, actions local Bureau agents did little to stop. Near Milton "from early dawn till after midnight—nothing can be heard save the drum and fife and words of military command." League members claimed portions of Caswell's roads and fields for their activities, actions that threatened the white monopoly of the landscape. Lotten concluded his letter with a sexually charged allusion that would flavor many a postwar white racial appeal, warning that black parades "endanger in a number of cases the lifes of ladies who may be traversing the public highways."[83]

Public rallies and political meetings in the county were prone to violent outbreaks. An 1868 Democratic Party barbecue in Yanceyville ended in an assault on handy Republicans, an attack that included the severe beating of one black farmer and the shooting of another. Two weeks later a riot erupted in Milton when black organizer George Bow tried to recruit freedpeople to join a political club devoted to the success of President Ulysses S. Grant. The threat of violence was so omnipresent that John Cook, a black minister who reported the Yanceyville riot to Governor William Holden, felt it unsafe to remain in the county. He informed Holden, "I had to mak my escapt out of the PlaCe for fear of them [white Democrats]." He concluded there could be no true freedom for black people in Caswell without a "forest of the union Army."[84]

Cook's belief in the need for a "forest" of troops proved prescient, as local chapters (or "dens") of the Ku Klux Klan, often operating under such sobriquets as the Constitutional Union Guard, White Brotherhood, or the Invisible Empire, sprang up throughout the county in 1868 and 1869.[85] First organized in Caswell by farmer John G. Lea, the local Klan rode the countryside at night pulling freedpeople from their homes, beating them, burning their dwellings, and threatening worse if they did not sign work contracts and stay away from the polls. Both Tourgée and John W. Stephens, Caswell's state senator, testified that the Klan all but ruled the county, having beaten hundreds of freedpeople and executed several prominent Union League members. Lea supported their statements, reveling in the way Klansmen had whipped Jim Jones, a justice of the peace, "and drove him out of the county."[86]

When interviewed by Work Progress Administration (WPA) employees during the Great Depression, former slave Ben Johnson recalled an incident involving the Caswell Klan that encapsulated the secret society's tactics. Though Johnson's tale (again in the exaggerated dialect common in WPA ex-slave narratives) contains a number of apocryphal elements, it captures the essence of the violence and intimidation employed by Lea's night riders during the late 1860s and the determined resistance of many of the county's former slaves:

Sam Allen in Caswell County wuz tol' ter move an' atter
a month de hundret Ku Klux come a-totin' his casket an' dey

tells him dat his time has come an' if'en he want ter tell his wife good bye an' say his prayers hurry up.

Dey set de coffin on two cheers [chairs] an' Sam kisses his ole oman who am a-cryin', den he kneels down side of his bed wid his haid on de piller an' his arms throwed out front of him.

He sets dar fer a minute an' when he riz he had a long knife in his hand. 'Fore he could be grabbed he done kill two of de Ku Kluxes wid de knife, an' he done gone out'n de do'. Dey ain't ketch him nother, an' de nex' night when dey comed back, 'termined ter git him dey shot ano'her nigger by accident.[87]

The rapid spread of Klan activities in Caswell and neighboring Alamance so impressed John Pool, one of North Carolina's federal senators, that he wrote to Governor Holden confiding that he had learned Nathan Bedford Forrest himself was riding the counties' roads at night, directing vigilante activities.[88] Although Forrest likely never rode with Caswell's white supremacists, Lea and his compatriots proved trouble enough.[89] In his posthumously published memoirs, Holden lamented the power of the local Klan in the late 1860s. Relating the difficulties of convicting Klansmen, he described the violence wrought by night riders who intimidated many of their opponents into silence. "Some of these victims were shot, some of them were whipped, some of them were hanged, some of them were drowned, some of them were tortured, some had their mouths lacerated with gags, one of them had his ear cropped, and others, of both sexes, were subjected to indignities which were disgraceful not merely to civilization but to humanity itself."[90] These night riders reigned supreme in Caswell, but they did face obstinate opposition from the remnants of Union Leagues and their champion, Caswell's state senator John Stephens.

A lawyer and tobacco trader originally from Rockingham County, North Carolina, Stephens moved to Yanceyville to pursue his tobacco business following Appomattox and became an active member of the local Republican Party. Stephens's character remains unclear. Although he seemed exceedingly ambitious, temperamental, and incapable of getting along with his neighbors—white opponents labeled Stephens a thief and even hinted at matricide—he served as one of the strongest supporters of Caswell freedpeople during the late 1860s.[91] He was instrumental, along with Tourgée, in establishing the Caswell Union League, and he volunteered to serve as a Freedmen's Bureau agent in the county, working as a justice of the peace after 1868. In that same year, Stephens ran for the North Carolina Senate against Democrat Bedford Brown. Brown won, but the state legislature refused to seat him due to his connections with the former Confederate government, and Stephens was victorious in the second election.[92]

Wearing all three of his hats—Union League organizer, Bureau justice of the peace, and senator—Stephens managed to anger the Klan and many of his other white neighbors. And he remained active as a tobacco agent; the likelihood that Stephens purchased tobacco directly from black growers contributed to his unpopularity with white farmers. Known for his "extensive popularity among the negroes," a tract sympathetic to the Klan claimed that "at the election, when he was chosen Senator, the colored vote of the County elected him. He did not receive the votes of twenty white men, and probably not more than half of that number."[93] George Anderson, a resident of Yanceyville, declared that Stephens was an official who "had a complete mastery over the negroes in the County & they were ready to follow his lead."[94] As justice of the peace, Stephens also provoked local landowners by attempting to strictly enforce Bureau-approved contracts, claiming that former slaves deserved a stake in the landscapes they helped create. In his early twentieth-century recollection of Reconstruction in Caswell, Lea remembered Stephens as a thorn in landowners' sides, constantly "persuading the darkies to warrant the farmer."[95]

The Klan also accused Stephens of inciting the local Union League to violence. In 1869, Governor Holden appointed Stephens as his personal investigator into the growing Klan activity in Caswell, a commission Stephens seemed to have interpreted as a mandate to directly combat night riders. According to several reports, Stephens and Union League members began to strike at the very heart of wealthy Klan leaders' power, their tobacco crops. Rumor swirled throughout the county that the Union League aimed to "put down" the Klan with the "torch," and over the course of several nights a number of tobacco fields and barns of prominent Klansmen went up in flames, including William Lea's "entire crop of tobacco."[96] Stephens's role in the fires was accepted among county Democrats. The common tale as relayed by a later reactionary historian accused Stephens of attending "a meeting at the home of his brother-in-law, one Jones, [where he] gave to each of the twenty negroes present a box of matches, still a rarity among them, and told each one to burn a barn."[97] Caswell's landowners found this agricultural warfare terrifying, as tobacco barns full of cured leaf were exceptionally vulnerable targets. Barns of bright leaf represented a year's labor, the culmination of the seasons, sweat, and expended soil; attacks on the crop were symbolic of attacks on white landowners' control of the rural landscape (fig. 8). Indeed, Caswell's tobacco producers responded to these arson attacks with the same ingrained fear displayed by William Faulkner's townspeople toward accused barn-burner Flem Snopes in *The Hamlet*.[98] By assaulting the bright leaf landscape and its products, Stephens and the Union League threatened the core of white control over local environments. The Klan responded with further acts of violence, beating and abusing league members. Local papers fueled the conflagrations, praising the work of the Klan, "those equestrian wonders of the times," and circulating rumors that Holden planned to send an army of black troops to occupy Caswell and adjoining Alamance County. In this frenzied atmosphere, Lea organized a Klan meeting to judge Stephens for the tobacco fires. Tried in absentia "before a jury of twelve men," the Klan sentenced the senator to death, and Lea and his men began devising an assassination plan.[99]

The Caswell Klan carried out the assassination on May 21, 1870. Stephens was in a court day crowd listening to Democratic Party political speeches, including oratory from Bedford Brown, when he was lured to the courthouse basement by a Klansman who feigned interest in running for the office of sheriff. A group of eight conspirators forced Stephens into a room used to store firewood. Lea pulled a cord around the senator's neck, strangling him, and then held his body upright while a fellow assassin stabbed him twice in the throat and once in the heart. After an aborted attempt to lay the blame for the murder on local freedpeople by moving the body to a nearby Freedmen's Bureau school, the Klansmen left the corpse locked in the basement room, threw the key into nearby Country Line Creek, and returned to their homes or melted back into the crowd.[100] Searchers led by Stephens's family and local freedpeople found the body the following day.[101]

FIGURE 8: The wooden construction of Southside barns, as seen here, made them ideal targets of arson during Reconstruction. Courtesy Lucinda L. Austin.

Stephens's assassination convinced Governor Holden that drastic measures were needed to combat the Klan in Caswell and Alamance. Drawing on the authority of a martial law bill spearheaded by Alamance Republican Senator T. M. Shoffner, Holden declared the two counties in a state of insurrection and suspended their citizens' rights of habeas corpus. The governor placed Colonel George W. Kirk (a former Union officer) in charge of quelling the Klan and authorized the use of troops recruited from the mountains of western North Carolina and eastern Tennessee. Kirk marched into Caswell on July 15 with a force of 670 black and white soldiers.[102] Kirk's troops arrested more than a hundred suspected Klansmen (including Lea and several of his co-conspirators) over the next two weeks and transported them to Raleigh to stand trial for the murder of Stephens and other acts of violence.[103]

Holden's victory over the Klan, soon known as the "Kirk-Holden War," proved fleeting. Local newspapers and conservative political leaders railed against the suspension of due process and Kirk's methods, accusing his troops of torturing their captives. Many state conservatives were incensed about accounts of black troops administering martial law over white North Carolinians. After a stillborn attempt to try the arrested Klansmen, the state remanded them to the Caswell and Alamance county courts, where they were promptly released. In a further blow to justice for Stephens and his fellow victims, in 1871 a "redeemed" North Carolina legislature voted to absolve the state's Klansmen of certain crimes committed up to that point.[104] Holden's enemies lost little time blaming the governor for launching what they believed was an illegal domestic war. A movement by conservative legislators for Holden's impeachment quickly gained steam, and after a lengthy trial the North Carolina Senate voted to remove the governor from office on March 22, 1871.[105] Holden's removal was the first successful impeachment of a governor in the history of the United States. Emboldened by their acquittal and Holden's impeachment, Lea and his compatriots resumed their campaign of terror in Caswell, driving the last vocal members of the local Union Leagues from the county.[106]

In April 1871 the United States Congress passed the Ku Klux Act, which legitimated Holden's actions after the fact by authorizing the use of federal troops and the suspension of habeas corpus to combat Klan violence throughout the South. Stephens's assassination, the Kirk-Holden War, and Holden's impeachment figured prominently in the congressional debates. Drawing on the power of the act, Tourgée spearheaded a campaign against the Klan in Alamance, the success of which intimidated the Caswell Klan. This breaking of the first Klan came too late to reverse the balance of power in Caswell, however. Lea and his fellow white supremacists had already accomplished most of their goals. The Union Leagues were broken, Stephens was dead, and white control over the county's land and black labor was as firm as at any point since the Civil War.[107]

Historians have often framed Stephens's assassination and the ensuing events as typical Reconstruction struggles over political and social power, just another example of violence that raged from the South Carolina Piedmont to Colfax, Louisiana. These characterizations contain more than a grain of truth; Lea and his fellow landowners were concerned about Union League activities, black voting, and racial deference, but they were equally concerned about the continuance of a demanding bright tobacco agriculture when the crop was so lucrative. Like the violence that took place around contract signing season, planting, and harvest, Stephens's death was linked to the rhythms of the tobacco season. The Klan, when "trying" Stephens, did not frame his crimes in terms of racial agitation or political usurpation; rather, the Klansmen argued that Stephens threatened farmers' livelihoods through the destruction of tobacco at a point when the crop was particularly vulnerable. The tension around Stephens's death was part of the tension that pervaded the countryside when flammable tobacco hung in wooden barns. When the local Union League, perhaps at Stephens's suggestion, burned barns of tobacco, it not only challenged white political and social dominance, it also shook the economic foundations of the Southside and drew on the natural properties of a valuable weed.

By the time Klan struggles in Caswell wound down, the labor situation in the Southside had largely stabilized. Despite the opposition of influential planters like Ott and Ragland, by the early 1870s sharecropping was competing with wage contracts as the most common form of labor agreement in the three counties, though it is unclear exactly when and how this shift occurred. Sharecropping certainly resulted in part from determined black resistance to the delayed-wage contracts that mimicked the terms of bonded labor, coupled with former slaves' desires to manage their own piece of ground. Perhaps enough African Americans echoed the desires of an anonymous Pittsylvania freedman who told northern traveler John Dennett that he and his friends wished local planters would let them work farmland on shares rather than wages.[108] Or, the move to sharecropping could have reflected revised calculations on the part of white landowners who began to see the benefits (for themselves) of a shares system: decreased management costs coupled with retention of control of the tobacco crop through sharecropper debt. William Hatchett, who turned to sharecropping contracts soon after the war, argued that the system worked best for both landowners and freedpeople: "The hands work very well, as they know if they do not make a good crop, they will get a small share."[109] Although the details remain fuzzy, the move to sharecropping was likely brought on by a combination of these factors.

Sharecropping terms had also stabilized somewhat by the early 1870s. Landowner A. E. Hairston's 1871 share agreement with cropper John Owen was a typical arrangement. Hairston promised to furnish the land, draft animals, and tools needed to raise a crop, along with money for half the necessary seed. In turn, Owen swore to take care of the remaining expenses and deliver half the crop to Hairston at the end of the season. Although the contract never mentions tobacco specifically, the details (especially splitting the cost of seed—tobacco seed was expensive and by the postwar period was often grown off the farm) suggest that Owen was promising to produce the crop.[110] Sharecropping arrangements eliminated some of the risk for white landowners, such as Elijah Hundley. By the early 1870s, Hundley rented his land to a number of sharecroppers. He received a portion of their tobacco at the end of the season and made additional profit selling the croppers such necessities as bacon, flour, thread, molasses, and sugar on credit. Although the sharecroppers sometimes settled their debts with cash, they more commonly turned over a greater portion of their crop to Hundley.[111]

The 1870 census did not include a separate tabulation of farms operated by owners, sharecroppers, and cash tenants, making hard figures about Reconstruction sharecropping difficult to obtain. The census does, however, provide the data needed to calculate the number of farms and average farm size. Throughout the three counties, the number of farms increased by roughly a third from 1860 to 1870, and the acreage per farm declined by approximately 100 acres. The statistics in Pittsylvania were typical of the three counties. From 1860 to 1870 the total number of county farms grew from 1,680 to 2,366, and the average farm shrank in size from 345 acres to 253 acres (tables 5.3 and 5.4).[112] Since census takers enumerated farms based on management rather than ownership (for example, a 1,000-acre farm separated among four tenants was recorded as four farms even if owned by one planter), an increase in total farms and shrinking farm size are strong indicators that large landowners were dividing their holdings among tenants.[113] Since the 1880 census did record sharecropper numbers and indicated that by that time most tenants were in fact farming on shares, it seems safe to say that shrinking farm size in 1870 indicated a growth in sharecropped acreage rather than cash rentals or freedpeople obtaining their own small farms.

Although the cause of the shift from wage labor to sharecropping in the Southside is unclear, the particularities of bright tobacco culture and the region's environment certainly played a role in postwar labor struggles. Landowners believed that bright leaf remained a profitable

Table 5.3.

Average Farm Size in Acres, 1850–1900

COUNTY	1850	1860	1870	1880	1890	1900
Caswell	339	374	264	180	205	140
Halifax	346	461	265	183	180	120
Pittsylvania	335	345	253	152	140	121

Source: Compiled and calculated from Historical Census Data Browser, 2004, retrieved December 19, 2008, from the University of Virginia, Geospatial and Statistical Data Center, http://fisher.lib.virginia.edu/collections/stats/histcensus/index.html.

Table 5.4.

Total Number of Farms, 1850–1900

COUNTY	1850	1860	1870	1880	1890	1900
Caswell	707	692	852	1,319	1,284	1,745
Halifax	1,309	1,029	1,294	2,463	2,426	4,092
Pittsylvania	1,524	1,680	2,366	3,502	4,177	4,885

Source: Compiled and calculated from Historical Census Data Browser, 2004, retrieved December 19, 2008, from the University of Virginia, Geospatial and Statistical Data Center, http://fisher.lib.virginia.edu/collections/stats/histcensus/index.html.

crop, one worth struggling to grow, a belief that encouraged their efforts to control black labor. Landowners also understood bright leaf to be a demanding crop, one that relied on an intimate understanding of soil, weather, and seed, and this belief coupled with racist assumptions about black abilities led many whites to believe that freedpeople were unable to produce quality tobacco without white supervision. These assumptions gradually broke down over the course of Reconstruction, due in part to a growing realization among white landowners that sharecropping arrangements entailed less effort on their part and meant greater profits. The former slaves themselves must also receive credit for their struggles to obtain land and independence. Despite difficult odds, they continued to seek land, challenged abusive labor contracts, and braved violence in pursuit of economic opportunities. Key to all these debates were conceptions of soils and environment that grew more and more entrenched through Reconstruction. Whether white or black, farmers increasingly believed that tobacco cultivation was the best use of local land, and they rooted this understanding in ideas about the natures of soils, plants, and people: Reconstruction struggles were about the devilish details rather than the viability of tobacco culture.

Landowners' control over sharecroppers grew in the 1870s with the passage of state crop lien laws. Passed by the Virginia General Assembly on April 2, 1873, "An Act to Secure Advances for Agricultural Purposes" gave landlords the first lien on crops tended by their tenants. In essence, the landowner was guaranteed his share of the tobacco crop before the sharecropper or any other lenders were entitled to a penny.[114] North Carolina passed a similar statute three years later.[115] These crop lien laws set in print the labor and land relationships hammered out through negotiation, legal action, and, frequently, bloody violence during Reconstruction, all but cementing labor relationships and patterns of land ownership in Caswell, Halifax, and Pittsylvania for decades to come.

SIX

A Barren and Fruitful Land

IT IS A TALE OF TWO FARMS. On the first, the year after General Lee's surrender at Appomattox, the young couple Henry and Julia Brumfield decided to raise bright tobacco for the first time on their 360 acre tract along Bearskin Creek, a stream near Chatham in Pittsylvania County. The decision paid off handsomely. The couple hired Beverly Halley, a freedman, on wages, and together the three—with the aid of one horse—raised bright leaf worth $1,800 in their first year of cultivation. The Brumfields, with the help of their children and perhaps continued assistance from Halley, continued to raise tobacco on Bearskin Creek for the next fifteen years. By 1880, the family had exhausted their farm's extensive timber resources on fuel wood, barn building, and field clearance. Forced to move, the Brumfields found a tenant to work their deforested farm and relocated to a new site in the northeastern corner of the county where the woods were thicker. At that point the Brumfields were employing advanced bright leaf production techniques: they used sheet iron flues in their curing barns, covered their plant beds with unbleached muslin to keep insect pests off the developing plants, and built a separate packing barn in which they stripped cured leaves from the stalks for grading. Despite these developments, the family remained relatively small producers of tobacco.[1]

In a second farm community, roughly fifteen miles to the south of Bearskin Creek, tobacco manufacturer William Sutherlin also raised bright leaf following the war, but he followed a different path than did the Brumfields. Sutherlin owned twelve substantial plantations on some of the best bright leaf land along the Dan River, and he hired dozens of freedpeople as wage laborers during Reconstruction. On his Danville-area properties for which records survive—Tar Heel, Sunny Side, Norwood, and Clover Dale—Sutherlin's farm managers supervised black workers in the production of thousands of pounds of bright leaf that served as the raw material for the landowner's manufacturing endeavors. Like many substantial landowners, Sutherlin continued to favor wage labor over sharecropping into the 1880s and 1890s, and his ledgers from these later periods resembled his industrial records; he carefully tallied farm inputs—fertilizer, wages, hardware, foodstuffs, and small loans to his workers—against tobacco production, and he kept a tight rein on his employees. One account of his agricultural enterprises declared that "all the appliances of labor-saving implements, fertilizers, &c., have been liberally provided, regardless of expense." Sutherlin's farms operated as tiny kingdoms; he advanced his workers food and other goods against their wages, and he kept running balances for all his employees, further binding them to his estates.[2] These farms fused the agricultural scale and discipline of antebellum plantations with the business calculations of the postwar industry of which Sutherlin was a part.

The differences between the Brumfields and Sutherlin highlight at least two of the paths of postwar bright tobacco production. The crop benefited from few economies of scale, and almost every farmer in possession of a tract of land was a potential bright tobacco grower, whether they could muster the financial resources of Sutherlin or only the more limited labor available to the Brumfields. Yellow tobacco offered the promise of prosperity for both the yeoman farmer so mythologized in American history and capital-intensive agribusinessmen like Sutherlin. Both paths to bright leaf production also illustrate the changing nature of cultivation on the ground. Certain cultivation techniques survived the war intact and would persist into the twentieth century, yet environmental challenges and new thinking altered other production methods.

The very land that promised bright tobacco sometimes shifted under people's feet, changing in response to continued cultivation and other manipulations of the environment. The timber shortage on the Brumfields' first farm was a common occurrence as more and more regional landowners raised bright leaf, and the soil erosion that often followed deforestation and intensive cultivation may have been a problem as well. While this wood crisis encouraged the family to abandon the property, their removal did not ease agricultural pressures on the land, as a tenant farmer moved into the place that they vacated. Sutherlin's large plantations would have consumed timber as well, and his records demonstrate the increasing use of commercial fertilizers in the post-Reconstruction era. Planters increasingly spent large sums on specially formulated tobacco amendments, often borrowing against their future crops, a trend that locked landowners into future tobacco cultivation as creditors demanded they plant more leaf to secure their debts. Soil amendments thus continued cultivation begun by ideas about soils.

The farm management of the Brumfields and Sutherlin also highlighted the continued difficulty of African Americans in acquiring land; it was no coincidence that black men (and perhaps women) worked as landless laborers in both situations. White landowners appreciated the profit potential of bright leaf and were thus less willing to part with land than planters in other regions of the cash-strapped South. The continued popularity of wage labor among landowners and the resulting debts incurred by many laborers reduced freedpeople's buying power. Rural blacks across the Southside were much more likely to hire out as wage workers for the Sutherlins and Brumfields of the world than to end up owning their own piece of soil.

Erosion, farmers' growing reliance on commercial fertilizers, and deforestation characterized broad swaths of the post–Civil War South, but the patterns these environmental problems took in the Southside owed their details to bright tobacco culture.[3] The speed at which agricultural soils eroded was the product of local geology, topography, and the cultivation techniques of bright leaf. Cotton or dark tobacco cultures, with their emphasis on fertile soils, would have taxed different portions of the Southside landscape and resulted in different erosion rates on different stretches of the countryside. Likewise, fertilizer usage was not a uniform agricultural practice across the landscape: application rates, fertilizer composition, and the fields subject to enrichment varied across the region and were tied to particular crops, terrain, and farmers' ideologies. Timbering was subject to similar variances. Expanding southern agriculture consumed woodlands through land clearance, but bright tobacco, with its fuel wood demands for curing, proved a particularly voracious consumer of forests. These environmental details specific to bright tobacco culture also shaped regional race relations. White landowners pointed to environmental problems on black-operated farms in defense of a virtual white monopoly of the landscape, while ignoring or minimizing similar issues on white-managed farms.

Although growing concerns about erosion, deforestation, and the cost of fertilizers began to trouble regional farmers following the Civil War, other antebellum aspects of bright tobacco culture endured. Bright tobacco boosters continued to tout the crop as an economic savior, and they devoted reams of paper to instructing farmers on how to produce quality bright leaf. Much as during the antebellum period, prominent tobacco manufacturers, seed producers, and fertilizer dealers pushed growers to raise more and more bright leaf, endorsements that often as not stemmed from their own economic interests. Bright tobacco also received valuable advertising from a new auction system that spread from its point of origin in Danville to cover the entire bright tobacco district. This "Danville System" sold loose tobacco in massive public warehouses, where labels and auctioneers placed a name and a monetary value on each farmer's bright leaf in public. Much like the newspaper articles and pamphlets of the antebellum era that celebrated lottery prices, these public auctions encouraged farmers to raise more and better tobacco and associated their agricultural skills with the price that their leaf brought.

As during earlier periods, local landowners remained willing to use or exploit violence to consolidate their control over labor, landownership, and tobacco marketing following the conclusion of Reconstruction. An 1883 riot in the city of Danville expressed black and white political frustrations as a racially diverse Readjuster Party seemed poised to gain power. But the violent event can also be seen as an expression of the racial tension explicit in the region's tobacco cultivation, marketing, and manufacturing. White tobacco manufacturers and black tobacco workers were involved in the riot, and local warehousemen and industrialists

touted the suppression of black political aspirations as evidence that
Danville remained the region's premier tobacco center—a message directed at white landowners who controlled the marketing of the majority
of tobacco raised in the countryside.

A new auction system, increasing demand, and the sometimes violent control of labor combined to continue bright tobacco's expansion
during and after Reconstruction. Danville, Milton, and Halifax warehouses teemed with loose piles of golden leaf; tobacco factories hummed
in almost every town and hamlet; and tobacco fields spread across the
landscape, popping up everywhere that the soil was suitable. By the late
1880s, farmers in the three counties raised more tobacco than ever
before, and their crops were almost exclusively bright leaf.

Regional tobacco marketing took a distinctive form following the Civil
War, one that connected bright tobacco with particular landscapes, those
landscapes with farmers' environmental knowledge, and then put that
knowledge on public display. Known as the auction sales system—or,
commonly, the Danville System—public marketing auctions became
entrenched during the postwar years. The region surrounding Danville
and Milton had pioneered this unique system of tobacco sales during
the antebellum era. At that time, Virginia and North Carolina law required that farmers bring their tobacco in hogsheads for state inspection at official warehouses, and these inspection warehouses also served
as the primary taxing point for the crop. Inspectors certified that the tobacco was of good quality and thus eligible for foreign export. Whereas
inspection may have been necessary to ensure the reputation of regional
tobacco for the export market, it was much less important for manufacturers who purchased their tobacco from local sources. These buyers
connected crop with grower and thus had little use for a middleman in
the form of a state inspector, especially when that middleman added to
the cost of tobacco.[4] Both growers and manufacturers asserted that tobacco quality was assured by their knowledge of local environments and
people rather than through the state officials whom more distant buyers used as proxies for these relationships. Danville-area growers and
buyers during the 1840s and 1850s increasingly exploited a loophole in
the law in order to sell and buy tobacco loose (not packed in hogsheads)
and thus bypass state inspection. They claimed that all the leaf that

passed through Danville was destined for local manufacturers rather than for foreign trade—a partial fiction, as at least some leaf found its way overseas—and hence the law mandating inspection and the use of hogsheads did not apply to their transactions. By one estimate, more than 90 percent of the Danville trade by 1850 was in loose form. Growers, buyers, and manufacturers all appreciated this loose sale system because it bypassed some state taxes, allowed for the sale of tobacco outside of the limited number of state warehouses, removed the need for thousands of costly wooden hogsheads, and provided greater flexibility in lot size.[5]

With these loose-leaf sales came an auction system that revolutionized tobacco sales and also worked to reinforce existing relationships between bright leaf growers, marketing, and environmental knowledge. At auction, growers arranged piles of their leaf in rows on the warehouse floor, and buyers moved down the lines of piles, personally inspecting the quality of the leaf as auctioneers solicited their bids. Legend has it that this process began in 1837, when the national financial crisis closed Danville's sole state inspection warehouse and growers began selling their leaf to buyers on the street. This practice remained popular after the state warehouse reopened, and the city's first facility dedicated to indoor auctioning—Neal's Warehouse—opened in the late antebellum period. For growers, this system stimulated competition among the various buyers and manufacturers' agents, an arrangement with the potential to drive up prices. There were benefits for buyers as well. They could visually inspect the tobacco rather than having to rely on the word of the state inspector or a small sample taken from a hogshead. Loose piles promised fewer cases of mold and bruised leaves than tightly packed hogsheads, and growers were unable to trick buyer with such practices as "nesting," in which growers packed poor-quality tobacco (or even stones) into the center of a hogshead where a casual inspection was unlikely to find it. By Reconstruction these auction practices came to be known as the "Danville System" of tobacco marketing.[6]

Following the Civil War, in an effort to reestablish some control over regional sales, the state placed inspectors in Danville's auction warehouses where they inspected loose tobacco destined for interstate trade. Both farmers and warehouse owners complained that the inspectors graded tobacco too low and charged high fees besides, and most warehouses insisted on having their own private inspector as well. In 1877, the Virginia legislature gave up attempting to supervise the Danville System and caved in to pressure from warehouse owners by eliminating the state inspection of tobacco, openly permitting loose-leaf auctions, and taxing the auction warehouses in more direct fashion. This action made the Danville System completely legal, but it also removed the remaining state oversight of loose-leaf auctions (limited as it was), and warehouse owners passed their new tax burden along to growers in the form of increased handling, storage, and commission fees. Despite the absence of state inspection, warehouses continued to employ their own inspectors to grade tobacco, and inspection fees actually tripled during the 1880s. These increases were nothing less than warehouse owners gouging farmers under the guise of eliminating unnecessary state inspectors.[7]

By 1880 the auction warehouses in Danville, Milton, Halifax Courthouse, and surrounding towns were well-oiled machines. Farmers or landowners brought their tobacco to the warehouse as loose piles in large wagons rather than in hogsheads. The warehouses often provided parking lots for the vehicles, stables for farmers' teams, and bunkhouses where men from the countryside could stay while selling their tobacco. Farmers who came to town were notorious for their rowdy escapades. One local history claimed, "If they didn't wake up with a hangover in the local jail, they woke up in a hotel room without their billfolds—victims of their own sins and disreputable women." Inside the warehouses, officials weighed and inspected each lot of tobacco and then placed the leaf on the auction floor with an attached tag that listed the grower's name and the weight of the pile. Buyers from regional manufacturers, Richmond or Petersburg operations, or speculators who planned to resell tobacco on other markets purchased the pile from an auctioneer, who roved the floor at set times selling the warehouse's current inventory in a rapid, sing-song lilt. The fastest auctioneers reportedly moved as many as 150 piles of tobacco each hour. Once the pile was sold, the name of the buyer and the price was written on the tag, and the farmer was given a receipt that he could cash at the warehouse office.[8]

This auction system put farmers' abilities—and, by extension, their worth as agriculturalists—on public display. Other growers wandered the auction floor, studying the quality of each pile and noting the name on the attached tag. The auction placed a monetary worth on tobacco quality, and it served as a measuring device of farmer expertise. Farmers who attended the auctions satisfied their most pressing questions: Who made the best tobacco on a regular basis? How much did they get for it? Underlying both questions was another: Who knew their land best? Auctions thus served to qualify a farmer's worth much as did the newspaper price announcements of the antebellum period, though the auctions reinforced these public claims to expertise in a much more visceral fashion.[9] Attendees could walk among the leaf, touch and smell the tobacco, listen to the auctioneer's chant as prices rose or stalled, and read for themselves the names and figures connected to each pile. The transition from state inspection of hogsheads to sale at public auction might seem minor, but this shift brought the physical leaf from the relative obscurity of the inspector's storeroom into the public purview. It cannot be coincidence that this shift occurred in a place and time when bright tobacco emerged as a crop characterized by the importance of quality above all else. Although it is an impossible impulse to quantify, it seems almost certain that many farmers returned to their farms after an auction weekend in Danville intent on improving both their profits and their public reputations as bright leaf producers.

What individual piles did for a single farmer's reputation, the impressive rows of golden tobacco did for the region's reputation as a whole. Reporters—whether locals with booster fever or more distant journalists investigating the region—invariably left the warehouses impressed by the quality of local crops, growers, and the distinctiveness of regional soils. Articles on the warehouse system almost always referred to both the color of the tobacco and to the sandy, thin soils of the surrounding countryside, both connecting and juxtaposing the two, expressing incredulity that such poor soil could make people so rich.

The Danville auction system's increased association of a staple crop with its growers is somewhat antithetical to environmental history's best-known analysis of late-nineteenth-century agricultural marketing. William Cronon has demonstrated that Chicago's wheat trade gradually removed the physical grain from the view of buyers and sellers. An increasingly stratified classification system defined wheat by type; farmers received receipts for their crops when they delivered them to silos across the Midwest, where the grain was commingled with like grades from all over the region. Chicago's buyers ultimately bought millions of bushels of grain sight unseen, relying on the perceived standardization of grading. On the market wheat became a tradable and somewhat abstract commodity rather than a tangible product of soil, agriculture, and science.[10] In contrast, on the tobacco warehouse floor every leaf was connected to individual farmers and specific farms, more so than under the previous inspection system. Bright tobacco would eventually be classified by specific grades; by 1930 the USDA's Bureau of Agricultural Economics defined an incredible 450 possible types of bright leaf. This government categorization divided yellow tobacco into six main groups, each subdivided into seven levels of quality, with five possible colors and eight additional special factors ranging from leaf size to storage potential (not all factors could be combined). If anything, this incredible parsing of tobacco made it less likely that standardized and abstract trading would dominate leaf sales. Despite official definitions, these grades still relied heavily on opinion, personal inspection, and the human eye.[11] The notion that bright leaf was an artisanal crop with limitless variation persisted well into the twentieth century, and it comfortably existed side by side with ever more complicated forms of marketing. Bright tobacco illustrates one exception to the general agricultural trend of anonymous commodification that Cronon records and the attention that environmental historians must pay specific agricultural systems when conceptualizing broader farm and market trends.[12]

A few years after Reconstruction's end, tremendous quantities of bright leaf flowed through Danville's warehouses, and the city was well on its way to becoming the largest tobacco market in the world.[13] Between the 1873–74 season and the 1879–80 season, Danville bright leaf sales expanded from 12 million pounds to more than 33 million pounds. By 1879, the city had eight major warehouses—substantial brick structures with as much as 28,000 square feet of auction floor—that each sold between 3 and 5 million pounds annually. An individual warehouse often had more than 100,000 pounds of bright tobacco on its floor at any given time during auction season. By 1885 there were ten warehouses and annual sales approached 41 million pounds, and in 1899 sales would exceed 54 million pounds of loose leaf. These clearing houses provided the raw materials to the city's tobacco manufacturing district, which grew to include at least thirty factories along the Dan River. These warehouses and factories soaked up almost all the tobacco grown in Caswell, Halifax, and Pittsylvania and began to draw from emerging bright leaf districts in adjoining counties, including eastern Henry and Franklin counties in Virginia and Stokes, Person, Rockingham, and Granville counties in North Carolina.[14]

The agricultural landscape of the rural Southside changed almost as quickly as the region's market system following Reconstruction. As noted above, the number of farms in the three counties increased following the war, and farm unit size shrank dramatically.[15] This was due in large part to the growth of tenancy in the form of sharecropping and farm rentals, as some landowners moved away from wage labor by force or necessity. In 1860 the three counties had contained 3,401 farm units, with an average size of 386 acres each. By 1880, Caswell, Halifax, and Pittsylvania had more than double the number of farms as two decades earlier ($n=7,284$) but the average farm size had shrank to just 168 acres. The percentage of farms under 100 acres in size grew from 38 percent ($n=1,290$) in 1860 to 42 percent ($n=3,086$) by 1880. A substantial number ($n=1,431$) of these smaller farms recorded in the 1880 census were managed by sharecroppers.[16]

If farm size shrank with tenancy arrangements, the size of tobacco plots under the control of individual families diminished as well, even as overall tobacco production increased. Because of its extreme labor demands, most antebellum growers had cultivated only a few acres of tobacco, but some of the largest planters, such as James Bruce and members of the Hairston family, raised well over a hundred acres of leaf each season.[17] By 1880, many large landowners had divested direct control over tobacco production to their sharecroppers and leasers, who in turn cultivated a few acres of tobacco on their small portion of the historic plantations. According to the 1880 agricultural census, the majority of regional farmers tended ten or fewer acres of tobacco during the 1879 season. A few larger operations persisted—Emanuel Girst, Sawyer Watkins, William Carrington, and E. B. Spencer tended fifty acres each in Halifax, where they were the county's largest growers; and John Chatten of Pittsylvania tended the largest regional tobacco plantation, where he raised seventy acres of leaf—but the vast majority of bright tobacco was produced on relatively small agricultural units.[18] The days of thousand-acre plantations with dozens of hands laboring in each field were past.

Small tobacco plots did not necessarily lead to more diversified or subsistence farming; indeed, agricultural records indicate that the Southside
became less self-sufficient in the decades following the Civil War.
Between 1860 and 1880 the population of the three counties swelled
from 74,839 to 104,002 people, but on the whole agricultural output
failed to keep pace with population growth.[19] Farmers' production of
corn, the staple of most southern rural families, increased slightly over
this period, from just under 1.5 million bushels to a little over 1.6 million
bushels, and the number of mules almost doubled, but every other major
agricultural crop (excepting tobacco) and livestock type decreased. In
1880, Caswell, Halifax, and Pittsylvania farms grew less wheat, oats, and
rye, and fewer sweet potatoes than they had on the eve of the war. Their
pastures held fewer horses, oxen, milk cows, and beef cattle, and less
than a third of the sheep than in 1860. Even the ubiquitous southern hog
had declined slightly in number; in 1880 there were 12,717 fewer pigs in
the three counties than twenty year previous. In fact, with more than two
people to every hog ($n = 49,107$) in the region, it could hardly have been
self-sufficient even in pork.[20] Although some of this agricultural simpli-
fication might be attributable to the destruction of the Civil War, these
basic census figures make clear an ever-increasing reliance on tobacco as
a cash crop to the detriment of diversified farming. The region that had
long been tobacco obsessed was all but dependent on the plant by the end
of Reconstruction as food crops and livestock lost ground to bright leaf.

Farm and field sizes grew smaller during the postwar decades, tobacco replaced grain and livestock, and tobacco culture fostered other,
more subtle rearrangements of the landscape as well. Tobacco barns had
long stood next to fields to facilitate the movement of leaf from cutting
to curing.[21] As farmers moved the bulk of Southside tobacco from bottomland to the sandy ridges, new barns appeared on the slopes as well.
An 1880 map of George Venable's Halifax farm along the Staunton
River provides an example of the changing built environment of regional farms (fig. 9). Venable's farm contained cleared bottomland, and
the presence of a tobacco barn just above the bottom, where the structure could serve the low fields yet was protected from freshets, strongly
suggests that the lowland was once the site of tobacco cultivation. In
1880, Venable's tract contained a substantial stretch of cleared upland
as well. His barn along the Staunton bluffs could serve the sandy land

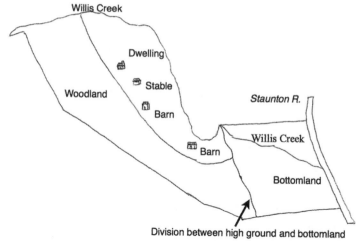

FIGURE 9: Map of the George C. Venable farm in Halifax County, 1880. Total farm size was 179¹/₄ acres, with all of the low ground and roughly half of the high ground in arable land. Note that both tobacco barns, which were almost always on the edges of the tobacco fields, were located on the high ground. Adapted from Halifax County Deed Book 68, 1880–1881, p. 302, Halifax County Courthouse, Halifax, Virginia. Map by the author.

above, but Venable also owned a barn higher along the ridge, indicating that his tobacco production was concentrated on the hills above the river by that date.[22] Venable's farm geography, when coupled with thousands of similar farmscapes across the region, presented environmental consequences. Landowners and tenants typically cut timber for new barns on site, and the construction of tobacco barns on the sandy ridges placed additional pressure on the most fragile portions of regional watersheds. Barn building combined with cordwood production and clearance for agricultural land to denude upland forests.

Tobacco culture's pressure on timber resources was a recurring concern for many local farmers. Wood served as curing fuel, barn timbers, hogshead staves, wagon frames, hoe handles, sticks for hanging cut plants, and as a source of ash for plant beds, among other uses. The 1880 census estimated that farmers in the three counties consumed or sold well over a quarter of a million cords of wood each year.[23] As one agricultural historian has concluded, "an abundance of woodlands was more essential to tobacco planting than any other type of farming."[24] Contemporaries agreed with this assessment. An article in the *Southern Workman* (a Hampton, Virginia, journal directed at freedpeople) warned readers against the wanton deforestation of watersheds, concluding, "The farmer or land owner can commit no greater mistake than the reckless cutting down of trees, or show no greater wisdom than in careful and systematic replanting."[25] The *Southern Workman* also advised that "hillside washings can be arrested by judicious tree-planting" on the "worn-out lands of Virginia and the Carolinas."[26] Likewise, a booster publication issued by the Virginia State Board of Agriculture in 1889 suggested that deforestation was a growing problem in the southern Piedmont. The book optimistically characterized regional wood resources as "considerable" but proceeded to caution readers that the region could ill afford to export any timber. (This admission appears more revealing considering that the same hyperbolic tract promised that the Piedmont's poor soils were in fact "among the most fertile known.")[27] Farm families had to husband their timber resources to ensure they had enough wood to raise and cure their tobacco crops. Miscalculations could mean the end of farming on a given site until the woods could recover, as seen in the experience of the Brumfield family.[28]

A casual survey of the census figures concerning farmland somewhat muddies these assessments of the Southside's forests. Improved farmland (land in fields and pastures) in the three counties actually declined by almost 200,000 acres between 1860 and 1870 as local farms recovered from the Civil War, adjusted to changed labor arrangements, and concentrated on tobacco at the expense of other crops. Improved acreage climbed slowly but steadily through the remainder of the century, and by 1900 improved acreage and total farm acreage once again stood near their 1860 levels. These numbers suggest a steady forest cover that actually grew during Reconstruction before receding to antebellum levels by the end of the century. It would be a mistake, however, to assume that all unimproved farmland was forest. A substantial portion of "unimproved" Southside farmland was not covered in timber. Rather, much of the land not in fields or pasture was abandoned cropland that had yet to regrow trees. These abandoned and galled stretches of landscape were especially prevalent in districts of bright tobacco cultivation, where farmers kept their fields in cultivation until the land was too weak to produce even yellow tobacco and thus was slow to revegetate when abandoned. Geographer Stanley Trimble estimates that between 15 and 35 percent of land along the states' border was deforested old field in 1880.[29] Census figures from ten years prior support Trimble's assessment. In 1870, only 73 percent of the unimproved land across the three counties was described as "woods" (unfortunately, this was the only year that census takers made such a distinction). The situation was particularly dire in Caswell, where 57 percent of unimproved lands were denuded of timber, leaving only a little over a quarter of the county's farmland in woods.[30] Lands so depleted were even slower to regrow shrubs and trees than the average abandoned ground. These old fields were erosive eyesores that threatened surrounding cropland with their runoff. Severe flooding in Pittsylvania in spring 1887 demonstrated the hazards of deforested watersheds. Rains that year led to freshets that damaged bottomland crops, and even small streams rose high enough to wash away "fencing, crops, &c., on bottoms" across the county.[31]

Contemporary land surveys in the three counties provide a more intimate glimpse into Southside forests. Surveyors delineated properties for sale by recording a series of markers around a tract's bounds; these points were usually trees, though they were sometimes rock piles or stakes—referred to as pointers—on open ground. Surveys listed marker trees by species, and thus the historian willing to wade through the courthouse survey books can compile a rudimentary list of tree species in any given year, and from these lists one can make some educated speculations regarding land clearance and ecological change. Such assumptions are fraught with uncertainty, however. Surveyors favored large but healthy trees that would be easy to spot in future surveys and preferred certain hardy species over others, survey lines often followed field margins and thus recorded trees in land that was essentially open, surveyors often ranged off the exact line to use trees as markers, and in any given year only a small percentage of land came under survey, and thus marker trees may not be entirely representative of the state or composition of a region's woodlands as a whole. Nevertheless, regional surveys tend to support contemporary observations concerning the changing regional landscape.[32]

For the purposes of this book, I examined all land surveys recorded in the three counties during 1880 and compared the marker data with surveys from a century earlier.[33] In total, the 1780 land surveys recorded exactly 1,000 line markers, and the 1880 surveys listed 1,151 markers.

The most dramatic difference between the 1780 and 1880 surveys was in the number of pointers used to mark land boundaries. Surveyors turned to trees as markers whenever possible, and as many property lines remained the same over multiple surveys, even swaths of land under cultivation often had line trees left from previous surveys. Only in completely deforested areas would surveyors resort to making pointers to mark the line. In the 1780 surveys, 12.5 percent of all markers were pointers. By 1880, pointers represented 23 percent (264 of 1,151) of survey markers. The use of tree species commonly associated with old fields increased as well, even though surveyors would likely have resisted selecting successional trees, which are generally short lived, as markers. Only three such trees were among the 1780 markers: one cedar, one persimmon, and one cherry. One hundred years later, 4.2 percent (48) of all line trees were old field species, including black and honey locusts, plums, and persimmons. White oaks and red oaks, species characteristic of mature Piedmont forests, dominated the 1780 line trees; 494 of the 1,000 markers were one of these two oaks. In 1880 these two species remained the most common markers, but they had become proportionally less prevalent. White and red oaks made up only 26.5 percent of markers at that date. In their place was a growing percentage of hickories and post oaks (which together made up 10.6 percent of markers, up from 5.6 percent in 1780), species often indicative of a younger forest or poorer, drier soil. All of this data remain suggestive rather than definitive, but these land surveys do seem to support the notion that Southside forests were markedly different in extent and composition in 1880 than they had been a century earlier, reflective of a century of tobacco culture.[34]

Along with deforestation, soil erosion continued apace. Antebellum bright leaf growers had worried that bright tobacco cultivation techniques were particularly erosive, but postwar tobacco agriculture made few changes to limit erosion. Popular cultivation and curing guides continued to advocate the use of new hillside ground for each season's plant beds, planting on ridge land, and shallow, frequent cultivation. These guides, and, it may be assumed, their techniques, were quite popular with regional farmers. E. L. and E. P. Love's *The Art of Curing Fancy Yellow Tobacco* went through several editions in the 1880s, and Robert Ragland's popular guide to raising and curing bright leaf (published under a variety of titles) appeared in six editions and had sold more than

FIGURE 10: By the early twentieth century, substantial portions of the Southside—like this old field in Caswell—suffered from sheet erosion and gullying. Such erosion would lead New Dealers to classify the old tobacco belt as an agricultural and environmental problem area. Courtesy Library of Congress.

100,000 copies (one for every man, woman, and child in the Southside) by 1885.[35] In a typical soil management passage, the Loves declared that "early, rapid and thorough cultivation is necessary to produce first class goods," and thus soil cover and retention took a backseat to frequent plowing, with obvious consequences.[36] They also cautioned farmers to not cultivate too deeply, as turning up the clay subsoil would reduce leaf quality. They warned readers, "Don't stir up the clay, which is hostile to the growth of fine tobacco. A lump of clay as large as a hen's egg at the root of a plant of tobacco will cause it to grow coarse and mature late. It matters not how fine you break the top of the soil, the finer the better, but do not plow deep enough to reach the clay."[37] Such advice was sure to produce a finely powdered upper soil horizon with a hardpan layer below, a recipe for rapid soil erosion during spring and summer deluges (fig. 10).

As during the antebellum period, these cultivation techniques and their erosive consequences provoked worry in a number of farmers, and these concerns proved difficult to reconcile with their enthusiasm for bright tobacco as a crop. From the early postwar days, bright leaf farmers made familiar complaints. A Granville County, North Carolina, farmer, living in a district that was beginning to experiment with the new crop, wrote that "farms every where [were] going down" as fields were "running away into galls and gullies."[38] Ten years later, a farmer wrote of similar concerns from the oldest of bright tobacco districts, the Pittsylvania-Henry county line, where he worried about the "murderous policy pursued of cultivating a piece of new land until it will produce nothing more, and abandoning it to the rains and hot sun until it is gullied often beyond redemption."[39] Even such enthusiastic bright leaf boosters as Ragland lamented the effects of soil erosion that often accompanied the crop. In 1879, he recorded that "the deterioration of tobacco lands is estimated as high as 15 per cent. per annum, especially the light soils on the rolling uplands and slopes. This damage is more than half the result of surface washing." Ragland went on to assert that careful soil conservation techniques could control this erosion, but much like George Jeffreys—who had lamented the damaging effects of his bright leaf experiments on the soil in the late 1840s—Ragland's claims seemed based more on hope than on practical evidence.[40] A few years later a Brosville (Pittsylvania) grower seconded Ragland's concerns. He wrote to the Southern Planter in 1887 that heavy spring rains had wreaked havoc in the bright tobacco district because the storms caught farmers in the middle of their intensive cultivation period. The rains caused the land "to wash much worse than if it had not been broken up. In a great many places the [top]soil is all gone and holes washed out in the clay."[41] The modern geographer Stanley Trimble also points to the 1880s as the time when soil erosion reached new heights in the Piedmont, a condition he attributed to a large extent to expanding tobacco cultivation.[42] Planters like Ragland and Jeffreys wanted to believe that soil erosion could be controlled, but at the same time they were unwilling to deviate from tillage methods that had proven successful at producing fine bright leaf, even as those techniques ate away at the region's sandy topsoil. This tension between the necessities of quality production and the need for soil conservation remained unresolved throughout the late nineteenth century and highlights the role of a particular crop culture in explaining abusive soil practices in one region of the postwar South.

Environmental concerns along the Border were never far removed from issues of race. Some white landowners found in the region's freed-people scapegoats for growing soil and forest crises, arguing that Emancipation had fomented a social erosion that became an agricultural problem. The first part of this argument labeled African Americans as inefficient farmers. A number of regional planters argued that black farmers did not have the energy or intellectual capacity to produce the highest-quality bright leaf without white direction, despite ample antebellum evidence to the contrary. These commentators stressed that African Americans were well suited for the menial aspects of tobacco culture, the endless hoeing and transplanting and cutting, but that white decision making at key points, during curing and topping, for example, was crucial to making bright leaf. Thus they argued against sharecropping arrangements with their degree of independence. Writing to the *Southern Planter* in 1867, "H" provided an example from Caswell County, which he declared representative. He outlined a case where a freedman and a white landowner controlled the same number of acres and hands, yet the landowner produced tobacco worth more than six times as much. "H" concluded from this result that the black laborer "is as yet only fitted for the place of a hireling."[43] Almost twenty years later, Charles Bruce echoed similar sentiments, declaring black workers "incomparable as a hired laborer" but only "when under constant and vigilant supervision and strict discipline."[44]

Even pundits who did not openly question freedpeople's abilities as independent farm laborers decried the labor situation in the tobacco district in terms that implied that African Americans were poor stewards of their own bodies and by extension of the land. William Sutherlin, speaking before an agricultural meeting, expressed his worry that black laborers would soon be too few to fulfill the needs of Virginia's landowners as "this limited supply was daily diminishing by reason of disease and death which has become incidental to that sudden abolition of slavery, which remitted the negro to a life of licentious liberty."[45] According to Sutherlin, killing the self and killing the soil marched hand in hand.

After characterizing freedpeople as inefficient farmers, critics of their farming abilities then linked soil depletion and erosion to tenancy, ignoring the voluminous literature instructing tobacco growers of all colors and ownership statuses to engage in methods bound to damage the land.[46] Opponents of tenancy argued that renters and sharecroppers had little long-term stake in the land, no financial incentive to conserve wood or soil, and little promise of changing their economic status. They believed that these conditions led to an ethic of waste, or an "extensive" rather than an "intensive system" of farming.[47] A Prince Edward County farmer's comments were representative of the sentiments of many Southside landowners when he explained that portions of his farm were "badly worn out, having been rented out since the close of the war, and nearly everything carried off the place and not much returned to the soil."[48] Although both white and black tenants existed in large numbers in the Southside, comments about tenants' destructive habits often bore racial overtones.[49] What these arguments ignored is the fact that tenants could hardly work the land harder than bright tobacco experts advised. Tenants who strived to make the best-quality bright leaf left behind eroded fields and timbered forests, just as did landowners who grew the crop. The problems that resulted from bright tobacco culture were no different in 1880 than in 1860, aside from scale. African Americans who focused on subsistence crop production rather than tobacco cultivation (as suggested in the example cited by "H," above) faced criticism as crude or inefficient farmers. Thus black tenants in particular faced an impossible situation; white critics claimed that freedpeople were not intelligent enough to raise high-quality tobacco, but if they managed to produce top bright leaf, the same critics declared that the resulting ravaged land proved that blacks were poor stewards of the soil. White racism thus combined with the dictates of bright leaf crop culture to create a stereotype of black farming that ignored universal soil abuse throughout the Southside. Freedpeople were, to employ a popular southern saying of the time, "damned if they did, and damned if they didn't."

This debate encapsulated much of the African American experience with bright leaf following the Civil War. Despite the fact that slave labor and knowledge had been so critical to the development of the crop, from cultivation to manufacturing, white landowners and tobacco experts increasingly defined quality farming in terms of professional guides and land ownership, almost exclusively white domains. In the eyes of Sutherlin, Bruce, and many other whites, black experience with bright leaf took a back seat to white stories about the crop and its culture constructed over the previous four decades. According to this body of cultural knowledge, white men—the most prominent of whom were Abisha Slade and Robert Ragland—developed and perfected bright tobacco with the inadvertent aid of their slaves, white boosters spread the crop's culture across the Southside, white auctioneers and warehouse-men built a new market system, and white experts created pamphlets and guides that systematized cultivation and curing. Absent from these stories were slave contributions to planting, topping, or curing practices drawn from their long experience working with the crop. Adding to the inadvertent irony, bright leaf was the most profitable agricultural option for black farmers, be they sharecroppers, renters, or landowners, but the crop's culture was a white-dominated system, and African American ag-ricultural efforts were doomed to white criticism, if not open hostility. The notion that black farmers were incapable of quality tobacco produc-tion or were particularly poor land stewards worked to further circum-scribe black landownership and social opportunity.

Working within a landscape increasingly characterized by soil de-pletion and deforestation, white and black farmers across the Southside turned to commercial fertilizers to increase the land's productivity. Al-though the impulse to apply these new fertilizers to old lands was com-mon across the South and the nation during the second half of the nineteenth century, bright tobacco farmers quickly developed an espe-cial reliance on tobacco "manures" that promised heavier crops and yel-lower leaf. Farmers across the Southside had long sought amendments that would increase bright tobacco production without damaging the quality of the leaf. The two most popular southern fertilizers, barnyard manure and pure Peruvian guano, often turned yellow tobacco harsh and made the crop difficult to cure. In contrast, new commercial prod-ucts that billed themselves as specifically formulated for bright leaf cultivation proved an unprecedented commercial success. By the 1880 agricultural census a report noted, "In all the tobacco districts of south-side Virginia the use of commercial fertilizers is general."[50] Tobacco growers in Virginia and North Carolina proved even more fertilizer-hungry than the typical southern farmer, a reliance fueled by their long-standing reluctance to use manures produced on the farm. Over the course of the 1879 crop season, farmers in the three counties spent roughly twice the Virginia and North Carolina average on fertilizer. Whereas the typical North Carolina farmer invested $13.40 in fertilizers and the Virginia farmer paid out $18.03, in the district around Danville the average annual fertilizer expenditure ranged from $32.00 in Halifax to $35.68 in Caswell. In all, landowners and tenants in Caswell, Halifax, and Pittsylvania spent almost a quarter of a million dollars annually fertilizing their fields.[51]

On its face, the spread of commercial bright tobacco fertilizer seems difficult to reconcile with the crop's nutritional demands. After all, it was poor agricultural soils and a general lack of humus and nitrogen that made Southside ridge land so suited to yellow tobacco production. Part of the answer to this seeming paradox lay in the continuous cultivation of bright leaf on the same fields for a decade or more. Bright tobacco only seemed to not deplete soils; on these overworked stretches of ground, land often suffering from moderate to severe soil erosion, the nutrients in fertilizers did add needed elements to tobacco ground. On these sites farmers had worked the land until it was too weak to support even bright leaf. Commercial tobacco fertilizers, many of which still bore the terms guano or manure on their labels, were also specially formulated to feed bright leaf. The most popular brands contained substantial doses of phosphorus and potassium, but a much lower percentage of nitrogen, especially when compared to such traditional amendments as barnyard manure or pure Peruvian guano.[52] A single 1886 issue of the *Halifax Record* contained advertisements for Bono Tobacco Fertilizer, British Mixture, Liebig Ammoniated Superphosphate, Flamingo Guano, Lister's Agricultural Chemical Works, Eagle Brand Guano, Capital Tobacco Fertilizer, Orchilla Guano, "Yellow Leaf" Tobacco Fertilizer, and Durham Bull Fertilizer.[53] As some of the brand names suggest, these amendments were formulated for bright leaf tobacco growers. Tobacco fertilizers encouraged plants to grow larger and faster, traits that had two main benefits for farmers. First, fertilized tobacco leaves were heavier than unfertilized ones, a characteristic that paid in a crop sold by weight. Second, supplemental phosphorus and potassium encouraged tobacco plants to mature early (nitrogen had the opposite effect), which in turn permitted growers to cut their tobacco well before the frost. Stiff doses of phosphorus and potassium also permitted tobacco plants to utilize additional nitrogen without developing a coarse texture or harsh taste.[54]

Furthermore, these amendments fertilized ever-changing plant varieties, as such horticulturalists as Ragland continued to develop new strains of bright leaf that matured earlier and grew larger.[55] Fertilizer popularity in the Southside thus grew at the intersection of environmental degradation and agricultural improvement as tobacco farmers sought to make better bright leaf while dealing with soil depletion and erosion.

The spread of commercial superphosphates in particular would create long-lasting, though long-hidden, environmental effects. As tobacco farmers shifted from guano phosphates to rock phosphates, particularly the apatite mined along the Carolina coast during the last two decades of the nineteenth century, they unwittingly created new health hazards for tobacco consumers. Much of the commercial rock phosphates used in North Carolina and Virginia fields contained high levels of radium-226, which eventually decayed to lead-210 and then to an even more toxic daughter isotope, polonium-210, one of the more radioactive substances known to science ("250 million times more toxic than cyanide [by weight]" and volatile enough to serve "the military as an initiator for nuclear weapons"). Year after year of superphosphate application built up soil levels of radium and lead, which tobacco plants drew from the ground through their roots or captured from the air with their fine, sticky leaf hairs. Lead-210 survived the manufacturing process, which in the twentieth century increasingly meant being rolled into cigarettes. Cigarette combustion in turn proved a most efficient means of transferring lead from tobacco into human lungs, where long-term exposure allowed the formation of polonium-210, a significant cause of lung cancer. Farmers who applied superphosphates in a search for renewed fertility and brighter tobacco knew nothing of radium, soil lead levels, or polonium; it was not until the 1960s that scientists and the tobacco industry developed an understanding of this chain of events, but the consequences for smokers became part of the long-term legacy of the tobacco belt's late-nineteenth-century fertilizer boom.[56]

The proliferation of commercial fertilizer use among tobacco farmers almost invariably entailed an expansion of credit and debt relationships. Landowners and tenants, black and white, often took out fertilizer loans against the upcoming year's crop. "As early as 1881 only 10 per cent of the fertilizers sold in North Carolina was purchased for cash."[57] Landowning farmers went in debt to fertilizer companies, such as Danville's

Patassco Guano Company or Baltimore's Piedmont Guano Company, which in turn placed liens on the farmers' tobacco crops, livestock, and other agricultural produce as surety on the loan. Merchants who retailed guano in smaller amounts often followed suit, as did local auction warehouses, which simultaneously profited from auctioning farmers' tobacco and selling them fertilizer for the following year's crop.[58] Both Virginia and North Carolina had homestead exemptions, laws that prevented creditors from seizing homes and land to fulfill debts, but fertilizer companies typically required that borrowers waive their exemptions in order to obtain credit. Though of dubious legality, this tactic was omnipresent.[59] Tenants were in an especially precarious position. Without a homestead to place at the mercy of the fertilizer companies or local merchants who served as middlemen, tenants, many of whom were black, were forced to find a landowner or patron to cosign their loan. To secure their own interests, landowners often demanded additional compensation from tenants. In a typical example, plantation owner Katherine Moses cosigned a fertilizer note for her tenants near Hurt, in Pittsylvania County. The merchants who extended the note reassured Moses that the tenants themselves took all the risk as she "could take it [the cost of the fertilizer] out of their part of the tobacco and run no risk."[60] In Pittsylvania and Halifax tenants' positions were further weakened by *Parrish v. The Commonwealth* (1884), in which the Virginia Supreme Court ruled that sharecroppers were mere employees who had no inherent interest in the land or its produce.[61] Bright tobacco culture following the war shaped land ownership patterns, largely encouraging whites to retain control of the countryside, and the spread of fertilizer dependence further constricted tenants' power over the land's productivity, perpetuating a dominance of landowner over landless that frequently equated to white over black.

Manufacturers of commercial tobacco fertilizers advertised the products of distant landscapes and at the same time sought to connect those exotic resources to local landscapes and people. Such ingredients as Peruvian guano and phosphate from mines along the southeastern coast of the United States traveled hundreds or even thousands of miles to reach the Southside, and manufacturers often stressed the origin, and hence the authenticity, of their ingredients. Baltimore Phosphate and British Mixture, for example, declared that their products contained only "very superior" Peruvian guano or "high grade phosphate" and questioned the quality of fertilizers produced by smaller companies.[62] Tobacco farmers often refused to take these claims at face value; they invested a great deal in commercial fertilizers and were concerned that they might not be getting their money's worth. At the encouragement of farmers, Virginia and North Carolina both established state experiment stations in the 1880s with the explicit aim of testing commercial tobacco fertilizers, guaranteeing farmers that their phosphates and guano were indeed legitimate.[63] Much as manufactured tobacco advertising linked bright tobacco products to the Piedmont landscape, fertilizer companies sought to connect their products to sites renowned for their fertility, be they Pacific islands or South Carolina mines.

While these fertilizer blends promised to bring the fertility of the globe to the Piedmont, companies also worked to assure farmers that their products were particularly suited to the regional environment. To do so, they solicited endorsements from prominent local bright leaf growers. Brand advertisements and instructional pamphlets produced by fertilizer companies invariably included several testimonial statements from Piedmont farmers assuring readers that commercial fertilizers helped maximize the productive potential of local tobacco soils. Like modern athletes who sign endorsement deals with particular companies, bright tobacco experts lent their names, and their assurances of quality, to fertilizer products. Advertisements touted a specific brand of fertilizer as superior to all others and encouraged a general belief that quality bright leaf could not be made without heavy applications of commercial amendments. Robert Ragland, the Halifax curing and seed expert, was by far the most popular spokesman; in fact, the Southern Fertilizing Company, makers of the popular Anchor Brand tobacco fertilizer, published Ragland's popular cultivation guides. In addition to Ragland, prominent cultivators William Sutherlin, Ed Pace, and John Lea (the Caswell Klan organizer) were among the local endorsers.[64] Enterprising companies also offered prizes for the best crops grown with their brand and used the winning growers as advertising fodder.[65] Growers who lent their names and reputations to fertilizer companies sold their experience and expertise, but they also linked commercial inputs to the local environment through agricultural activities. Through these advertising campaigns, bright leaf cultivators came to see the products of Peru and the southern Tidewater as integral components of local agro-ecosystems.

Heavy investments in expensive fertilizers posed obvious risks as small growers were often only a bad thunderstorm or drought from going into debt. New bright leaf producers, often struggling to master the crop's cultivation and curing, also faced the specter of higher input costs coupled with uncertain tobacco quality. Results at market remained highly variable, as grower Langhorn Scruggs experienced in 1871, when he sold lots of tobacco ranging from $1 to $20 per hundredweight.[66] A poor crop often led to financial hardships. Leatherwood Valley resident William Sours wrote his brother John of a neighbor who suffered from an all-too-common situation. The farmer had sold a poor crop of tobacco, the proceeds of which were $15 short of covering his fertilizer note.[67] Protesting the growing dependency on tobacco fertilizers, planter Charles Bruce wrote that a tobacco grower who relied on commercial amendments was "traveling a road which will surely land him in bankruptcy."[68] Margins were typically tightest for African American farmers, who rarely had much in the way of savings as insurance. Despite these pessimistic assessments, bright tobacco grown with the aid of commercial fertilizers could still turn strong profits. According to one estimate, good bright leaf land sold for around $40 per acre in 1880, tobacco hands demanded $150 per year, and fertilizer cost as much as $70 per ton, but if everything went well (a big assumption in farming), experienced farmers on the best soil could still clear more than $130 per acre annually, a profit unrivaled by any regional alternative. While bright leaf growers had the potential to turn healthy profits, the few remaining dark tobacco producers "received prices that were 'scarcely sufficient to meet the cost of production.'"[69]

Fertilizers could temporarily reverse losses in productivity created by soil erosion or exhaustion (the two were often effectively the same in the Southside), but they could do little to address other environmental challenges to bright leaf culture. Insects and plant diseases threatened all Southside farmers with economic disaster, threats that grew during the postwar years. Certain insect pests had long attacked Mid-Atlantic tobacco. Flea beetles cut small holes in tobacco leaves, damaging a crop's value; wireworms bored into young transplants, withering the plants or, on occasion, cutting them off entirely at ground level; hornworms munched mature leaves with astonishing rapidity; and budworms attacked the new foliage at the apex of maturing plants. These pests had been present in regional tobacco fields for decades, but bright leaf agriculture exacerbated their threat. The habit of bright tobacco farmers to plant their crops in the same field year after year promoted booming insect populations. All four species produce pupae or grubs that overwinter in the soil, thus the numbers of worms and beetles present in a tobacco crop owed a great deal to how long tobacco had been planted on a particular piece of ground. In addition, wireworms and hornworms also fed on oxeye daisy and horse nettle, common "weedy" species associated with the margins of tobacco land.[70] A number of tobacco growers recognized the prevalence of these insects in long-farmed fields, but the necessity of specific soils in the production of quality bright leaf meant few farmers were willing to abandon productive land.[71] Treatment methods could also exacerbate other agricultural problems. One of the most commonly prescribed insect control methods—fall plowing for the purpose of crushing pupae in the soil—actually promoted further erosion by leaving land bare of plant cover at the start of winter.[72]

If insect pests remained a constant threat, a number of persistent and new tobacco diseases gave Southside farmers even greater causes for concern. Root knot had long been a problem in Virginia and North Carolina. The product of nematodes, microscopic soil-dwelling worms, root knot caused tobacco plants to wither and dry in the field. Tobacco mosaic also likely attacked bright leaf fields in the three counties, although the disease was inconsistently identified and its history somewhat uncertain. The viral disease stunted plant growth and produced leaves with patchy, mottled surfaces; both results led to low prices for infected tobacco. In the 1880s a new disease, Granville wilt (named for the North Carolina county where it first appeared in 1881), afflicted tobacco crops. The wilt was a bacterial infection that led to root decay, causing plants to slowly wither and die. Although farmers did not understand the immediate causes of these three diseases, they did correctly associate them with particular tracts of land. As with insect infestations, continuous cropping worsened all three plant diseases. Nematodes are present in almost all soils but often build to problematic levels only when coexisting with the same crop year after year, and the bacteria that caused Granville wilt and the virus that led to tobacco mosaic both overwintered in the soil. Tobacco farmers typically responded to these threats much as they did to insect infestations: they tolerated diminished returns and experimented with various cultivation techniques but refused to take fields out of production unless they suffered catastrophic losses.[73] As was the case with so many contradictions between traditional land management and bright leaf culture, farmers chose, sometimes willingly and other times with great reluctance, to make the most valuable tobacco possible despite the economic and ecological hazards involved. They remained convinced that because they understood local soils they knew the best use of the land, equating mastery of a crop culture with mastery of an environment.

If environmental and economic challenges in Southside fields brought increasing difficulties for tobacco farmers, struggles over exactly who would farm the landscape and work in regional tobacco factories, and under what terms they would do so, remained contentious even as Reconstruction ended. Violence, long a tool used by white landowners to keep black hands at work in fields and factories, continued to plague the Southside, connecting seasons, field routines, and ballot boxes through tobacco. Even when farmers and manufacturers did not initiate racial violence, they were quick to exploit bloodshed for the benefit of the existing tobacco culture.

Racial tensions endemic to the region since the start of Reconstruction came to a head once again in Danville in November 1883, in an event that has become a seminal marker in the racial politics of the emerging New South. Conflict between the city's black and white residents left four black men and one white dead and a number wounded, and along with the body count the riot shifted the course of Virginia politics for decades. The Danville riot took place three days before a crucial state election pitting the Readjuster Party—with its coalition of white liberals and black voters—against the conservative Democrats.[74] The previous year the Readjusters, headed by Governor William Cameron and party boss William Mahone, had gained control of the Virginia General Assembly, and as part of statewide reforms, they had divided Democrat-dominated Danville into new, racially proportionate, wards. The city's black majority proceeded to elect a number of black officials, including "a majority in the twelve-man council, and four of the nine policemen."[75]

Speeches by members of each party concerning the new political land-scape stirred local tensions throughout 1883, and a month before the riot a number of prominent white businessmen signed a petition decrying the "misrule of the radical or negro party." *Coalition Rule in Danville*, popularly known as the Danville Circular, declared that black Readjust-ers were harming Danville's economic and social fabric, especially its tobacco industry, by drawing "large numbers of idle and filthy negroes" to the city.[76] The circular served to mobilize Democrats across the com-monwealth and raised racial animosity in the city to dangerous levels.

In her influential interpretation of the riot, Jane Dailey character-izes the violence as an example of Readjuster-era political tensions in Virginia, brought to a head by struggles over racial deference and the control of public space. In particular, Danville freedpeople claimed the right to use the city's sidewalks without stepping off to let whites pass, a public assertion of self-worth that the city's white population, long used to black subservience, found repugnant. Black and white, male and female, Danville's residents strode city sidewalks determined to preserve their rights-of-way, and failures to yield ground led on occasion to verbal and physical abuse. Dailey describes these sidewalk battles as "the open and public actions of an enfranchised and politically empowered people" and the countering resistance of white townsfolk intent on reestablishing their social hegemony. It was just such an encounter between a white and a black man that initiated the November 3 riot, which began as a fist fight, drew an angry crowd, and turned from punches to pistols.[77]

As Dailey asserts, the Danville riot was almost certainly influenced by Southside politics and racial animosity, but the event, like regional Reconstruction violence, was also shaped by local tobacco culture. The Danville Circular mentioned black political control of the city and a gen-eral lack of black deference in social relations as factors contributing to white anger, but it also made another key assertion. Eleven of the twenty-eight businessmen who signed the circular were tobacco manufacturers or warehouse owners, and six more were general merchants who may have bought and sold tobacco on occasion, and they laid out their con-cern in the pamphlet's text: "It is well known that hundreds of the North Carolina tobacco raisers who live within a few miles of Danville, and used to sell their tobacco in our market, now go five times as far to a market in their own State, *on account of the negro rule in our town*." The circular's authors asserted that North Carolina (and, by implication, Virginia) farmers would avoid spending the money they made selling tobacco in a town where black sheriffs patrolled the streets and black vendors sold food and other goods on the squares.[78] These Democratic businessmen worried that local politics would drive away the white farmers and landowners who marketed the tobacco raised by their black tenants. Compounding these worries, new warehouses in North Carolina advertised Danville's "Negro rule" in an attempt to draw business away from the preeminent bright tobacco market.[79] Racial politics in Danville threatened Democratic control of the polls, and, perhaps of equal importance, tobacconists feared that it threatened profits generated from bright tobacco sales.

Scattered comments suggested that local African Americans—from political leaders to factory workers—understood their potential leverage over the city's white businessmen. In a speech a week before the riot, Squire Taliaferro, a local black politician, exhorted black laborers to flex their political muscle. He warned white Danvillians that they needed the muscle and expertise of black workers to man their tobacco factories and fields. Whites would have to accept black political rights to maintain economic stability, according to Taliaferro: "If negro rule would cause Danville to sell a few more pounds of tobacco, they were going to have it."[80] These assertions of the vitality of black factory labor mirrored Reconstruction struggles over the importance of black labor in tobacco fields, and like those earlier struggles, the possibility of violence was never far removed from public debates. Rumors swirled about the city that blacks involved in tobacco work were planning some sort of demonstration, and local farmers and factory owners may have feared that this action would lead to the sabotage of tobacco manufacturing or cultivation if African Americans were denied access to the polls.[81] The specter of the barn burner grew to immense proportions as industrialists imagined factory burners. Coupled with these fears was the realization that no matter how much white landowners denigrated the capabilities of freedpeople as farmers or factory workers, they remained integral in every step of producing and finishing tobacco.

The actions of one tobacconist is suggestive of the connections between the riot and the tobacco economy. William P. Graves, a prominent tobacco warehouse owner, was among the white rioters. When he shot at and wounded the brother of a black Danville policeman, even as his victim was proclaiming his peaceful intentions, he may have been expressing frustration with black social assertions, as Dailey claims. But perhaps Graves was also asserting his mastery over African Americans as workers. Earlier in the year Graves had a run-in with one of his black warehouse workers, striking the man when he bumped him with a to-bacco basket. The injured employee took Graves to court, where the em-barrassed employer was forced to pay a fine.[82] On the factory floor and in the city's streets, Graves was a white man worried about social bounds, but he was also a tobacconist concerned with his power over laborers; he was thus intent on preserving his economic interests as well as his racial prerogatives.

Danville's Democratic leadership, writing as the Danville Committee of Forty, issued a pamphlet defending the results of the riot, in which they blamed the violence on armed and insolent blacks. Claiming that the white crowd acted in self-defense, the Committee of Forty declared that the riot had little effect on the subsequent elections, which proceeded in "peace and good order."[83] As with the Danville Circular signees, a num-ber of the Danville Committee of Forty were involved with tobacco marketing and manufacturing. Of the thirty-four committee members who appear in the 1880 census, fourteen were tobacco businessmen of some sort. These industrialists included the committee chairman, Wil-liam Sutherlin, as well as the prominent manufacturers John Pace, J. M. Neal, and John Rison.[84] Among the ten witnesses interviewed by the committee who also appear in the census, six were tobacco farmers, dealers, or factory workers. That a large percentage of the committee and witnesses were involved with tobacco is no surprise; most Southside resi-dents had some tie to tobacco culture. But an acknowledgment that these people were farmers and tobacconists serves as a reminder that tobacco permeated almost every facet of regional life. Both the riot and its subse-quent explications were the product of men entwined in the cultivation, manufacturing, and sale of bright tobacco.

Whatever the rioters' intent, the results of the violence and ensuing debates were brutally clear. Democrats won local elections just days after the riot and swept back into statewide power. The violence proved supremely intimidating within Danville, where armed whites patrolled election-day streets. Of the city's 1,301 registered black voters, only 31 cast

ballots. In the bordering communities of North Danville and New Design the figures were just as stark; only 4 of 700 potential black voters
appeared at the polls.[85] Similar actions took place in Halifax, where a
Readjuster observer wrote that local conservatives "have carried the election here by fraud, intimidation, shooting, and cutting the negroes."[86]
The *Southern Workman*, assessing the election from the remove of
Hampton, decried the Democratic victories as the "results of terrorism,
backed by actual violence at Danville, South Boston, and elsewhere."[87]
Conservatives had "redeemed" the Old Dominion, and regional tobacco
producers, manufacturers, and warehousemen worked to exploit the political change.

In the aftermath of the riot, two warehousemen who claimed to
have participated in the violence made clear their thoughts on the aims
of white action. Their Banner Warehouse circular framed the riot as an
action taken to preserve tobacco interests. Its text read:

> Virginia is safe
> And her people are free
> Danville is redeemed
> No more negro rule
> Banner Warehouse takes the lead in high price, as in everything
> else. Come to Danville with your tobacco and to Banner Warehouse
> go.
> Your interest will be protected.[88]

These warehousemen captured the complex relationship between power,
violence, and control of the tobacco countryside. The *Independent*, a New
York periodical, blamed the violence in Danville on racial tensions that
plagued the South and lamented the power of "the low white, with the
shot-gun in his hand."[89] Several years after the riot, a writer for *Harper's
Weekly* attributed the incident to a struggle over the shape and form of
the local tobacco industry, though the author mistakenly believed the
city's tobacco workers to be white.[90] These three assessments of the Danville riot need not be mutually exclusive. Political and social struggles
linked to race were certainly factors in the bloodshed, but the white and
black riot participants were by and large people who lived in and depended on tobacco culture. Even if the tensions of tobacco production
and manufacturing did not inspire the riot, they shaped how some local

citizens and distant commentators perceived events. Danville tobacco manufacturers and warehousemen framed the riot as a struggle that ensured that their city would continue to serve as the bright tobacco hub of Virginia and North Carolina for at least a few more years.

The spread of bright tobacco following the Civil War continued many antebellum themes. Farmers worked to master the intricacies of bright leaf cultivation, from soil management to curing techniques, and successful production was still far from guaranteed. Raising yellow tobacco remained an exercise in tacit, experiential knowledge despite the thousands of pamphlets, guides, and articles that flooded the Southside with professional advice. Bright leaf was also still a crop that was hard on both land and people. Yet despite all its hardships, yellow tobacco continued to hold a great deal of economic promise. No other crop could pay as well in the Southside, but no other crop carried the same risks. Fertilizer loans combined with a poor season could drain a farmer's bank account just as quickly as the crop's intensive culture could erode a farm's soil and deplete its forests. And for freedpeople, bright tobacco offered economic promise yet encouraged racial violence and economic oppression.

Regional growers did not enter tobacco cultivation blindly—to varying extents they understood the hazards of bright leaf agriculture—but time and again, farmers chose to raise bright tobacco. It would be easy to argue that this decision was the result of greed, to say that tobacco farmers put profits before all else and steadily ruined the land and their future prospects. But this would ignore a number of constraints critical in farmers' decisions to cultivate bright leaf. A long local history of tobacco farming encouraged farmers to plant the crop they knew best, experts assured growers that the crop was sustainable and the best use of Piedmont land, and fears of outmigration and agricultural decline created an atmosphere in which rural residents regarded a lucrative staple as an economic and societal savior. In addition, the erosion that came with several crops of bright leaf limited agricultural alternatives. Farmers who placed marginal land in tobacco fields could hardly be expected to go back to corn or wheat on those sites once they were even less productive. Once bright tobacco was well established in the Southside many farmers must have believed that their options were more bright tobacco or nothing at all.

As during the antebellum period and Reconstruction, racial violence tied to the tobacco seasons and infrastructural patterns remained one of the weapons in white landowners' arsenal for retaining control over black workers and the tobacco landscape. The violence inherent in slavery had secured labor for antebellum plantation owners; Reconstruction-era violence had stifled black dreams of widespread land ownership and ensured white landowners a regional farm workforce at low monetary cost; and continued post-Reconstruction violence, typified by the Danville riot, worked to maintain white dominance of polls, fields, and factories. Although these bloody events were political and social as well as agricultural and industrial struggles, the culture of tobacco, from its cultivation to its manufacturing, was never far from the minds of the blacks and whites involved. This violence was firmly rooted in the material realities of the region, an expression of the social and environmental framework of the Southside.

Despite erosion problems, deforestation, increasing fertilizer debt, and continued violence, bright tobacco remained the Southside's economic engine, and many farmers and businessmen were optimistic that the crop could fuel regional growth indefinitely. A correspondent from the *Richmond Dispatch,* traveling through Halifax in the summer of 1885, captured this optimism. He wrote that the energy and industry of the countryside was attributable to "the magic of gold, the product of their golden yellow tobacco that you may now see piled in bright heaps, or hanging like yellow clouds in every barn you pass, or meet twenty times a day on the market roads, neatly packed high in covered wagons going to Danville, or South Boston, or Clover, where the auctioneer finds quick sales and large profits for the well-pleased planters."[91] Ragland, though he had elsewhere admitted the difficulties and hazards of raising bright leaf, made a similar glowing assessment in an article for the Baltimore journal, *American Farmer.* Describing the impact of yellow tobacco on the Southside, he wrote of a crop that came out of nature and at the same time bettered it: "Twenty-five years ago, some of the poorest regions in what now constitute the yellow tobacco belt of North Carolina and Virginia, offered a scant living to the poor inhabitants dwelling in huts amid uninviting surroundings. But yellow tobacco came, and lo! what a change! The log houses have given place to neat and substantial, comfortable dwellings; commodious school houses and imposing churches erected; the very face of nature and all of the surroundings changed, bettered and beautified."[92] Absent from Ragland's idyll are the gullies and rills of the sandy hills, the expanses of stumps where oak forests once stood, the poverty of poor black and white sharecroppers and wage workers, and the increasing debt of many Southside tobacco growers.

Hyperbole aside, the growing environmental, social, and economic difficulties associated with bright tobacco culture had done little to dampen Southsiders' enthusiasm for the crop by the 1880s. Key to this continued faith in yellow tobacco was its durable potential to create wealth. Even as fertilizer debts, soil depletion, and continued labor struggles made success far from a sure thing, bright tobacco continued to offer the chance of getting rich. A tobacco exposition held in Danville in 1887 reassured local farmers that yellow tobacco's potential remained limitless. The first prize for a bright wrapper at the exposition garnered $200, and that lucky farmer sold his entire lot for an astonishing $660 per hundredweight (a rate that, given an average yield, translated to roughly $3,300 per acre).[93] Farmers retained at least a little skepticism of news of extraordinarily high prices, recognizing them as advertising gimmicks on the part of warehouses. Noting reports of fantastic sales in Lynchburg, the *Milton Chronicle* (perhaps in defense of local warehouses) declared, "That bait wont [sic] take, in these parts."[94]

Despite the profit potential, during the late 1870s and 1880s, bright tobacco cultivation in the three counties became a riskier venture; the economic margins grew tighter and the landscape became more and more degraded, yet yellow tobacco remained the most promising option for most farmers. As one commentator pointed out, "Of late years tobacco has certainly not been a profitable crop, except to those who have the proper soil for 'bright' tobacco, and have learned the art of making and curing it, which few have. If tobacco culture is to be continued, planters should turn their attention more to 'bright.'"[95] As it was during the emergence of the crop despite (or, in part, because of) the message of antebellum agricultural reformers, the centrality of tobacco seemed only natural.

SEVEN

The Decline of the Border

BY THE END OF THE NINETEENTH CENTURY, after six decades of bright tobacco cultivation, the crop permeated almost every corner of Southside life. Although yellow tobacco dominated the region as never before, it held less promise for regional farmers. This bright leaf malaise was the product of four concurrent forces. Caswell, Halifax, and Pittsylvania faced increasing competition from new bright tobacco districts; the American Tobacco Company (ATC), a trust formed by the largest cigarette manufacturers, achieved a monopsony over local markets that farmer organizations were unable to break; the rise of the cigarette as the dominant consumer bright leaf product limited top market prices even as it stimulated national tobacco addiction; and environmental problems stemming from standard cultivation practices made growing yellow tobacco more difficult and expensive while at the same time limiting farmers' ability to turn to alternate crops. The turn of the century did not spell the end of bright tobacco cultivation along the border, but it did signal the conclusion of the crop's most promising era.

Bright tobacco, long the staple of the sandy hill farms along the Dan and Roanoke, spread across the Southeast during the 1890s. Fields of yellow leaf appeared in places as distant as northeast Florida and coastal North Carolina. This diffusion was in part the result of efforts by Southside boosters to spread their crop, as seed producers, warehouse managers, tobacco manufacturers, and fertilizer agents all envisioned expanding profits. The agronomic work of Southside experts contributed to the adaption of bright tobacco to new landscapes. By the century's end, more bright leaf was grown outside of the Southside than within the region, and the three counties would never regain their predominance.

Southside tobacco growers attempted to combat competition, environmental problems, and economic difficulties through organization. The Grange and then local chapters of the Farmers' Alliance supported growers' demands against warehouse owners, fertilizer suppliers, equipment manufacturers, and the ATC. Farmers also turned to these organizations for cheap fertilizer, cash loans, agricultural advice, and better plows. In the end these organizing efforts failed; tobacco prices declined and erosion continued. Organization did not address the root of the region's environmental problems—farming on fragile land—and it could not break the power of warehouse owners and tobacco manufacturers.

When the Grange and Farmers' Alliance failed to significantly better tobacco growers' situations, farmers turned to state and federal agricultural scientists in the hope of making tobacco culture more profitable. Bright tobacco growers were key in the creation of government soil surveys, which were an outgrowth of earlier cultivation literature produced by private tobacco "experts." Farmers believed that these surveys would help them locate productive land for bright tobacco cultivation and combat competition from new tobacco producing regions. Soil surveys did benefit tobacco cultivation, but the bulk of the rewards went to the expanding tobacco districts south and east of Caswell, Halifax, and Pittsylvania. The first county-level soil surveys in the nation mapped potential bright tobacco land on the North Carolina coastal plain and hastened the ascendance of the Southside's competitors.

The rise of the cigarette as the dominant form of tobacco product, though incomplete at the turn of the century, transformed tobacco culture as well. Bright tobacco became a critical ingredient in American cigarettes. Smokers enjoyed the variety's mild flavor, and the chemistry of flue curing meant that bright tobacco transferred nicotine to the human body more efficiently than competing forms of smoking tobacco. The cigarette ensured an expanding market for bright tobacco growers, but it also undermined the economic position of the three counties that first produced the crop. Cigarette manufacturers were less concerned about premium quality than were plug producers, and an expanding market encouraged further competition from new tobacco districts in eastern North Carolina and South Carolina. Cigarette producers, synonymous with the ATC monopsony after 1890, refused to pay premium prices for the best tobacco and thus removed much of farmers' incentive to focus on quality rather than quantity.

The environmental problems that had plagued bright leaf culture since the antebellum era peaked at the end of the nineteenth century as the crop spread to its greatest extent yet along the border. Erosion continued to plague the region's hillsides, stripping topsoil from ridge fields and choking bottomland in silty runoff. The deforestation that accompanied widespread tobacco cultivation proceeded apace. Farmers worried about competition from new tobacco districts and decreasing crop prices planted more of their land in bright leaf and bought more and more commercial fertilizer, securing loans that required them to continue producing tobacco even as the crop became less profitable. Regional farmers' long history of cultivating tobacco also worked to weaken their willingness to turn to alternative crops, and decades of soil abuse made the success of such alternatives unlikely. Yellow tobacco had caught its cultivators on the horns of a dilemma; many farmers had turned to the crop because they believed that their land was unsuited to grow anything else, and years of diligently following tobacco cultivation advice made their fears reality.

Competition, monopsony, cigarettes, and erosion changed the bright tobacco calculus. The lottery prices of yellow tobacco's golden age faded away at the very moment that commercial fertilizer became a necessary addition to weakening land, and the ATC all but guaranteed that crop shortages would do little to elevate market prices. Southside tobacco growers quickly found themselves in a situation similar to that of other southern commodity crop producers—on a treadmill running faster and faster just to keep up. Tobacco cultivation would continue in the Southside into the twenty-first century, but it would never realize the economic or agronomic promise it held in the nineteenth century.

The last decade of the nineteenth century was a period of tremendous expansion of bright tobacco culture as cultivation and curing techniques became more systematic and predictable; eventually the crop appeared on farms from southern Virginia to northeastern Florida. During the late 1880s, bright tobacco cultivation had spread beyond the ridge land surrounding Danville into adjoining counties in North Carolina and Virginia, where there were pockets of suitable soil. Farmers in other parts of the impoverished South took note of the profitability of the crop and attempted its cultivation in their own locales with mixed success. Only sites with sufficiently sandy soil stood a chance of success, and with some notable exceptions, growers in other regions who procured the proper seed and followed proven curing methods still failed to produce bright leaf of as high a quality as that made in the heart of the old belt. Despite a less than perfect record, planters in other parts of the South persisted in their bright leaf experiments, and they gradually developed seed types and cultivation methods that worked well in their regions. Their success flooded the bright tobacco market with leaf and helped depress prices throughout the South.

Southside producers and tobacconists were in part responsible for this increased competition. Key to the success of bright tobacco expansion were experts from Caswell, Halifax, and Pittsylvania who had spurred the original development of bright leaf during the antebellum and postwar period. Boosters and businessmen who had a stake in the crop's success, including seed producers, such as Ragland, and warehousemen, such as Danville's Ed Pace, believed that the expansion of yellow tobacco cultivation could only benefit their bottom lines and the growth of Southside business. Farmers in the new regions bought seed, barn flues, and fertilizer from established Southside companies, and tobacco from the new districts moved through the manufacturing facilities sprouting up in Virginia and North Carolina. This expansion would eventually challenge and then eclipse the Southside's bright leaf dominance as the epicenter of yellow tobacco production moved south and east. Durham, Raleigh, Winston, Greenville, Kinston, Oxford, and a host of smaller towns in the North Carolina Piedmont and coastal plain developed into warehousing and manufacturing centers, drawing on growing local bright leaf supplies as well as the railroad expansion that followed Reconstruction, often supported by northern capital, for example, the spur of the Richmond and Danville Railroad that connected Winston to Greensboro in the 1880s, further stimulating the growth of R. J. Reynolds's formidable Winston operations.[1]

The two most significant new areas of cultivation were eastern North Carolina and the northeastern coastal plain of South Carolina.[2] By 1890, bright tobacco began to displace corn and cotton as the dominant cash crop in the flat lands around Greenville and Wilson, North Carolina, where weak, sandy earth was the predominant topsoil.[3] Ed Pace opened the first regional bright leaf warehouse in 1890, an enterprise began at the urging of Ragland.[4] When farmers in the Pee Dee River basin of South Carolina looked for advice on how to begin bright leaf cultivation, they too consulted Ragland, who mailed instructions (probably copies of his popular cultivation pamphlet). Pace appeared in the Pee Dee district as well, serving as an outside judge in an 1886 tobacco-growing contest.[5] Other border tobacconists attempted to spread bright leaf culture on a smaller scale. Representative were Halifax's R. B. Davis, who took up production in Catawba County, North Carolina, and Pittsylvania planters W. T. Dickinson and J. D. Wilder, who experimented with bright leaf culture near Asheville.[6] Likewise, a small bright tobacco district modeled on the Southside system developed along the border of Georgia and Florida in the years bracketing 1900.[7]

Bright tobacco boosters could not export the Southside soil that was so suited to bright leaf cultivation, but they could export seed varieties and curing systems along with a degree of agricultural knowledge rooted in a particular southern environment that farmers in other regions could then modify to fit their locales. More than any other factor, the work of seed producers like Robert Ragland made this expansion possible. Perhaps no farmer had done more work since the Civil War to promote bright tobacco, and by 1890 Ragland produced twenty-nine varieties of bright leaf tobacco, five dark types, two burley lines, and five cigar types of tobacco. Such varieties as Yellow Oronoko and Yellow Pryor predated the war, but Ragland also developed new varieties adapted to the soil types found in eastern North Carolina (at least according to his descriptions). Gold Finder, Bullion, and Oak Hill Yellow were among the varieties that gave potential bright tobacco farmers options matched to local soil conditions. Ragland made it clear that he sold environmental knowledge and experience as a plant breeder in addition to seed. Ragland's seed company had moved beyond basic seed selection; many of his seed lines were early experiments with vigorous hybrids created by crossing two older, inbred lines of yellow tobacco. Ragland combined old advice with his new plant material, cautioning would-be growers that despite plants "greatly improved by propagation of selections and judicious crossing," "the soil must be adapted to the type, or failure is certain."[8]

There was certainly a good deal of economic stimulus behind expansion. Bright tobacco prices that seemed poor to Pittsylvania or Caswell farmers, jaded by accounts of prime wrapper sales, seemed quite a windfall to eastern North Carolina cotton farmers suffering through their traditional staple's price depression. New seed types coupled with commercial fertilizers made tobacco a viable alternative across vast swaths of the coastal plain. In locations like Wilson County, North Carolina, sandy soils that were too deficient in nitrogen and phosphorus to produce even bright leaf became quite productive with the regular addition of guano and phosphates.[9] B. W. Arnold, a contemporary tobacco expert, described this eastward spread as the result of the new amendments: "The border counties of Virginia and North Carolina, covered for years with dwarf oaks, broomsedge and pines, were reclaimed by the use of commercial fertilizers, and were converted into the choicest lands for growing a particular variety of bright yellow tobacco."[10] Thus the same developments that allowed Southside farms to remain productive in the face of soil erosion and depletion brought on by decades of continuous cultivation also stimulated robust competition in the form of new bright tobacco districts.

The spread of bright leaf into North and South Carolina, along with the continued growth of burley tobacco production (another distinct variety that found its way into plugs and cigarettes) in Kentucky and Tennessee, provided stiff competition for Southside tobacco growers. In 1880, South Carolina growers produced only 45,678 pounds of tobacco; by 1900, they raised almost 20 million pounds of bright leaf. Georgia farmers increased their production from 228,590 pounds to 1,105,600 pounds over the same period. The increases were even more dramatic in eastern North Carolina, where Wilson and Pitt counties were representative of this bright tobacco boom. In 1880, Wilson growers raised 8,745 pounds of leaf, but by 1900 the county produced more than 7 million pounds. And in Pitt, farmers raised their production from a meager 598 pounds, grown on only three acres, to almost 11 million pounds.[11] Partly as a result of this competition, bright leaf prices plateaued or even gradually declined through the 1880s, although bright tobacco prices remained well above those paid for dark tobacco on the Danville market. This price stagnation is more meaningful when one considers the added costs of the commercial fertilizers that had become a ubiquitous part of cultivation. As farmers' profits declined, the acreage they planted in bright leaf in the Southside did not decrease. Rather, farmers engaged in a common agricultural paradox; as each acre of bright leaf became less profitable, growers planted more and more ridge land in the crop in an attempt to maintain their income. During the 1879 season, Caswell, Halifax, and Pittsylvania farmers sold a total of 24,162,039 pounds of tobacco. By 1899, when bright leaf had spread across widely scattered portions of the South, the three counties produced 35,799,650 pounds of leaf. Pittsylvania and Halifax remained the two leading tobacco counties in Virginia, but Caswell had dropped from the second to the seventh most productive county in North Carolina.[12]

These increasing production figures reflected continued expansion of the number of tobacco farmers in the three counties. Over the same twenty-year period, the number of farmers raising tobacco increased. This increase, even as the profitability of yellow tobacco slowly waned, reflected the popular—and well-published—perception that bright tobacco remained the most profitable regional crop and a lack of promising agricultural alternatives. Industrial labor in the region offered little alternative employment for blacks or whites. Most of Danville's growth revolved around selling, storing, and manufacturing bright tobacco, with the sole exception of the Riverside Cotton Mills, a textile concern organized in 1882 (and Riverside Mills hired only white workers during its initial decades).[13] Many tobacco opponents, such as William Sours, a transplanted Pennsylvanian long critical of the Southside's tobacco dependency, had given in and begun to plant bright leaf by the mid-1880s. Sours rejected his former trade, blacksmithing, in favor of what he believed to be the most reliable way of making money in the Southside.[14]

By the turn of the century, bright tobacco was no longer centered on the counties ringing Danville; it had become a southeastern crop, with farmers from Virginia to Georgia planting golden leaf and selling their produce in auction warehouses modeled on the Danville system. Bright tobacco fields appeared in most southern districts where the soil was generally sandy and weak, or where it could be adapted to tobacco culture with the appropriate combination of chemical fertilizers. Indicative of the growing bright belt to the south and east of the crop's cradle was the movement of the *Southern Tobacco Journal,* a trade publication catering to manufacturers and founded in Danville in 1887, to Winston, home of R. J. Reynolds's growing plug operation, in 1891.[15] This crop diffusion did little to lessen the importance of bright leaf in the Southside, however. Tobacco raised around Danville, Milton, and South Boston continued to bring some of the highest prices in the nation (though lower than before) even as competitors entered the marketplace; Southside soils remained peculiarly suited to bright leaf and the region's growers had the most experience producing the finicky crop. In addition, the growing popularity of a relatively new form of consumer tobacco—the cigarette—ensured a continued market for the increasing glut of yellow tobacco.

Historian Barbara Hahn has pointed to this expansion of bright leaf into coastal North and South Carolina and the diffusion of the crop during the twentieth century to places as remote as Southern Rhodesia (now Zimbabwe), China, and Brazil as evidence that environment mattered little in the development of bright tobacco. She argues that technology and culture created conceptions of tobacco more reflective of institutions and intellectual constructs than any biological reality.[16] There is another way to understand this expansion however, one that places soil at center stage. It seems too unlikely to be coincidence that bright tobacco culture first emerged in a landscape combining weak sandy soil with a historic tobacco culture situated on adjacent richer lands. The successful culture of bright leaf was rooted in these soil relationships, and it was the search for soils with similar possibilities that presented the opportunity for and then drove expansion. In other words, boosters like Pace and Ragland did not travel to South Carolina and Georgia because tobacco would grow anywhere, they evangelized in those places because they saw potential in specific soils. And the failure of bright leaf to spread to other areas of tobacco culture, despite its profit potential, indicated the crucial importance of soil and environment in successful cultivation. Boosters who tried to raise bright leaf in the mountain valleys of southern Appalachia, underlain with limestone or rich with humus washed from surrounding slopes, invariably failed, and yellow tobacco made no significant inroads in the western burley districts or the fields of Connecticut or Wisconsin. It is also important to note that tobacco varieties changed over time through selective breeding as growers continually adapted to new environmental conditions and bred plants with an eye toward increased use of commercial fertilizers. In this regard Hahn makes a crucial observation: like all crops bright tobacco was a moving target, ever-changing to suit human desires as well as environmental conditions. But for all this variability, tobacco's characteristics and the soil in which it grew were always intimately connected, and the fine margin between success and failure often lay in a farmer's knowledge of the land beneath his feet.

Many of the cultivation practices that had characterized bright tobacco production, especially since the end of the Civil War, continued largely unchanged at the turn of the century. Farmers still sought new ground for plant beds, burned the soil, clean tilled their crops shallowly and often, and planted on ridge land. Contemporary manuals that attempted to systematize bright tobacco culture emphasized all of these practices—and even advised new methods that promised to further damage the landscape. A common early twentieth-century prescription called for farmers to plow their tobacco lands in the early winter. This tillage was supposed to expose the soil to freezing and thawing which would improve the land's tilth and make it easier to work in the spring. Experts also believed that winter plowing worked to crush insect larvae and expose them to birds and the elements, lessening infestations the following season. Winter plowing may have accomplished these goals, but it also exposed the vulnerable soil to additional months of wind and rain, accelerating the erosion that already plagued Southside farms. Taken together, the continuation of older techniques and the new advice of tobacco professionals served to maintain or even increase erosion rates that had reached disastrous levels in the 1880s. By the turn of the century, erosion and deforestation in the Southside had never been more severe.[17]

Erosion problems remained a grave concern as the century drew to a close, but Southside farmers' willingness to criticize bright tobacco practices seemed to decline. Tobacco had long had its opponents, farmers who argued in the regional papers and agricultural journals that the crop was a destructive force that consumed the soil as it hampered farm diversification. These voices grew noticeably quieter during the 1890s. The boosterist fawning over tobacco was such that a commentator in a Caswell newspaper could blame grain farming for regional erosion problems without a hint of irony. The writer accused "continuous grain cropping without manures" of ruining the county's land, without mentioning the even more intense cultivation that took place on Caswell's ubiquitous tobacco patches (or the fact that local cereal cultivation was in the midst of a long decline).[18] Decreasing criticisms of bright tobacco culture did not reflect improved agriculture; rather, the opposite. Regional tobacco dependence had grown to a point where life without bright leaf was hardly imaginable.

Rather than focusing on the visible soil destruction wrought by tobacco, most commentators on the bright belt wrote of the crop as an improving force. Much like their antebellum counterparts who argued that slavery served as a tool of improvement rather than a hindrance, they pointed to increasing land values and the economic stimulus of tobacco manufacturing as fruits of tobacco culture and evidence of the crop's value. The old tale of the alchemical transformation of poor, abandoned piney fields and broomstraw land into the most valuable soil along the border remained a popular topic.[19] To be sure, some of this was booster rhetoric (and places with modest prospects always seem to contain the loudest boosters), but assertions of tobacco's positive influence also stemmed from decades of reoccurring claims. Since at least the 1850s, tobacco experts, warehousemen, manufacturers, and not a few farmers had been declaring bright tobacco the future of the Southside. The crop made money while it seemed to defy agricultural conventions, and it took farmers a long time to accept that bright tobacco culture, as practiced in the late nineteenth century, was fundamentally unsustainable.[20]

The use of commercial fertilizers both encouraged and complicated existing tobacco culture and its boosters by temporarily obscuring tobacco unsustainability. Phosphorus-rich commercial amendments kept Southside soils viable in the face of continual erosion, but they also helped open vast southern and eastern lands for bright tobacco, competition that slowly drove down the price of yellow leaf. Fertilizers also wrought a subtle intellectual transformation among bright tobacco farmers. Whereas a slavish attention to the minute properties of local soils had long characterized successful growers, a universal use of fertilizers diminished the importance of this expertise. By masking the deficiencies of weak or depleted ground, or by making nitrogen-rich land more suitable to tobacco cultivation through the addition of phosphorus, these amendments transformed local soils from living, nuanced ground to a mere substrate amenable to the addition and subtraction of bagged nutrients.[21] The new calculus called on farmers to memorize fertilization rates rather than to study what sort of trees grew on the land or what the earth felt like underfoot. And the enormous quantities of fertilizers advocated by some manuals surely strained farm budgets: an 1898 USDA Farmers' Bulletin called for the application of as much as two tons of phosphorus-rich commercial fertilizer per acre on bright tobacco land.[22] Granted, the growing disconnect between soil type and quality tobacco was far from complete; bright leaf from the sandy ridges surrounding Danville remained the highest-priced tobacco in the Southeast due in part to local soil qualities, but fertilizers dramatically narrowed the gap between the birthplace of yellow tobacco and the new districts of the eastern Carolinas.

Southside tobacco farmers seemed to believe that they had little control over the environmental conditions that troubled their livelihoods, and their efforts to challenge political and economic orders that threatened their profit margins were equally unsuccessful. From the 1870s to the 1890s, regional farmers formed several organizations to challenge the economic dominance of warehouse owners, fertilizer companies, and merchants. The first and most conservative of these organizations was the Grange, or "Patrons of Husbandry." The Grange was a national organization that promoted fraternal bonds among farm families and worked to advance rural interests.[23] Although the Grange was never as strong in the Mid-Atlantic South as it was in the Midwest, it was particularly popular among the tobacco farmers of Halifax and Pittsylvania counties. Each county had more local chapters than any other Virginia county, with the exception of Augusta.[24] Grangers were hardly radicals intent on overthrowing the existing agricultural and social order; instead, they advocated moderate reform of existing systems, and Border Grange governance reflected this moderation. Among the leadership of the local Grange were influential farmers and businessmen, such as William Sutherlin, who had a vested interest in much of the status quo. The Border Grange lobbied to regulate railroad rates and warehouse fees, two sources of economic annoyance to most Southside tobacco growers (and such regulation happened to coincide with the interests of tobacco manufacturers, who would also benefit from lower shipping and handling costs).[25]

The Border Grange's primary target of reform by the late 1870s was the Danville warehouse system. The city's warehouses had united to form the Danville Tobacco Association (DTA) in 1869. Ostensibly, the DTA's purpose was to cut down on private tobacco buyers, or "pinhookers," and to ensure uniform weighing and sales practices. Pinhooking involved the speculative purchase and quick resale of tobacco, often accompanied by false information or some other sort of trickery, and the trade had existed on the edges of the warehouses since the earliest days of the auction system. A pinhooker might meet a farmer outside the warehouse, regale the man (and sellers were almost always men) with tales of poor wrapper sales, and offer to buy his load for a modest price. If the farmer was gullible enough to agree to the sale, the pinhooker would then turn around and sell the leaf on what was, in reality, a strong day for yellow wrappers. In another popular tactic, a pinhooker purchased several poorly graded piles on the auction floor, sorted them into piles of top leaf and poorer-quality tobacco, and generally made a profit on the resale. This trick took advantage of buyers' tendency to assess a pile based on the lowest grade of tobacco it contained. Although farmers almost universally acknowledged that pinhooking was an irritant, the DTA was less effective at controlling pinhookers than it was at achieving its unstated goal. In reality, the organization's most important function was to guarantee that all of the city's warehouses charged farmers the same fees. The association also apparently encouraged the major tobacco buyers to consult on their bids, removing a source of competition between the cooperating warehouses. The Grange characterized the DTA as a prime example of a trust that subjected regional growers to unfair market conditions. In an effort to combat the power of the DTA, the Grange went so far as to establish its own warehouse in Danville in the 1870s, but poor management forced its closure after only a few seasons.[26]

In the 1880s another farmer organization, the Farmers' Alliance, took up the Grange's agenda with renewed vigor. Like the Grange, the Farmers' Alliance was never as active in Virginia and upper North Carolina as it was in other portions of the country, but the organization did engage more directly in political activity than its predecessor.[27] The

Farmers' Alliance was a national organization, and Alliance members on both sides of the border had many desires similar to those of farmers across the South and the Midwest. Like the Grangers, Southside Alliancemen were particularly troubled by the political and economic power of the large businesses with which they regularly dealt, especially the railroads and the tobacco warehouses. Alliance members believed that high shipping costs lowered the prices that tobacco buyers would pay for raw leaf, in addition to imposing a financial burden on the growing number of farmers who shipped their tobacco to market by rail, and they constantly complained of the costs associated with selling tobacco in auction warehouses and of the power of the DTA in particular. Among their demands of the Virginia and North Carolina legislatures were a commission to regulate railroad freight rates, lower litigation costs that would enable poor farmers to take their cases to court, and an end to taxes placed on tobacco. Unlike midwestern Alliances, Southside farmers had little need for Alliance-owned storage facilities; they could store their tobacco in their barns for almost a year, waiting for an opportune time to sell. They did, however, clamor for an Alliance-owned auction warehouse where they could market their tobacco without paying the various fees and commissions charged by the public warehouses, and nine of these facilities eventually operated across the Virginia and North Carolina Piedmont.[28] This idea did not emerge sui generis; Alliancemen drew on the Grange's unsuccessful warehouse of the previous decade for inspiration.[29]

White farmers in Caswell, Halifax, and Pittsylvania were willing to challenge the railroads and warehouse owners, but they were unwilling to engage in biracial politics (or organization) to do so. Local Alliance chapters were dominated by wealthy and influential farmers, and Alliancemen worked hard to assure their fellow white farmers that theirs was a thoroughly white organization. A Halifax paper sympathetic to the movement informed readers, "There is not a principle announced in this platform [of the Virginia Farmers' Assembly] that is not thoroughly Democratic."[30] A Caswell newspaper that served as an "Alliance Advocate" was more explicit. An editorial explained that all county Alliance members were white Democrats because "for years in Caswell county the Democracy has been fighting almost without hope of success against a heavy Radical majority, composed almost entirely of negroes."[31] Local

Alliance chapters thus positioned their struggle against white business owners as a question separate from the postwar battles over black labor and land ownership. According to William Link, when the Alliance attempted to engage in Populist politics in 1892 and 1893, racial issues quickly doomed the organization in the Upper South.[32] As the Alliance made clear, the same racial animosity that fueled the assassination of John Stephens and the Danville Riot continued to influence local politics.

Although the DTA remained a foe of Southside Alliancemen, regional tobacco farmers had a new, more imposing opponent beginning in 1890: the ATC. The ATC was one of the nation's first mega-corporations, along with Standard Oil and U.S. Steel, and it held a near monopoly on a product that would revolutionize consumer tobacco, the cigarette. American cigarettes first appeared in the 1850s, but they remained uncommon until the 1880s. The man most closely associated with the rise of the cigarette was James B. Duke, who owned W. Duke, Sons and Company of Durham, North Carolina. Duke made his first cigarettes in 1881, and through mechanization, effective advertising, and aggressive business practices he came to dominate the early regional market. Chief among Duke's innovations was his early adoption of cigarette-rolling machines, developed by Virginian James Bonsack, which sped up production and lowered the cost of making each cigarette. These machines were at first delicate and unreliable, but Duke persisted with the new technology until it was perfected.[33] W. Duke, Sons and Company was so successful that Duke was able to arrange a merger of the leading American cigarette producers into a trust with himself as president. At its creation, the ATC brought together five firms—Duke's company; Allan and Ginter of Richmond, Virginia; the Kinney Company of New York City; W. S. Kimball and Company of Rochester, New York; and Goodwin and Company of New York City—that combined produced nine out of every ten cigarettes sold in the United States.[34]

Despite a virtual monopoly on cigarette sales, the early ATC did not dominate the American tobacco industry to the degree that most historians assume. Indeed, the cigarette was a relatively minor form of manufactured tobacco well into the twentieth century. Cigarettes remained a novelty in the United States until the 1880s, when hand-rolled Egyptian and Turkish products became somewhat fashionable. The mechanized production of Duke and his competitors quickly overwhelmed these imported luxuries, but cigarettes remained a niche good among American consumers. As late as the 1910s, only 5 percent of tobacco products sold in the United States by weight were cigarettes. Plug chewing tobacco, snuff, and cigars continued to be the most popular forms of tobacco.[35] Historian Allan Brandt summarizes the slow rise of cigarettes succinctly: "The cigarette seems such a ubiquitous part of American Culture that it is difficult to imagine that it is really a twentieth-century phenomenon."[36]

Despite the monopoly's growing power, local buyers and manufacturers persisted in the years immediately following its formation. Illustrative is the example of the small town of Milton, along the Dan River in northeastern Caswell County. In 1893, the Sanborn-Perris Company prepared a town map for fire insurance purposes, which displayed the location and function of each structure in Milton. More a village than town, composed as it was of just a few residential blocks with some businesses mixed in and a population of 975 (according to the mapmakers), 1893 Milton was home to fourteen companies devoted to warehousing, manufacturing, or prizing tobacco, as well as to three guano and fertilizer warehouses catering to tobacco farmers. A walk down Main Street from east to west passed Captain Irvin's tobacco warehouse, the Winstead Tobacco Factory with its steam heating system and "sweating" rooms, and then the Public Warehouse. A turn south on Bridge Street skirted J. A. Hurdle's prizery, the N. E. Oliver plug factory, E. Hunt's tobacco prizery, the Liberty Tobacco Warehouse, and the American Tobacco prizing facility. A turn east again to intersect Lee Street led to R. L. Walker's prizery and the Golden Belt Tobacco Company. Although this insurance map does not specify the exact activities that took place in each structure, nor their financial health, businesses in search of local tobacco stood, seemingly, on every corner in this market town.[37]

If the ATC did not dominate the national tobacco industry, or even a town like Milton, during the early 1890s, the new trust quickly came to wield inordinate power in the bright tobacco belt. As Duke and his fellow manufacturers pioneered mass-produced cigarettes, they found that bright leaf made an ideal substitute for the smooth-smoking Turkish and Egyptian tobaccos used in the first cigarettes. Yellow tobacco was both milder and more abundant than the imported varieties, and many cigarette manufacturers soon made their new product almost entirely from bright leaf. Poor-quality manufacturing tobacco could be hidden in the middle of a plug, where it might go unnoticed by the average consumer, but smooth tobacco became essential to the reputation of a cigarette brand.[38] Advertising for the brand Dixie Darlin' emphasized the importance of bright tobacco in smoking products, relating the tale of a consumer who "had poisoned his lungs, destroyed his nerves and ruined his temper by smoking" harsher tobaccos. Upon switching to the bright leaf in Dixie Darlin' he supposedly broke out in jubilant doggerel:

> Tobacco is a glorious plant,
> No matter how I used to rant;
> I've quit my everlastin' snarlin' —
> I found the "cure" in "Dixie Darlin'."[39]

With the popularity of bright leaf as a cigarette ingredient ensured by customer preference and advertising muscle, manufacturers sought as much yellow tobacco as they could get. Buyers for cigarette manufacturers worked the Danville markets and the new sales centers opening in North Carolina towns and quickly became the most important buyers of high-quality yellow tobacco. Compounding this monopsony, by the turn of the century the ATC bought its way into plug manufacturing, most notably through its acquisition of the R. J. Reynolds company of Winston (then the world's largest chewing tobacco manufacturer), and became the largest purchaser of plug wrappers in the country, increasing its chokehold on yellow tobacco markets. By 1910 the ATC sold 85 percent of American chewing tobacco, monopolizing plug sales to almost as great a degree as cigarettes.[40]

Turning again to Milton hints at the effectiveness of ATC consolidation and its impact on local townscapes as well as local economic relationships. The Sanborn mapmakers returned to Milton in 1908 and found a town of just 450 residents. Where fourteen tobacco businesses had operated in 1893, just six remained, most of them dedicated to storing or packing tobacco rather than manufacturing leaf. The mapmaker noted that the Winstead Manufacturing Company, once Milton's most impressive facility, stood "now vacant," with its brick "ch[imney] falling down," symbol of a once busy manufacturing district fallen on hard times.[41]

Southside farmers were ultimately less worried about who was buying their tobacco than they were about the prices that buyers paid for their leaf. Growers might have been content to sell their crops to the

ATC if they felt adequately compensated, but such was not the case. The ATC thoroughly exploited its control of the market in an attempt to force cigarette prices higher and raw leaf sales lower, with the aim of absorbing their small competitors and then completely controlling both ends of the market. When the trust artificially lowered cigarette prices to drive their competitors out of business, their buyers paid less for bright leaf on the auction floor. When the ATC then restricted cigarette production to drive up the sale price, buyers further depressed the market by taking less leaf. The trust also absorbed or forced out of business small tobacco manufacturers who offered growers alternative buyers. In the two decades following the creation of the ATC, the trust gobbled up approximately 250 competing tobacco manufacturers. Even many of the pinhookers who bought tobacco outside of the warehouses saw the writing on the wall and went to work as buyers for the ATC.[42] The ATC's buying and sales tactics angered Southside growers, who felt as though the monopsony was working with the alliance of warehousemen to attack their profits from all sides. Frustrated farmers in Caswell complained of being "oppressed and in bondage to the tobacco combine for the price of our tobacco" and organized a small county Tobacco Growers' Association in a futile attempt to combat the ATC.[43] Reviewing the trust's tactics, another opponent declared the ATC akin to a natural disaster of epic proportions. It was, he wrote, "by far the worst enemy the farmer has ever encountered. Far-reaching and disastrous to the tobacco grower, in its aim and results, its creation was a calamity to him in comparison with which drought, hail-storm and frost, all combined, sink into insignificance."[44]

In Pittsylvania, Alliance members organized one of Virginia's most ambitious cooperatives in the early 1890s, in an effort to break away from the tentacles of credit relationships with merchants and fertilizer manufacturers. On August 16, 1890, members of twenty-three local Alliance chapters met in Chatham and determined to form the Pittsylvania Central Alliance Trade Union (PCATU). The new cooperative rented an office and a warehouse, and it established a general store in Chatham. County farmers could purchase a share of the cooperative for ten dollars, which capitalized the venture and granted them access to the store's goods, cooperative discounts on fertilizer and farm implements, and agricultural meetings. By June of the following year the PCATU had enrolled several hundred farmers and was an active force in rural Alliance chapters scattered across the county, including those in Green Rock, Dry Fork, Hollywood, Swansonville, Ashbury, Caddo, Harpin Creek, Chatham, High Point, Callands, Lolo, and Wayside.[45]

Farmers in the cooperative sought a number of benefits. The PCATU store sold goods—from coffee and oranges to nails and Brazil nuts—to members for 10 percent less than most local merchants.[46] More important for tobacco producers (and almost all county farmers raised tobacco), the PCATU bought fertilizer in large quantities at wholesale prices and passed these savings along to members. Between March 1891 and April 1892, the PCATU purchased well over 200 tons of fertilizer. Even more significant than the reduced cost of goods was the cooperative's policy of extending credit to its members. According to its records, unlike most merchants or fertilizer companies, the PCATU apparently did not require borrowers to sign a crop lien.[47] In addition, the cooperative store also provided small, unsecured cash loans to needy members.[48] The cooperative planned an even more direct attack on the mechanisms of farm credit and debt; the union intended to build a fertilizer factory and a machine shop to produce agricultural implements in Chatham, in the hope of bettering local agriculture while further lowering the prices of farm inputs.[49]

Cooperative records are relatively silent on the organization's larger political aims, though the PCATU seemed largely to have reflected the purposes of the state Farmers' Alliance. A cooperative ledger urged "the brothers to unite together in business" but did not explicitly challenge the Southside's political or social status quo.[50] As with other branches of the local Alliance, this call for brotherhood did not extend to the region's black farmers and tobacco workers. In short, the union's primary goal seems to have been to provide its white members with less expensive fertilizer and general merchandise, though the challenge to large merchants who controlled regional agriculture was implicit. Undercapitalization prevented the full realization of even these limited goals. The PCATU was never able to construct either the fertilizer plant or the implement factory. Instead, the cooperative purchased farm machinery from outside sources and entered into agreements with the Durham Fertilizer Company and Venable's Fertilizers to furnish amendments at bulk rates.[51]

Although it was ultimately unsuccessful, the PCATU, like the Grange and Farmers' Alliance chapters before it, attempted to address the agricultural and environmental issues that made farmers' lives difficult. The cooperative's overwhelming focus on affordable fertilizer reflected growers' belief that the best way to combat soil depletion and erosion was through purchased fertility. Likewise, the organization's attempt to build implements suited to tobacco culture reflected a historic local interest in modern machinery that might make farming more efficient (and in the case of better plows, less erosive). Cooperative meetings were a place to discuss strategies for combatting the ATC and Danville warehouse owners, but they must also have been a forum for conversations about tobacco cultivation techniques, just as agricultural club gatherings had long served to disseminate farm knowledge. The PCATU was an organization for combating Pittsylvania farmers' economic and political difficulties, but it also worked (however briefly and ineffectively) to improve farmers' relation to the land.

Ambitious as it was, the PCATU was short-lived. The cooperative collapsed in the fall of 1892. The PCATU's demise was the product of a failure of its members to meet their debts to the organization. Farmers who promised to pay their subscription fees often failed to follow through, and tobacco growers who purchased fertilizer on credit just as often failed to settle their accounts. In some cases these defaults may have represented poor faith on the farmers' parts, but it seems likely that for many debtors the realities of a poor crop caught up with them. By September 1892, the union was almost $8,000 in debt to its creditors, the largest being the Durham and Venable fertilizer companies. Ironically, or perhaps tellingly, the PCATU succumbed to the same credit structure and market depression that plagued Southside tobacco farmers.[52]

Declining bright tobacco prices made farmers' concerns all the more pressing. Demand for yellow leaf was such that Danville market prices had muscled through the Panic of 1873 without a decline; in fact, bright tobacco prices increased steadily through economic turbulence of 1873 and 1874. Such was not the case with the more severe financial crisis in 1893. The ATC's efforts to restrict cigarette production combined with the crisis to thoroughly depress bright tobacco markets. In 1889–90, bright tobacco moving through Danville's warehouses averaged $12 to $13 per hundredweight, and this price was low enough to stimulate Alliance organization. In 1893 the price fell to $6.46 per hundredweight.

Although this figure was still roughly double what a farmer could expect for dark tobacco, it made the margins tighter than ever for the average grower. Prices rose slowly toward the end of the decade, but farmers remained suspicious that the ATC had engineered the decline.[53]

Growers disenchanted by the failures of farmers' organizations held out hope that a new ally, the federal government, would improve their situation. The last decade of the nineteenth century marked a federalization of tobacco expertise as government agencies gradually assumed control of tobacco studies and the dissemination of agricultural knowledge. Government tobacco research did not necessarily indicate the top-down power of the state through experts at the Department of Agriculture; rather, it was in part a reflection of popular farmer demand.[54] Farmer requests for government aid came out of the economic troubles of the 1890s, but they were also a continuation of historic tobacco culture. Bright leaf growers had long turned to tobacco "professionals" for advice, from the correspondents in antebellum agricultural journals to the mass-produced guides of the post–Civil War years. The tobacco research conducted by the Department of Agriculture (USDA) and other government agencies produced at the solicitation of growers were a continuation of a decades-old tradition.

Many farmers found the few government publications of the early 1890s to be of little use. Although copies of Robert Ragland's and the Love Brothers' detailed bright tobacco guides circulated the Border, state and federal tobacco publications tended to be general and relatively unhelpful to the bright leaf grower.[55] John Estes's *Tobacco: Instructions for Its Cultivation and Curing* (1892), the sixth number in the USDA's new Farmers' Bulletin series, was typical. Estes's brief guide furnished general advice on cultivating tobacco, but it failed to differentiate between varieties until it covered curing. At that point the author admitted that curing bright tobacco was "a somewhat difficult process, requiring practice to insure the best results," and left it at that. Estes also confessed that aspiring farmers could do little unless they matched tobacco varieties to appropriate soil types and climate, but he offered no instructions on how to do so.[56] These shortcomings meant that *Tobacco* was a poor guide for the practical Southside farmer. Along these lines, Otto Carl Butterweck's *The Culture of Tobacco* (Farmers' Bulletin No. 82, 1898) requested a comprehensive federal study of the requirements of various tobacco types, the sort of practical document that might aid aspiring growers.[57]

Commonwealth agricultural boosters took the lead in promoting more practical government studies. At the behest of growers, the Virginia State Board of Agriculture, anxious to promote and improve bright tobacco agriculture, approached USDA officials concerning a systematic soil survey of the tobacco belt. Milton Whitney, head of the Division of Soil Investigations, agreed to undertake such a survey if the state board would foot the bill, and Virginia officials agreed.[58] Whitney's 1898 report attempted to provide farmers with a rational, systematic way of identifying soil types. Instead of defining quality tobacco soil by vegetative cover or topography, Whitney outlined a mechanical method of separating soil into its constituent particles. The report provided an illustration of the proper ratios of gravel, sand, clay, and silt in ideal bright tobacco soil. Whitney also furnished farmers with moisture readings from forty-three locations scattered across the bright belt and suggested that farmers use these figures as guidelines when searching for appropriate tobacco land. Whitney repeated a Southside truism—"the flavor and quality of the leaf are greatly influenced by the conditions of climate and soil"—but his methods of identifying these microenvironments differed greatly from the inherited techniques of regional farmers. Whitney's report was indicative of the professionalization of agricultural knowledge that had come to dominate bright leaf production.[59] Whitney's findings achieved mass circulation in a USDA Farmers' Bulletin (No. 83) the same year, a more concise version of his preliminary report.[60] This bulletin received immediate attention in the Southside, where, for example, one newspaper editor encouraged his readers to take advantage of the systematic study to improve regional tobacco culture.[61]

Whitney's studies of tobacco soils and farmer agitation for useful scientific guides eventually led to one of the USDA's most ambitious early agricultural programs: the county-level soil surveys. Whitney's preliminary report called for mapping soils on a detailed, local level, asserting that such studies would benefit tobacco farming to a great degree.[62] After two years of looking at small, well-defined shade tobacco districts in Pennsylvania and Connecticut, the Bureau of Soils turned its attention once again to the bright belt of Virginia and North Carolina.

Whitney pointed to these surveys of eastern North Carolina land, much of which the bureau touted as prime bright leaf land, as a perfect example of the value of soil survey work.[63] Along with the work of Danville-area bright leaf boosters, the general soil survey and the county-level soil surveys that followed in the early 1900s were key stimulants in the rapid growth of bright tobacco culture in the North Carolina coastal plain. This was not the result that Southside growers had sought.[64]

In addition to the accumulating studies of Southside soils, a new tobacco experiment station in Pittsylvania County conducted studies of the crop and reported its findings to interested farmers. Organized in the mid-1890s, the station near Chatham was funded by the USDA's Committee of Analytical and Agricultural Chemistry.[65] The Hatch Act (1887), which guaranteed state agricultural experiment stations federal funding, had stimulated some bright tobacco work at Virginia Polytechnic Institute at Blacksburg and the North Carolina experiment station at Raleigh, though the early trials were almost exclusively focused on fertilizer use and application rates. The Chatham experimental farm was the first station in the nation exclusively devoted to the study of tobacco (though it too was very interested in fertilizer studies and guano purity tests). All indications are that the experiment station concentrated on inviting regional farmers to view tobacco cultivation and fertilization techniques, a method of operation that likely appealed to bright leaf growers who held hands-on experience in high esteem.[66]

Although these efforts were encouraged by farmers, this federal tobacco work was indicative of an increasingly powerful and ambitious USDA, an agency on its way to becoming "the most dynamic portion of the national state in the early twentieth century."[67] James Wilson, the secretary of agriculture at the turn of the century, pointed to the linked work of the Bureau of Soils and tobacco field experiments as a prime example of the potential of the USDA's nascent farm programs. Wilson declared that tobacco studies showed the USDA to be "a bureau well organized and well equipped to carry on the work [of agricultural improvement] in a larger and broader field than has ever before been possible."[68] The tobacco soil surveys were a rigorous scientific way of understanding a crop culture and transferring that knowledge to farmers, and "the success of their [field researchers'] work is attracting trained men, who look to this as one of the most promising fields for research work along economic lines."[69] Wilson's vision of the tobacco soil survey work as the launching pad for more general soil studies came to fruition in 1900, when the Bureau of Soils adapted the tobacco survey format to address soil suitability for a wide range of crops.[70]

As soil surveys and the work of experiment stations strove to systematize and regiment bright tobacco production through the application of scientific methods, the growing association of bright tobacco with cigarettes served to make the crop more predictable in a way growers disliked. The prominence of cigarettes gradually circumscribed tobacco prices and reduced crop variability. Whereas plug manufacturers could afford to pay lucrative prices for the best wrappers, which served as advertisement for (but minor ingredient in) top-dollar plugs, cigarette manufacturers were much more concerned with uniformity in tobacco. They wanted to produce consistent, identical cigarettes. Many cigarette brands were made entirely of bright leaf, and manufacturers could hardly pay lottery prices for their main ingredient in a mass-produced product. Besides, a cigarette's paper wrapper hid the tobacco inside, making consistent taste the most important quality. The cigarette-consuming public was less interested in a product that contained exceptional-looking tobacco than they were in purchasing a reliably mild product. As a consequence, ATC buyers sought abundant middle grades of tobacco rather than competitively bidding on the very best lots. This shift in purchasing contributed to a decline in the average prices paid for bright leaf on the Danville market—the 1880s average price of $11.10 per hundredweight fell to $7.89 per hundredweight in the 1890s—but, perhaps just as important, it removed the potential for extremely high prices for select lots.[71] Farmers knew there was little chance of striking it rich with an exceptionally yellow crop of tobacco by 1900. This reality lessened the association of the best tobacco with the best farming practices, and it encouraged standardized production methods that led to consistent crops. As one farmer wrote in resignation, "The day of the white [extremely light yellow] wrapper and cutter is passed, so far as fancy prices are concerned. Let us, therefore, produce what is wanted."[72]

Manufacturers' reliance on bright tobacco in cigarettes drew on some of the same physical and chemical qualities that made the variety so successful in plugs. Cigarette advertising's use of brand names often continued the plug trends of evoking the color or taste of bright tobacco.

Companies like Lorillard manufactured cigarettes and smoking tobacco meant for cigarette rolling under such brand names as Golden Floss and Old Gold, and other brands alluded to their sweet flavor with names that included fruits or desserts.[73] Although this old-fashioned advertising remained a part of cigarette marketing, it was quickly surpassed by more sophisticated efforts, including names that evoked the exotic Middle East (R. J. Reynolds's Camel cigarettes being the classic example), cash giveaways, endorsements by foreign actresses, and campaigns touting a particular brand's health benefits or allure. James Duke even sponsored a roller-skating polo team named after his Crosscut brand of cigarettes.[74] A number of manufacturers also placed cards featuring attractive women, historical figures, or baseball stars in each pack of cigarettes to promote brand loyalty.[75] Cigarettes advertising embraced the modern consumer world that emerged in the late nineteenth century, a more elaborate and nuanced public arena than the world of Reconstruction plug tobacco advertising, and early cigarette advertising was well ahead of its time. Advertising appeals to bright tobacco's country of origin and its botanical qualities carried some appeal, but tobacco advertising had moved beyond the simple methods of antebellum- and Reconstruction-era plug manufacturers.

Bright tobacco imagery played a role in advertising success, but the crop contributed perhaps as much to the rise of the cigarette through biochemical means. Since the early nineteenth century, scientists and some farmers understood that tobacco contained a stimulating chemical substance, eventually labeled nicotine, and that the alkaloid had some effect on the human body. This understanding did not extend to a consistent conception of nicotine as a physiologically addictive substance, scientific confirmation of which would not come until the mid-twentieth century, though a few experts went so far as to classify it as both stimulating and "poisonous."[76] Despite the absence of a clinical understanding of addiction, nicotine undeniably served to lure consumers back for more, proving more powerful than any advertisement. Some of the same qualities that made yellow tobacco popular for chewing plugs and pipe tobacco—a smooth taste and relatively low levels of nitrogenous compounds, such as nicotine—made the crop ideal for use in cigarettes as well, but thanks to the chemistry of smoke and the human lung, low absolute levels of nicotine in bright leaf did not equal a lower likelihood of addiction; in fact, the opposite proved to be the case.

Bright leaf plants had wrought dramatic, even alchemical changes, to the Southside environment by the end of the nineteenth century. The crop had transformed poor land into the Piedmont's most valuable soil, it had turned the curse of little nitrogen into a blessing, and it had changed a region weary of tobacco cultivation into one crazed by yellow leaf. But bright tobacco had one last bit of biological and chemical magic in its leaves, tied directly to the technology and practices of curing and manufacturing—and to the rise of the cigarette. Thanks to the nature of bright leaf, cigarette users were more likely to become addicted to nicotine than consumers who used other forms of tobacco, even though nicotine levels in bright leaf were lower than in competing tobacco varieties.

The human body can absorb nicotine, and thus become addicted to the compound, in a number of ways. People who chew tobacco plugs and twists take nicotine into their bloodstreams through the linings in their mouths. Cigar and pipe smoke contains nicotine as well, which enters consumers' bodies through the mucous membranes of the mouth, throat, and nasal passages but is difficult to inhale deep into the lungs. This difficulty comes from the chemistry of the smoke: smoke from cigars and pipes is basic (or alkaline on the pH scale) and thus hard for lungs to inhale; deep inhalation often results in a choking sensation or coughing fit, even in habitual smokers. Cigarette smoke, however, proved to be slightly acidic. This property derived from the way growers cured their bright tobacco; the high heat of charcoal fires and then flues, developed in attempts to make tobacco yellower, acidified the leaf as it fixed color and sugar content, whereas the low heat of traditional curing had left dark tobacco products basic. As a consequence, smokers could pull the acidic vapors of a burning cigarette much more fully into their lungs, where the elaborate folds of millions of tiny alveoli provided many times the surface area of the mouth or throat through which nicotine could enter the bloodstream.[77]

Bright leaf's potency was thus "due to the extremely rapid absorption of nicotine, which produces an intravenous-like effect following inhalation," a particularly fast and powerful high.[78] Bright tobacco growers, manufacturers, and boosters did not realize they were creating a more addictive product as they built flues, installed cigarette-rolling ma-

chines, and advertised new product lines; indeed, the modern conception of addiction had yet to fully form, but the effects of cigarettes and
their acidic smoke were real nonetheless. Epidemiologists would later
conclude that death rates from cancer and other chronic tobacco-related
diseases were much higher in smokers who regularly inhaled than in
those who did not (in no small part because of the effects of substances
such as polonium-210, discussed in chapter 6), and that cigarette smokers were much more likely to inhale than consumers of other forms of
tobacco.[79]

As with the biochemical interactions that happened to fix high
sugar content in extremely yellow bright tobacco, mating smooth flavor
with an attractive appearance while appealing to a consumer's sweet
tooth, flue curing drew on plant biology and practical technology to create an exceptionally attractive product. There was contingency involved:
tobacco growers and manufacturers did not realize that cigarettes were
more addictive than other tobacco products until well into the twentieth
century (and then they actively worked to hide this information). Nor did
early cigarette manufacturers understand that bright leaf as an ingredient made cigarettes not only appealing to consumers but virtually irre-
sistible. Yet it would be remiss to fail to include the importance of nature
in America's smoking habit. Culture and chance intersected in critical
ways as the chemical makeup of organic material and the physiology of
human bodies interacted with the nutrient content of Piedmont soils
and the combustion of second-growth pines within sheet-metal flues.
Cigarettes ultimately surpassed chewing tobacco and cigars in part
because the biology of yellow tobacco varieties and the operation of
people's lungs and brains combined with the cultural realm epitomized
in the clever advertising dreamed up by James Duke. The nature of
bright leaf intersected the nature of human bodies through the mechanisms of addiction, ultimately with grave consequences for the health of
people and soils.

It is important here to pause for a moment, to take a deep breath,
and to again acknowledge that the chemistry of addiction was hardly
deterministic in the growing American obsession with tobacco. After
all, even chewing tobacco and cigars are significantly addictive, suggest-
ing that the consumer shift to cigarettes entailed more than just a
quicker, easier nicotine fix. Other forces were at work in the spaces

between barns and human bodies, encouraging the rise of the cigarette and a cultural as well as physical dependence on a particular nicotine-delivery device. On factory floors, cigarette-rolling machines, most closely associated with James Bonsack and the Duke tobacco company, made the production of cigarettes cheaper, faster, and less labor intensive, though their introduction and implementation were likely less inevitable and more gradual and variable than myth has long held.[80] Tobacco companies played on consumer desires in some of the earliest sophisticated mass-advertising campaigns as well, positing cigarettes in various ads as exotic, modern, convenient, safe, sanitary, and even sexy. The advertising power of the ATC exemplified this new marketing acumen in its tobacco trading cards featuring popular baseball players and scantily clad women, touting cigarettes as an indispensable element of Progressive Era consumption and attracting impressionable youths.[81] And World War I spurred cigarette consumption to an unprecedented degree: cigarettes became part of every soldier's kit, forming habits in camps and on the fronts that would follow troops home. It was only in the years following the Great War that cigarettes surpassed chewing products as the most common way Americans took their tobacco.[82]

And yet, despite being largely absent historians' explanations of the rise of big tobacco, the intersection of curing, chemistry, and cigarette consumption has some real explanatory power. The properties of inhalation forged in southern soils and barns did little to attract consumers to their first puffs, but they might, at least in part, explain why they took a second draw. Addiction remains a bit of a black box; science and culture accept its reality, in many cases there are satisfactory explanations of the various mechanisms of addictions, and yet much remains unclear. To exactly what degree do genetics, personal willpower, health, and habit shape addiction? Debates remain even in a substance as studied as nicotine.[83] And how might we historicize addiction as a causative agent or as a power limiting human agency?[84] Addiction might go far to explain broad trends—while threatening to rob any particular historical actor of agency if uncritically applied.

Tobacco curing and the chemistry of cigarettes can only suggest possible avenues into thinking about addictive crops, their histories, and their roles in broader culture and economy. As the case of soils, curing, and chemistry illustrates, even a phenomenon long defined as fundamentally biological—nicotine addiction—was and remains deeply enmeshed within technological systems, agricultural practice, economic structures, and chance. But it is also a relationship between the nature of bodies, plants, and places. It is as much a political ecology as a political economy.

If the nature of physiological addiction (or the hazards of radioactive polonium, for that matter) was little understood at the close of the nineteenth century, tobacco growers' economic and cultural addiction to raising yellow tobacco was all too clear. The cigarette captured tobacco growers as neatly and completely as it did the consumer tobacco market. The ascendency of the cigarette, as produced by the ATC and the large companies that followed the trust's dismemberment, constricted bright tobacco markets and lowered the prices that buyers paid for leaf. More significant, cigarettes stripped away the lucrative lottery prices of which growers once dreamed, transforming an artisanal crop (at least in farmers' imaginations) into an industrial commodity. The cigarette relied on bright tobacco, but it repackaged the agricultural product in a way that further distanced field and consumer. The yellow leaf as plug wrapper kept the tobacco variety and its quality at the forefront of manufacturers' and consumers' minds, and both were willing to pay handsomely for the best article. Cigarettes, in contrast, stressed consistency and hid tobacco inside a paper wrapper. Inside their packaging, all cigarettes looked the same. America's cigarette obsession would ensure bright tobacco a place in today's Southside fields, but it spelled the end of the crop's future as early tobacco boosters had envisioned it.

If the situation was bad for white farmers, it was particularly dire for the region's African American population. Almost four decades after Emancipation, black residents of the three counties had made relatively little progress toward landownership. In 1900, African Americans made up roughly half the population of Halifax and Pittsylvania, yet they owned only 4 percent of Halifax's acreage and only 2 percent of the land in Pittsylvania.[85] The situation was equally unbalanced in Caswell. Despite the fact that almost 55 percent of the population was black, African Americans owned only 60 of the 1,745 farms in Caswell.[86] In his study of agriculture and society in a Piedmont county, Crandall Shifflett argues that "above all, the man-land relationship stood at the center of

FIGURE 11: Halifax County's Black Walnut plantation, fallen into disrepair by the 1930s, symbolized the decline of bright tobacco's regional promise. Courtesy Library of Congress.

the social order."[87] This may have been so, but all such relationships were not equal, for as C. S. Lewis so eloquently put it, "What we call Man's power over Nature, turns out to be a power exercised by some men over other men with Nature as its instrument."[88] In the Southside tobacco kingdom the limits of black access to the land were part of an exercise in social power, but they also came from particular ideas about the relationships between people, race, and landscapes. Declining tobacco prices and increasing input costs made tobacco less profitable for black sharecroppers and renters, lessening African Americans' ability to save money for land purchases, even as decreased profitability might have encouraged some white landowners to finally consider selling. The conditions that worked to impoverish white tobacco growers—erosion, fertilizer debt, market monopsony, and the rise of cigarettes—further limited the economic opportunities available to African Americans. A few black families achieved landownership only to lose their farms as the tobacco economy and Jim Crow both grew tighter, but most never owned their own ground.[89]

The conditions that would characterize Southside bright tobacco production until the Great Depression were present in full force by 1900 (fig. 11). Ideas about soils had given rise to a new crop, and that crop had in turn fundamentally changed the regional environment. Tobacco culture dominated almost every farm in the three counties, competition was fierce thanks to the expanding yellow tobacco districts of eastern North Carolina and South Carolina, commercial fertilizers were an expensive and ubiquitous farm input, market prices remained lower than during earlier decades, there were no more lottery prices for the best tobacco, and racial inequities existed in the shadows and margins of every day's actions. Farmers would make continued efforts to better their situation; they tried new organization tactics, they continued to publish and read instructions on making better tobacco, and they sought new and better government research into tobacco cultivation.[90] Despite these efforts, the promising bright tobacco culture first created on the Southside's sandy ridges was effectively dead as the twentieth century began. In its place was an agricultural system that closely mirrored the cotton empire of the Deep South, with impoverished farmers in a land of gullies, beset by crop liens, increasing tenancy, low prices, and few alternatives.

Epilogue

In those parts of North Carolina where tobacco is grown almost exclusively, there is the most apparent poverty, not poverty of money alone, but poverty of culture, poverty of soil; poverty of good homes and social environment; poverty of health and of everything that goes to make rural life that ideal mode of living

—F. H. JETER, in Charles E. Landon, "The Tobacco Growing Industry of North Carolina" (1934)

DESPITE THE GROWING PROBLEMS of Southside tobacco culture, the year 1900 hardly marked the end of bright tobacco in the region. Piedmont farmers during the Great Depression—and even later—worked at tasks their tobacco farming grandparents would have readily recognized: hoeing weeds, firing log barns, and piling stones and brush in erosion gullies deep enough to swallow a mule (and later a tractor). Most regional farmers labored on, working to make a living from the crop they knew so well. Weather conditions, wars, and advertising worked to swing tobacco prices up and down, with brief pinnacles such as the spike that accompanied U.S. entry into World War I, but the general declining trend of the 1890s continued until the late 1920s, when tobacco markets fell into the abyss of the Great Depression. Although the environmental concerns associated with regional agriculture were full-blown by the 1890s, farmers locked into bright tobacco production by a lack of alternatives did little to move away from the crop. Environmental and economic troubles had arrived to stay; farmers would continue to battle soil erosion and depletion and the power of large tobacco companies for the entire twentieth century.

Although labor on the land remained much as it was before the Civil War, tobacco manufacturing continued to rapidly change. The cigarette underwent a transformation, from a faddish product with an uncertain future to being the dominant form of consumer tobacco in the first decades in the twentieth century, culminating in the entrenchment of cigarettes during World War I, when they became part of the essential kit of American soldiers. By the early 1930s, four major brands of cigarettes dominated the U.S. market. Despite this transformation of consumption, tobacco manufacturing remained connected to the early decades of bright leaf production in certain ways. Bright leaf remained an important component of cigarette filler, and though the ATC trust was dissolved in 1911, the companies spun off by court order remained collusive in tobacco buying, ensuring that truly competitive auction bidding remained more fiction than fact. Two of the four dominant cigarette brands drew on historic forms of bright tobacco advertising: the ATC's Lucky Strike resurrected the historic plug name, and Lorillard's Old Gold recalled the tradition of naming plugs after high-quality ingredients.[1]

Ecological troubles continued to plague the foundations of tobacco agriculture. Erosion and deforestation issues grew worse in the Southside during the early decades of the twentieth century as farmers persisted in their reliance on bright leaf as a staple crop. According to one geographer, by the 1930s the average plot of ground in the three counties had lost between seven and nine inches of topsoil to sheet erosion, among the highest rates in the Upper South, and particularly disastrous in a region with a thin topsoil layer to start.[2] A series of soil erosion surveys undertaken by the Soil Conservation Service (SCS) graphically illustrated the damage to the region's topsoil; according to the surveys, the vast majority of Caswell, Halifax, and Pittsylvania land suffered from moderate to severe sheet erosion and gullying. Erosion was the worst on the ridges and uplands paralleling river and stream courses—prime bright tobacco land that had been under cultivation for more than half a century in most cases.[3] Southside forests were similarly imperiled. One 1927 study declared that little woodland along the border between the states had any value as timber, and USDA soil surveys of the three counties made similar assessments. The majority of nonagricultural land was "cut-over land, abandoned crop land and fire-damaged woodland," the legacy of decades of intensive tobacco farming.[4] It was a landscape far removed from the impressive forests and clear streams William Byrd had so admired two centuries earlier.

The identification of the Southside as an environmental and agricultural "problem" area during the New Deal brought the region into the national purview, but the label merely highlighted conditions that had long existed in the old tobacco belt. An article in the 1936 volume of *Soil Conservation,* written by P. F. Keil, directed the attention of farmers and conservationists across the nation to the Southside's erosion issues. The piece "Two Centuries of Accruing Tragedy along the Dan River" blamed poor agricultural practices for the sorry state of regional farmlands, declaring that "until recently, little thought had been given to the idea of holding the soil and plant food on the farms; consequently the soils have washed away."[5] Keil used the Schoolfield Reservoir on the Dan River in Danville as his central example of the destructive potential of local erosion. Built in 1904, the 1,150-foot-long dam impounded a 540-acre reservoir that stretched more than three miles west of the city, providing drinking water and hydropower for the city's textile mills. By the time of Keil's article, 400 acres had disappeared, filled by silt from upstream runoff. In fact, most of the reservoir had silted so quickly that by the 1930s moderate-sized trees grew where seventeen feet of water existed at the dam's closing.[6]

Although Keil's descriptions of local erosion were accurate enough, he was much less precise about the relation between tobacco agriculture, the nature of local soils, and environmental degradation. He did note that tobacco was the dominant local crop and that its culture had existed in the Piedmont from the early years of white settlement. But rather than connect local erosion conditions to environmental and agricultural particularities, in keeping with the SCS's agenda of tackling rural poverty and land use issues across the nation, he associated local problems with poor farming practices up and down the Eastern Seaboard.[7] Keil's representation of local soil and farm conditions ignored how and why bright tobacco had become the dominant crop and the intellectual and agronomic struggles that had long dominated local discussion of how people could and should farm. His article removed the region's problems from their material and social contexts and made them, instead, part of an amorphous idea of laggard southern agriculture, which served as a trope in national conservation conversations.

FIGURE 12: During the Great Depression, CCC workers labored at a variety of conservation tasks in the Southside, including seeding meadow strips to slow sheet erosion, as seen here in rural Caswell County. Courtesy Library of Congress.

Efforts to address the region's problems were not limited to surveys and written critiques. The Southside was so eroded that the SCS selected a portion of the Piedmont landscape along the Bannister River in Pittsylvania as one of twenty-six agricultural demonstration districts scattered across the southeastern states.[8] Along with sites in North Carolina, South Carolina, Georgia, and Alabama, the Bannister River district was supposed to serve as a test bed where SCS scientists could demonstrate proper soil management techniques to southern farmers. With the resources and manpower of the SCS (which had Civilian Conservation Corps [CCC] workers at its disposal), the Bannister River project invested a great deal of labor and money in slowing Southside erosion (fig. 12). Within the first year, farmers and agency workers constructed more than 160 miles of contour terraces and 4,000 check dams designed to stem gully erosion (though officials estimated that this one small river basin alone still had 25,000 unfilled gullies).[9] The district also oversaw the planting of approximately half a million trees in gullies and on worn fields.[10] Descriptions of these efforts along with SCS advice for Piedmont farmers appeared in a local publication, the *Bannister River Banner*, that touted agricultural reform through resource conservation.

The journal must have had a familiar ring to Southside tobacco farmers. Although some of the advice that appeared in the *Bannister River Banner* reflected new soil conservation thinking, many of the articles would hardly have seemed out of place in antebellum agricultural improvement journals. Twentieth-century science was evident in articles advocating the use of kudzu—an imported Japanese legume soon to become an iconic pest across much of the South—to stabilize eroded hillsides; plans for the planting of contour grass strips dividing blocks of grain and hay; and in the large-scale transplanting of shortleaf pine, loblolly, and black locust saplings on marginal lands.[11] Much more common, however, were proscriptions that would have sounded quite familiar to James Bruce or John Edmunds. Writers for the SCS called for Pittsylvania farmers to conserve local wood resources, to build terraces that would slow runoff, to plow deeply and along the contour, and to diversify their farms' production. Also akin to antebellum reformers, SCS officials were reluctant to advocate an end to bright tobacco culture. Publications, such as the USDA Farmers' Bulletin No. 1767, *Soil Defense in the Piedmont*, admitted that regional land could no longer grow much besides yellow tobacco and devised cropping plans for the old staple that excluded legumes to ensure best leaf quality.[12] A contemporary geographer, summarizing bright tobacco cultivation in North Carolina, breathed new life into old tropes concerning the magic of a crop that could thrive where little else could grow: "The finest sort of tobacco seems to grow on old worn-out fields that have been abandoned for several years. These fields grow up to trees and bushes and may lie for fifteen or twenty years without being used. They are so poor that nothing else grows in their soil."[13] Challenging the agricultural paradigm was difficult indeed with this sort of expertise circulating.

If the advice and criticisms coming from SCS officials were nothing new, the agricultural patterns they faced seemed equally entrenched. At the start of the Depression, most regional growers cultivated bright leaf in ways almost identical to those of the 1880s. Tobacco dominated the poorer upland soils, shallow plowing was the most common cultivation method, and commercial phosphate fertilizers remained a necessary input on most tobacco farms. Experts from the SCS could offer no solution to a fundamental problem: bright leaf faced little competition for economic dominance in the Southside. Farmers who wanted to make a living turned to tobacco as had their fathers and grandfathers, for, as a Virginia Polytechnic Institute study admitted, "Good bright leaf soils are somewhat deficient in nitrogen and it is usually not profitable to produce other crops on bright tobacco land."[14] Farmers recognized the ecological problems tobacco culture caused, but they could see no way around them. The *Bannister River Banner* and similar publications also did nothing to address racial and social relationships that cemented local agricultural practices. The SCS did not critique the dominance of white landownership or the credit relationships between sharecroppers and tenants, their landlords, and fertilizer companies. Soil scientists failed to tackle the inextricability of social networks, environmental conditions, and agricultural practices, a shortcoming that would continue to hamper the SCS's local demonstration district until its demise.

The New Deal's most lasting impact on the Southside lay not in its soil conservation programs but in a market management plan first implemented in 1933 as part of the Agricultural Adjustment Act (AAA). The AAA Marketing Agreement for Flue-Cured Tobacco, commonly called the "quota system" by Southside residents, set a limit on the acreage a farmer could plant in bright tobacco and tied this tobacco allotment to farms where the crop was already raised. Tobacco growers agreed to these restrictions because the quota system also established a government-supported price floor for tobacco, using funds exacted through taxes on tobacco manufacturers. The program set this price floor at the average figure paid for bright leaf between 1919 and 1929, a period of decent sales, and buyers from the major tobacco companies promised to buy at least as much leaf as they had during the 1932–33 season. The Supreme Court ruled the AAA unconstitutional in 1936, but the Soil and Conservation Domestic Allotment Act resurrected the quota system's main features almost immediately. The program proved popular with landowning farmers and, with modifications, remained in place into the twenty-first century, finally ending in 2004.[15]

The quota system proved effective in keeping bright tobacco supplies lower than they had been in the 1920s and thus stabilized prices, but the program dramatically reduced the number of sharecroppers and renters (both disproportionately African American) in the old tobacco belt. Reduced and fixed tobacco acreage meant that landowners needed fewer laborers, and the cost of transferring quota along with land meant that landless farmers who wanted to raise tobacco—the crop with which most local people had the most experience—had to come up with even more money than before to purchase a farm. In the Southside, the quota system thus solidified a situation that dated to the Civil War: landowners had a great incentive to retain their property, and it became more and more difficult for the landless to become property owners. As Pete Daniel writes, in practice the quota system "swept people from the land."[16] The Southside's surviving tobacco growers were disproportionately white and landowners, helping create the myth of tobacco as the heritage and last bastion of white yeoman farmers.

Farmers in Caswell, Halifax, and Pittsylvania counties still raise flue-cured tobacco (as bright leaf is now most commonly called) today, though there is less acreage than in past decades, and the decline in tobacco's prospects is only accelerating. Between 2002 and 2007 alone, years spanning the elimination of the quota system, the number of tobacco farms in the three counties declined from 761 to 299, though acreage shrank less dramatically, from 13,786 to 11,817, reflecting larger farm units. Farmers produce more tobacco per acre than ever before, thanks to enormous quantities of fertilizer and new plant varieties, but even so there is significantly less tobacco produced in the Southside than a century ago.[17] Although the quota system and tobacco's relatively high value per acre kept Southside small farms viable much longer than in most regions, flue-cured tobacco is today undergoing the same trend of farm consolidation that has swept through every corner of American agriculture.[18] With tobacco's association with numerous health problems, culminating in the mid- to late-twentieth-century identification of smoking as a leading cause of lung cancer, American demand for tobacco has lessened. Other developments were almost as critical: the quota system no longer supports prices, the old auction houses have closed and direct contracts with the tobacco companies have taken their place, and Hispanic laborers harvest the crop instead of African Ameri-

cans and white family labor.[19] Beef cattle and large-scale poultry operations are the region's other main agricultural enterprises; most of the
land is too poor to raise grain for a global agricultural marketplace.

Bright tobacco's current decline is in keeping with the history of its
first sixty years as a Southside staple. The crop experienced a meteoric
rise followed by a long and steady declension. Bright tobacco culture
worked to impoverish the land; there is no denying this fact. A tour of
the Southside today reveals gullies hidden in the woods, vast stretches
of red clay land where there is literally no topsoil left, and river and creek
bottoms choked with sand that has run off the uplands. Pine and hardwood scrub forest covers much of the historic tobacco land, slowly working to rebuild topsoil and slow erosion, though it will likely take decades
or centuries of unmolested woods to restore the land to its pre-tobacco
qualities. In other former fields cattle pastures cover the red clay, old terraces remain visible as green humps, and farmers wield lime, fertilizer,
and red clover seed in an effort to make the former tobacco kingdom produce beef.

The rural economy has suffered as well. Bright tobacco made landowners money at first, but around the turn of the century, environmental
problems, overproduction, competition, and marketing forces stripped
much of the profitability from the crop. Correctly or not, most Southside
farmers believed that it was too late to shift to a different form of agriculture after so much had been invested in bright leaf, and they soldiered
on until they lost their land or went to work off the farm part time. Today
a few hundred farmers raise tobacco where tens of thousands once labored. It is a fate these tobacco farmers share with multiple commodity-
producing landscapes increasingly tied to the global marketplace;
tobacco farms worked by transient Hispanic laborers find their parallels
in the Great Plains populated by as many combines as people, the Mississippi Delta where mechanical cotton pickers strip floss once picked
by African American sharecroppers, and reaches of Appalachia where
a few nonunion equipment operators remove mountaintops to get at
coal seams once worked by hundreds of deep-shaft miners.

Despite a reliance on tobacco from the Southside's initial Euro-
American settlement in the mid-eighteenth century until the present
day, the regional environment did not dictate or guarantee that the region would become a tobacco kingdom. The three counties did contain

lowlands suitable to dark tobacco cultivation and numerous waterways for transporting hogsheads of tobacco to eastern markets, and their soils and climate were among the best in the South for producing yellow tobacco, but it took the arrival and persistence of a particular human culture to create the land of the bright leaf. The region's early farmers grew tobacco in part because William Byrd and the settlers who followed him emphasized the land's potential for tobacco culture, and because many of them came from an eastern Virginia society that had depended on the crop for more than a century. That the Southside's thin, low-nitrogen, sandy soil produced some of the best-quality yellow tobacco in the world appeared providential, but it was a providence based on consumers' and manufacturers' cultural preferences for tobacco of a certain color and flavor. Nature, in the form of the biochemistry of plants and lungs, combined with culture, in the form of curing technology and cigarette manufacturing, also worked together to ensure bright leaf's popularity by creating a tobacco product that delivered addictive nicotine more efficiently than its competitors. But it was a foundation built from cultural preferences and habits that first made the crop regionally popular.

The region's dominant culture combined with the Southside environment to form an agriculture that had harsh consequences for a substantial portion of the region's population. That so much of tobacco's local history makes sense only when considered within the context of race relations is part of what makes this a "southern" story. African Americans' situation in the Southside in the early twentieth century was similar to their plight throughout the rural South in general. Black residents of Caswell, Halifax, and Pittsylvania faced limited economic opportunities, they encountered racism and stereotypes about their abilities as farmers, and they were less likely than whites to own their own land. The rise of bright tobacco illustrates the ways in which ideas about both soils and plants often drew on ideas about people—and vice versa.

This social and economic oppression was more than the result of amorphous southern attitudes or ideologies. In the Southside these realities reflected, in part, the demands of bright tobacco culture. White landowners and tobacco growers believed that their expertise was necessary to produce the best-quality tobacco (despite black experience with the crop), and they favored the close supervision of wage labor over sharecropping, rentals, or land sales whenever possible. In addition to racial prejudice, the lucrative nature of early bright leaf sales also made landowners extremely reluctant to sell land to African Americans, as whites feared giving away the goose that laid the golden egg. Universal environmental problems associated with bright tobacco cultivation also became racial characteristics in many whites' minds. White experts argued that soil erosion, deforestation, and farm monocultures were the result of African American tenants' poor stewardship when in fact these problems plagued all Southside tobacco farms. As bright tobacco profits declined around the turn of the century, these obstacles to black land-ownership eased, but new challenges emerged, and becoming an inde-pendent tobacco farmer no longer held the same promise that it had twenty years earlier.[20] Bright tobacco and the environmental conditions that underlay the crop were not the sole forces shaping African Ameri-can antebellum and post–Civil War experiences in the Southside, but they were important components of the region's racial legacy at the dawn of the twentieth century.

Of course the pursuit of yellow tobacco ultimately proved a dysto-pian experience for many of the region's white landowners as well. Histo-rians' tendency has been to chalk these problems up to either ignorance or greed: southern farmers did not know proper agricultural techniques, or they did know but cared more about profit than conservation. Bright tobacco challenges both assumptions. If this book has stressed any-thing, it is that Southside residents constantly thought about perma-nence and the world they were creating. They wanted to profit from farming, but just as strongly they wanted to be able to continue farming, and thus greed seems far too simple a description of their collective be-havior. Bright tobacco illustrates the way in which ideas of stability and permanence did not always equate with agricultural or environmental sustainability. Southside farmers turned to bright leaf in part because they sought an answer to agricultural decline; they wanted to remain on their farms and fashion a continuous living from their land rather than emigrate.

Yellow tobacco's tale equally refutes the stereotype of the ignorant tobacco farmer. Many cultivators put great emphasis on the technical nature of tobacco production and devoted their careers to advancing ag-ronomic knowledge. Experts—from Abisha Slade and his fellow pio-neers to USDA bureaucrats—promised that yellow tobacco could help landowners achieve better farming through bettering themselves as farmers. Agricultural improvement appealed to bright tobacco growers in part because of its emphasis on modern farming methods, and reams of agricultural literature assured growers that bright tobacco was the crop best suited to the Southside environment. Despite a veneer of professionalism, advice from these expert sources often exacerbated regional erosion, deforestation, and soil depletion. Farmers wanted to believe this advice because they wanted to remain on their farms. There is a sad irony here: although they did not think or write in terms of sustainability, farmers engaged in unsustainable practices in part because of their search for sustainability. The environmentally destructive form that bright tobacco culture took was not simply the product of lazy farming or unusually greedy farmers; it came out of a desire to preserve a familiar lifestyle in a certain place, but it was no less destructive for those admirable aims. We lose sight of this history when we couple agricultural knowledge with conservation, when in fact the two are not necessary linked.

Bright tobacco's story counters traditional accounts of tobacco as a staple crop and highlights the vital importance of histories that pay attention to specific crop cultures. Despite a general belief that "tobacco exhausted the soil rapidly," yellow tobacco in the Southside (and elsewhere) was not an unusually exhaustive crop per se.[21] The plant's nutrient demands, especially for soil nitrogen, were relatively light. And clean tillage crop cultivation as employed by tobacco growers was a practice that produced erosion in wheat and corn fields as fast as in tobacco plots. Rather, the damage caused by tobacco farming came from the composition and location of the soils best suited to the crop's cultivation and in the practices that made high-quality leaf. The perennial shallow cultivation of sandy topsoil with little organic matter on steep slopes in a region prone to thunderstorms was a recipe for disastrous erosion. This destruction was compounded thanks to watersheds stripped of trees to fuel tobacco barns, winter plowing to kill insects and soften the soil, and an absence of systematic crop rotation. Thus it was not tobacco that was hard on the land but the form of agriculture employed in making high-quality tobacco that led to gullied fields and sand-choked streams. The distinction is a subtle yet key one. To blame the tobacco plant for erosion and exhaustion is to blame nature, to make these troubles acts of God.

To place the burden of environmental problems stemming from bright tobacco cultivation more appropriately on people's ways of thinking about and managing nature offers hope of a path to constructive solutions.

Tobacco proves a troubling and yet illuminating place to think about such solutions, to search for lessons in American agriculture. In many regards tobacco seems to epitomize the negative side of the national experience working the land, from the colonial era to the present. As one of the first New World staples, tobacco was an early step on the road to plantation agriculture and monocultures, it fueled the rise of Chesapeake slavery, and it encouraged oppressive labor relations following Emancipation in the form of sharecropping and tenancy. And for all the harm tobacco cultivation caused to landscapes and workers, it wrought more immediate and visceral destruction on its consumers. Although there is no way to calculate its total health effects through history, the weed has killed millions of people and contributed to a poorer quality of life for countless others. Tobacco-related diseases continue to be one of the world's leading causes of death. Centuries of blood and sweat flowed into producing a crop that wasted rather than nourished bodies. And yet there is also some hope to be found in bright leaf's story. That debates over agronomy, farmers' moral worth, permanence, and conservation took place even with a crop like tobacco suggests just how seriously many farmers took the idea of stewardship—and makes a strong case that we should also take these understandings seriously when thinking about agriculture and nature both past and present.

As appropriate a place to leave this history as any is with a New Deal picture and what it represents. It is a photograph at once mundane and striking. Young Bobby Willis, tow-headed and a bit on the stringy side, in clothes obviously over-large, stares at a wall of canned goods (fig. 13). Row upon row of gleaming glass jars full of produce fill the frame: corn, green beans, pickles, and peaches, along with other unidentifiable but attractive food. The wooden shelves appear both sturdy and new, and even in black and white it is easy to imagine the kaleidoscope of colors confronting the boy. Bobby's left hand holds a narrow jar that he has just selected from the shelves, and though I cannot be sure, I am convinced that it contains jelly, for my mother and grandmother put up preserves in just such jars.

FIGURE 13: Bobby Willis is shown standing in his family's pantry, stocked with food preserved under the direction of FSA experts. Willis's family was part of the Farm Security Administration's New Deal efforts to better the condition of poor Southside farmers. Courtesy Library of Congress.

The photograph was taken sometime in October 1940, near Yanceyville in Caswell County, by Farm Security Administration (FSA) photographer Marion Post (later Wolcott).[22] Post was an employee of the FSA's Historical Division, tasked with roaming and photographing the rural portions of America during the Great Depression. Post and her colleagues bore two burdens: Roy Stryker, head of the Historical Division, charged them with providing visual testimony of agrarian America's hardships, and with furnishing evidence of the labors and successes of various FSA programs. Put another way, the Historical Division assured Americans that their tax money was both needed and being well spent. The images recorded totaled more than 270,000 between 1935 and 1943. The South, with its disproportionate rural poverty, racial turbulence, ex-

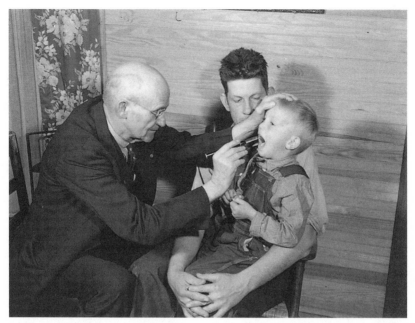

FIGURE 14: The FSA's work included free medical care for Southside families, among other programs. Here a FSA doctor examines Bobby Willis in Caswell County. Courtesy Library of Congress.

otic "otherness," and high rates of tenancy, attracted a substantial share of the agency's reform and documentary work.[23]

Post's pictures of Bobby Willis were part of the Historical Division's second mission. The Willis family was the subject of a series of "project" photographs, a documentary record of one family or community assisted, and in theory bettered, by the work of the FSA. Along these lines, Post recorded not only the Willis's canning successes under the agency's tutelage but also visits by a FSA-funded doctor who inspected the family's children (fig. 14), the new clapboard house built with an agency loan, efforts to install contour strips in community fields, and a happy father working with his tobacco crop. Although no FSA photographs exist of the Willis family before the agency's intervention, the photographic narrative implies that the various FSA initiatives, from subsidized loans to agricultural advice to modern healthcare, had bettered the family's condition.[24] One can hardly see the "after" pictures without conjuring the scenes that must have come "before."

The FSA rehabilitation programs worked in concert with the other New Deal agencies laboring to change regional society as Post snapped pictures across Caswell. The SCS's efforts to fill gullies and plant trees, agricultural demonstration districts along the Bannister River, and young men searching for a living in CCC camps all played a part in efforts to alter relationships between people and land during the 1930s. And most federal officials, as well as many local residents, agreed that there was a great deal to reform about Piedmont land use.

Ultimately, FSA efforts to improve the Willises' lives were but a new version of agricultural reform. The methods and messages of reform had changed over time: beginning in the 1830s, agricultural reformers of various stripes questioned tobacco, asserted its importance, stressed intensive cultivation methods, encouraged federal expertise, and, finally, turned the might of the New Deal's agrarian policy toward an impoverished countryside. What remained consistent was the failure of reformers to fully and effectively challenge locals' belief in the centrality of bright tobacco to agriculture, as well as reformers' inability to analyze the environmental relationships that ungirded that belief. Rare indeed was the observer willing to declare, "Where tobacco is grown almost exclusively, there is the most apparent poverty, not poverty of money alone, but poverty of culture, poverty of soil; poverty of good homes and social environment; poverty of health and of everything that goes to make rural life that ideal mode of living."[25] Piedmont farmers believed self was allied to a sense of place, and that sense of place stemmed in part from an understanding of soils, streams, and woods. At the center of this cosmology lay perceived relationships between tobacco and people.

Which brings us back to Bobby Willis. It would take more than canned goods to nourish land and body in the tobacco Piedmont. As in other portions of the southern agricultural economy, efforts to reform relationships between people and land usually amounted to little because they consistently refused to acknowledge what local people saw in the land and what they wanted from it. The FSA's efforts to better the Willis family's situation failed to account for why they and their neighbors believed so adamantly in the worth, the "rightness," of tobacco, even as their houses crumbled and their fields washed away, and could not explain why they continued to plant tobacco during the New Deal reform campaigns and long afterward. Government experts measured local precipitation rates, calculated proper sterilization temperatures in canners, and debated the costs of children's healthcare and subsidized farm loans. But they consistently failed to ask the Willises what they saw in landscapes and plants, how they envisioned the world. They failed to examine the connections between mental and physical environments. Ultimately, the FSA provided the right answers to the wrong questions.

True transformation of the rural South came not out of misguided general reform efforts but from the New Deal's authoritarian economic regulation in the form of production regulations and farm subsidies— which in the case of tobacco entailed quota allotments for landowners— and the economic and mechanical revolutions following World War II. Federal limits on tobacco poundage, coupled with the mechanization of certain tobacco tasks, pushed tenants, sharecroppers, and later small landowners out of tobacco farming, usually without immediate alternatives. These changes, however, amounted to a purge of the agrarian South rather than a reform. They were one route to the birth of modern American agriculture and the slow death of the nation's agrarian life.[26]

C. Vann Woodward famously defined the "burden" of southern history as the region's existence outside the boundaries of the mythical national narrative, its exception to American exceptionalism. In a nation built upon stories of success, on variations on the theme of Manifest Destiny, the South was long a land defined, at least in part, by loss, guilt, overt racial turbulence, economic inequities, and a self-conscious and defensive otherness. These characteristics could be found in the Southside, to be sure, but part of the region's burden also came from the very specific details of place intersecting with regional and national ideas. Nature ignored, bent too far to suit human dreams, sometimes broke. The Southside's burden stemmed as much from sandy hillsides, emerald green tobacco fields, summer thunderstorms, smoking burns in second-growth forest, hungry hornworms, and people's intellectual constructions of nature's elements as from the Lost Cause or the political rhetoric of the Redeemers. Nineteenth-century tobacco farmers sought to remake their world, and they did. That the results proved unexpected, that the Southside's unique combination of environmental conditions and human desires ultimately contributed to erosion, racism, agricultural unsustainability, and the persistence of a deadly crop is not, as Woodward might have suggested, ironic—it is tragic.[27]

ANTEBELLUM TOBACCO PRICES

THERE HAVE BEEN no reliable calculations of price differences between bright tobacco and dark tobacco in the Southside before 1869. References to the lucrative nature of the crop tended to be anecdotal and took the form of newspaper articles profiling successful farmers, agricultural journal pieces, or warehouse flyers. For the purposes of this book, I sought a more reliable—if still inexact—calculation of antebellum prices. By sifting through the financial papers of planters and farmers from the three counties I was able to locate records for the sale of nearly one thousand individual hogsheads and loose lots of tobacco between 1840 and 1860. These records were typically in the form of loose sales slips, small scraps of paper furnished by the merchant or warehouse that purchased the tobacco, though in some cases farm ledgers listing sales have survived. Although these sales were not always clearly labeled as bright or dark leaf, on the instances that tobacco type was noted, a pattern became clear. Before the Civil War bright tobacco sold for two to three times as much as the traditional staple, and it often went for much higher amounts.

The papers and sources that I consulted follow: Receipt of George Hairston, July 2, 1852, George Hairston Papers, Southern Historical Collection, Louis Round Wilson Library, University of North Carolina–Chapel Hill; William and Carrington to William Bailey, June 17, 1859,

Receipts of William Bailey, September 29, 1858; May 19, 1859, Account Ledger of William Sims, 1828–1867, William Bailey Papers, *Records of Antebellum Southern Plantations* (microfilm), Kenneth M. Stampp, gen. ed. (Frederick, MD: University Publications of America, 1987), series E, part 1, reels 1–2; Receipts of James C. Bruce, July 15, August 16, 1841, August 4, 1846, May 3, 7, July 21, August 6, 1847, March 11, April 11, 1861, Bruce Family Papers, *Records of Antebellum Southern Plantations*, series E, part 3, reels 7–30; Receipts of George Clement, January 12, February 1, April 20, May 16, 1843, May 30, December 12, 1844, June 14, August 8, 22, 1845, November 12, December 19, 1846, April 15, 22, September 4, 1847, April 21, May 9, July 16, August 2, November 24, 1848, June 8, June 21, September 13, 19, 1849, August 20, October 15, n.d., 1850, August 18, 30, 1851, September 18, 1852, August 8, 15, September 7, 1854, July 28, 1859; Receipts of Elisha Barksdale, September 1, 5, 1841, July 1, 7, September 15, 1842, July 6 (2), 7, 1843, May 2, 16, 1844, June 4, November 19, 1845, August 27, September 15, 1846, May 4, June 16, 24, 1847, July 22, 1848, March 2, 1850, September 10, 1852; Receipts of William Barksdale, n.d., 1850, February 7, 1851; Receipts of Rebecca Barksdale, July 14 (2), August 5, 1852; Receipts of Thomas Jones, March 11, April 13, 1840, August 25, 1856, July 28, 1860; Receipts of Thomas Jones and John Jones, April 13, 1840 (2); Receipts of Agnes Jones, August 20, September 15, 1855; Receipts of Martha Jones, April 13, 1840 (2); Receipt of John Jones, April 13, 1840; Receipts of Richard Jones, July 7, 1841, March 18, September 17, October 27, 1842, July 10, 1845, February 13, 1846, August 10, 1853, July 4, July 16, 1856, August 9, 1858, August 18, 1859, August 18, 1860; Receipts of Rebecca Jones, January 26, March 10, June 26, 1843, December 2, 1848, June 20, 1849; Receipts of Joel Hubbard, April 16, May 8, August 12, 1840, February 12, 20, April 13, 1841, September 28, 1842, August 1, 14, 1844, April 9, 28, 1845, July 29, October 2, 1846, March 3, 12, June 3, 1847, January 5, June 9, 1848, March 10, May 25, 1849, February 14, March 31, April 2, June 4, 1850, August 28, 1851, April 16, 1852, July 19, 21, 1854, March 30, July 20, August 6, October 24, 1855, July 18, 31, 1856, February 7, March 28, June 16, July 10, 1857, April 12, June 23, August 12, 17, 1858, January 5, June 16, 1859, October 14, 1865, two receipts with no date, 1850s; Receipt of J. Clayton, April 13, 1840; Receipt of P. Owen, December 14, 1840; Account of Ragsdale and Carrington, February 19–May 14, 1857; John Tyree to Joel Hubbard, May 9, 1857, all found in the Southside Virginia Family Papers, *Records of Antebellum Southern Plantations*, series E, part 3, reels 1–6; Receipt of William Bailey, September 23, 1852, in Bailey Family Papers, Virginia Historical Society, Richmond; Receipts of Nathaniel Ragsdale, October 12, 1855, May 5, September 2, 1856, February 25, 1859, Receipts of Ann Ragsdale, June 25, August 9, 1860, all in Ragsdale Family Papers, Virginia Historical Society; Receipt of John T. Garland, September 3, 1847, in Caswell County Historical Association Collection, Southern Historical Collection; Receipt of George Hairston, July 20, 1854, in George Hairston Papers, Southern Historical Collection; J. S. Totten's Account Book, 1832–1858, 96–97, 125, 170, in Caswell County Historical Association Collection, Southern Historical Collection; Account Book of Robert Wilson, 1848–1861, and Account Book of Robert Wilson, 1861–1865, both in Robert Wilson Account Books, Southern Historical Collection; Receipt of Marshall Hairston, August 10, 1858 (2), Hairston and Wilson Family Papers, Southern Historical Collection; Receipt of William Long, May 10, 1849, William Thomas Sutherlin Papers, Southern Historical Collection; and Receipts of William Thomas Sutherlin, June 16, December 30, 1859, William Thomas Sutherlin Papers, Southern Historical Collection.

NOTES

INTRODUCTION

1. Throughout the book the terms "bright," "yellow," and "light" are used interchangeably, as they were in the region until well into the twentieth century.

2. A note on terminology: the geographical term "Southside" is often used in Virginia history to refer to the counties of the Tidewater (or eastern) portion of the commonwealth south of the James River. The residents of Virginia and North Carolina in the district surrounding Danville, the principal town of Pittsylvania County, also referred to their immediate region as the Southside during the nineteenth century, and they continue to do so. In this book the term "Southside" serves as shorthand for the counties in question. This text also occasionally refers to the region as the "Border," another common nineteenth-century label.

3. Orville Vernon Burton, "Reaping What We Sow: Community and Rural History," *Agricultural History* 76, 4 (Autumn 2002): 634.

4. Wendell Berry, *The Unsettling of America: Culture and Agriculture,* 3rd ed. (San Francisco: Sierra Club, 1996), 31.

5. Some of the most influential or interesting general histories of American tobacco cultivation over the past century and a half include Robert L. De Coin, *History and Cultivation of Cotton and Tobacco* (London: Chapman and Hall, 1864); E. R. Billings, *Tobacco: Its History, Varieties, Culture, Manufacture, and Commerce, with an Account of Its Various Modes of Use, from Its First Discovery until Now* (Hartford: American, 1875); Paul Avery Werner, *Tobaccoland: A Book about Tobacco; Its History, Legends, Literature, Cultivation, Social and Hygienic Influences, Commercial Development, Industrial Processes*

and Governmental Regulation (New York: Tobacco Leaf, 1922); Joseph Clarke Robert, *The Tobacco Kingdom: Plantation, Market, and Factory in Virginia and North Carolina, 1800–1860* (Durham: Duke University Press, 1938); Joseph Clarke Robert, *The Story of Tobacco in America* (New York: Alfred A. Knopf, 1949); Wrightman W. Garner, *The Production of Tobacco* (New York: Blakiston, 1951); Jerome E. Brooks, *The Mighty Leaf* (Boston: Little, Brown, 1952); Pete Daniel, *Breaking the Land: The Transformation of Cotton, Tobacco, and Rice Cultures since 1880* (Urbana: University of Illinois Press, 1985); Jordan Goodman, *Tobacco in History: The Cultures of Dependence* (New York: Routledge, 1993); and Iain Gately, *Tobacco: A Cultural History of How an Exotic Plant Seduced Civilization* (New York: Grove, 2001).

6. For representative examples of the first group of histories, see Edmund S. Morgan, *American Slavery, American Freedom: The Ordeal of Colonial Virginia* (New York: W. W. Norton, 1975); Paul Clemens, *The Atlantic Economy and Colonial Maryland's Eastern Shore: From Tobacco to Grain* (Ithaca: Cornell University Press, 1980); Gloria L. Main, *Tobacco Colony: Life in Early Maryland, 1650–1720* (Princeton: Princeton University Press, 1982); T. H. Breen, *Tobacco Culture: The Mentality of the Great Planters on the Eve of the Revolution* (Princeton: Princeton University Press, 1985); Russell R. Menard, *Economy and Society in Early Colonial Maryland* (New York: Garland, 1985); Allan Kulikoff, *Tobacco and Slaves: The Development of Southern Cultures in the Chesapeake, 1680–1800* (Chapel Hill: University of North Carolina Press, 1986); Lorena S. Walsh, "Plantation Management in the Chesapeake, 1620–1820," *Journal of Economic History* 49, 2 (June 1989): 393–406; Lois Green Carr, Russell R. Menard, and Lorena Walsh, *Robert Cole's World: Agriculture and Society in Early Maryland* (Chapel Hill: University of North Carolina Press, 1991); Philip D. Morgan, *Slave Counterpoint: Black Culture in the Eighteenth-Century Chesapeake and Lowcountry* (Chapel Hill: University of North Carolina Press, 1998); Lorena S. Walsh, *Motives of Honor, Pleasure and Profit: Plantation Management in the Colonial Chesapeake, 1607–1763* (Chapel Hill: University of North Carolina Press, 2010), and for the second, see Benjamin W. Arnold, "History of the Tobacco Industry in Virginia, 1860–1894," in Herbert Adams, ed., *Johns Hopkins University Studies in Historical and Political Science* (Baltimore: Johns Hopkins University Press, 1897): 9–86; Richard B. Tennant, *The American Cigarette Industry: A Study in Economic Analysis and Public Policy* (New Haven: Yale University Press, 1971); Nannie May Tilley, *The R. J. Reynolds Tobacco Company* (Chapel Hill: University of North Carolina Press, 1985); Robert F. Durden, *Bold Entrepreneur: A Life of James B. Duke* (Durham: Carolina Academic Press, 2003); Robert R. Korstad, *Civil Rights Unionism: Tobacco Workers and the Struggle for Democracy in the Mid-Twentieth Century South* (Chapel Hill: University of North Carolina Press, 2003); Allan M. Brandt, *The Cigarette Century: The Rise, Fall, and Deadly Persistence of the Product That Defined America* (New York: Basic

Books, 2007); Peter Benson, *Tobacco Capitalism: Growers, Migrant Workers, and the Changing Face of a Global Industry* (Princeton: Princeton University Press, 2011).

7. Crandall Shifflett, *Patronage and Poverty in the Tobacco South: Louisa County, Virginia, 1860–1900* (Knoxville: University of Tennessee Press, 1982); Lynda Morgan, *Emancipation in Virginia's Tobacco Belt, 1850–1870* (Athens: University of Georgia Press, 1992); Jeffrey Kerr-Ritchie, *Freedpeople in the Tobacco South: Virginia, 1860–1900* (Chapel Hill: University of North Carolina Press, 1999); Evan Bennett, "King Bacca's Throne: Land, Life, and Labor in the Old Bright Belt since 1880" (Ph.D. diss., College of William and Mary, 2005).

8. Frederick F. Siegel, *The Roots of Southern Distinctiveness: Tobacco and Society in Danville, Virginia, 1780–1865* (Chapel Hill: University of North Carolina Press, 1987).

9. Barbara Hahn, *Making Tobacco Bright: Creating an American Commodity, 1617–1937* (Baltimore: Johns Hopkins University Press, 2011); Nannie May Tilley, *The Bright-Tobacco Industry, 1860–1929* (Chapel Hill: University of North Carolina Press, 1949). See also Hahn, "Paradox of Precision: Bright Tobacco as Technology Transfer, 1880–1937," *Agricultural History* 82, 2 (2008): 220–235.

10. Selected examples of works that to some extent treat major southern staple crops as monolithic entities include Shifflett, *Patronage and Poverty;* Bertha S. Dodge, *Cotton: The Plant That Would Be King* (Austin: University of Texas Press, 1984); Daniel, *Breaking the Land;* Goodman, *Tobacco in History;* Charles S. Aiken, *The Cotton Plantation South Since the Civil War* (Baltimore: John Hopkins University Press, 1998); Kerr-Ritchie, *Freedpeople in the Tobacco South;* Gately, *Tobacco;* and Stephen Yafa, *Cotton: The Biography of a Revolutionary Fiber* (New York: Penguin, 2006).

11. Hahn, *Making Tobacco Bright,* 223.

12. Some exceptions include Timothy Silver, *A New Face on the Countryside: Indians, Colonists, and Slaves in South Atlantic Forests, 1500–1800* (New York: Cambridge University Press, 1990); Mart Stewart, *"What Nature Suffers to Groe": Life, Labor, and Landscape on the Georgia Coast, 1680–1920* (Athens: University of Georgia Press, 1996); David S. Cecelski, *The Waterman's Song: Slavery and Freedom in Maritime North Carolina* (Chapel Hill: University of North Carolina Press, 2001); Dianne Glave and Mark Stoll, eds., *"To Love the Wind and the Rain": African Americans and Environmental History* (Pittsburgh: University of Pittsburgh Press, 2005); Scott E. Giltner, *Hunting and Fishing in the New South: Black Labor and White Leisure after the Civil War* (Baltimore: Johns Hopkins University Press, 2008); Mark Hersey, *My Work Is That of Conservation: An Environmental Biography of George Washington Carver* (Athens: University of Georgia Press, 2011); James C. Giesen, *Boll Weevil Blues: Cotton, Myth, and Power in the American South* (Chicago: University of Chicago Press, 2011).

13. On African slaves' contributions to rice culture, see Peter H. Wood, *Black Majority: Negroes in Colonial South Carolina from 1670 through the Stono Rebellion* (New York: Alfred A. Knopf, 1974); Daniel C. Littlefield, *Rice and Slaves: Ethnicity and the Slave Trade in Colonial South Carolina* (Baton Rouge: Louisiana State University Press, 1981); Judith Carney, *Black Rice: The African Origins of Rice Cultivation in the Americas* (Cambridge: Harvard University Press, 2001); Judith Carney, "Out of Africa: Colonial Rice History in the Black Atlantic," in *Colonial Botany: Science, Commerce, and Politics in the Early Modern World,* ed. Londa Schiebinger and Claudia Swan (Philadelphia: University of Pennsylvania Press, 2005): 187–203; and Judith A. Carney and Richard Nicholas Rosomoff, *In the Shadow of Slavery: Africa's Botanical Legacy in the Atlantic World* (Berkeley: University of California Press, 2009), esp. 150–154. For a challenge to these arguments, see David Eltis, Philip Morgan, and David Richardson, "Agency and Diaspora in Atlantic History: Reassessing the African Contribution to Rice Cultivation in the Americas," *American Historical Review* 112, 5 (December 2007): 1329–1358.

14. Drew Swanson, "Black Hands, Bright Leaf: African American Environmental and Agricultural Knowledge in the Making of a New Crop System," paper presented at the Graduate Association for African American History Conference, Memphis, TN, November 11, 2010. Paper in possession of the author.

15. Some examples include Ronald Davis, *Good and Faithful Labor: From Slavery to Sharecropping in the Natchez District, 1860–1890* (Westport, CT: Greenwood, 1982); Joseph P. Reidy, *From Slavery to Agrarian Capitalism in the Cotton Plantation South: Central Georgia, 1800–1880* (Chapel Hill: University of North Carolina Press, 1992); Lynda Morgan, *Emancipation in Virginia's Tobacco Belt;* Julie Saville, *The Work of Reconstruction: From Slave to Wage Laborer in South Carolina, 1860–1870* (New York: Cambridge University Press, 1994); Kerr-Ritchie, *Freedpeople in the Tobacco South;* J. William Harris, *Deep Souths: Delta, Piedmont, and Sea Island Society in the Age of Segregation* (Baltimore: Johns Hopkins University Press, 2001); and John C. Rodrigue, *Reconstruction in the Cane Fields: From Slavery to Free Labor in Louisiana's Sugar Parishes, 1862–1880* (Baton Rouge: Louisiana State University Press, 2001).

16. For a full discussion of Phillips's determinism, see Mart A. Stewart, "'Let Us Begin with the Weather': Climate, Race and Cultural Distinctiveness in the American South," in *Nature and Society in Historical Context,* ed. Mikulas Teich et al. (New York: Cambridge University Press, 1997).

17. For selected examples of calls for increased attention to the importance of agriculture in southern environmental history, see Stewart, *"What Nature Suffers to Groe";* Mart A. Stewart, "If John Muir Had Been an Agrarian: Environmental History West and South," *Environment and History* 11, 2 (2005): 139–162; Otis Graham, "Again the Backward Region? Environmental His-

tory in and of the American South," *Southern Cultures* 6, 2 (Summer 2000): 50–72; Lynn Nelson, *Pharsalia: An Environmental Biography of a Southern Plantation, 1780–1880* (Athens: University of Georgia Press, 2007); Paul S. Sutter and Christopher Manganiello, eds., *Environmental History and the American South: A Reader* (Athens: University of Georgia Press, 2009), 1–4, 18–19; Jack Temple Kirby, *Mockingbird Song: Ecological Landscapes of the South* (Chapel Hill: University of North Carolina Press, 2006); and Christopher Morris, "A More Southern Environmental History," *Journal of Southern History* 75, 3 (August 2009): 591–594.

18. Stewart, "If John Muir Had Been," 147.

19. Benjamin Cohen, *Notes from the Ground: Science, Soil, and Society in the American Countryside* (New Haven: Yale University Press, 2009), 24–25.

20. For a sensitive treatment of these sorts of peopled landscapes, see James McCann, *Maize and Grace: Africa's Encounter with a New World Crop, 1500–2000* (Cambridge: Harvard University Press, 2005).

21. Jack Temple Kirby, "Virginia's Environmental History: A Prospectus," *Virginia Magazine of History and Biography* 99, 4 (October 1991): 449.

22. Some influential interpretations over the years include: Hugh H. Bennett, *Soils and Southern Agriculture* (New York: Macmillan, 1921); Avery Craven, *Soil Exhaustion as a Factor in the Agricultural History of Virginia and Maryland, 1606–1860* (Urbana: University of Illinois Press, 1926) (though Craven does argue that the antebellum Chesapeake temporarily reversed this decline); Lewis C. Gray, *The History of Agriculture in the Southern United States to 1860*, 2 vols. (Washington, DC: Carnegie Institution, 1933); W. J. Cash, *The Mind of the South*, reprint (New York: Vintage, 1991), esp. 147–178; Clement Eaton, *A History of the Old South*, 2nd ed. (New York: Macmillan, 1966), 207–229; Eugene D. Genovese, *The Political Economy of Slavery: Studies in the Economy and Society of the Slave South*, 2nd ed. (Middletown, CT: Wesleyan University Press, 1989); Shifflett, *Patronage and Poverty*; Steven Hahn, *The Roots of Southern Populism: Yeoman Farmers and the Transformation of the Georgia Upcountry, 1850–1890* (New York: Oxford University Press, 1983); Daniel, *Breaking the Land*; Jack Temple Kirby, *Rural Worlds Lost: The American South, 1920–1960* (Baton Rouge: Louisiana State University Press, 1985). There have been a few studies that challenge the notion of southern agriculture as particularly inefficient and exploitative, including Carville Earle, "The Myth of the Southern Soil Miner: Macrohistory, Agricultural Innovation, and Environmental Change," in *The Ends of the Earth: Perspectives on Modern Environmental History*, ed. Donald Worster (New York: Cambridge University Press, 1988), 175–210; Jack Temple Kirby, "Virginia's Environmental History: A Prospectus"; Jack Temple Kirby, *Poquosin: A Study of Rural Landscape and Society* (Chapel Hill: University of North Carolina Press, 1995), 109–114; and Jack Temple Kirby, *Mockingbird Song: Ecological Landscapes of the South* (Chapel Hill: University of North Carolina Press, 2006), esp. chap. 3, though these works are disproportionately focused

on the soundness and sustainability of early swidden agriculture and free-ranging stock systems.

23. Richard White, "'Are You an Environmentalist or Do You Work for a Living?': Work and Nature," in William Cronon, ed., *Uncommon Ground: Rethinking the Human Place in Nature* (New York: W. W. Norton, 1995), 181.

24. Hahn, *Making Tobacco Bright.*

25. Perhaps the best place to start exploring this vast literature is with Brandt, *Cigarette Century.*

CHAPTER 1. ON THE BACK OF TOBACCO

1. Byrd's accounts of the survey work have been reprinted numerous times. This account draws on an early collection of Byrd's Southside writings: William Byrd, *The Westover Manuscripts: Containing the History of the Dividing Line betwixt Virginia and North Carolina; a Journey to the Land of Eden, A.D. 1733; and a Progress to the Mines,* ed. Edmund Ruffin (Petersburg, VA: Edmund and Julian C. Ruffin, 1841). For a history of the account's drafting and publication, see Pierre Marambaud, "William Byrd of Westover: Cavalier, Diarist, and Chronicler," *Virginia Magazine of History and Biography* 78, 2 (April 1970): 144–183. For an engaging modern tour of the line that explores the regional environment in comparison to Byrd's day, see Stephen Conrad Ausband, *Byrd's Line: A Natural History* (Charlottesville: University of Virginia Press, 2002).

2. Byrd, *Westover Manuscripts,* 39–84, passim.

3. Ibid., 75.

4. Ibid., 75–76.

5. Ibid., 2.

6. Maud Carter Clement, *The History of Pittsylvania County, Virginia* (Lynchburg, VA: J. P. Bell, 1929), 37. The naturalization clause was likely directed at William Byrd's plan to settle Swiss emigrants on his extensive Eden tract in present-day southern Halifax County, but it might also have been influenced by the growing number of German settlers moving from Pennsylvania to Virginia.

7. Ibid., 35–36, and Allan Kulikoff, *Tobacco and Slaves: The Development of Southern Cultures in the Chesapeake, 1680–1800* (Chapel Hill: University of North Carolina Press, 1986), 141–144.

8. Beeman, *Evolution of the Southern Backcountry,* 21–23; and Clement, *History of Pittsylvania,* 43, 46–47.

9. J. F. D. Smyth [John Ferdinand Smyth Stuart], *A Tour in the United States of America: Containing an Account of the Present Situation of That Country* (London: G. Robinson, J. Robson, and J. Sewell, 1784), 236.

10. Jessie Wilson, "Genealogy of the Wilson Family of Pittsylvania County, Virginia," unpublished typescript, Robert Hairston Papers, *Records of Antebellum Southern Plantations* (microfilm), Kenneth M. Stampp, general ed. (Frederick, MD: University Publications of America, 1987) (hereafter cited

as *RASP*), series J, part 11, reel, 9. The practice of early settlers monopolizing backcountry bottomland through surveying "water-hogging" tracts was a common one throughout Virginia. See James D. Rice, *Nature and History in the Potomac Country: From Hunter-Gatherers to the Age of Jefferson* (Baltimore: Johns Hopkins University Press, 2009), 222–224. A particularly illustrative regional example can be found in the plats of tracts settled along the junction of the Stinking and Bannister rivers in northeastern Pittsylvania County. See Land Plat of John Hawkins, November 8, 1751, Crenshaw and Miller Family Papers, folder 1, Southern Historical Collection (hereafter cited as SHC), Louis Round Wilson Library, University of North Carolina at Chapel Hill; and Land Survey of Hawkins, Crenshaw, Roberts, and Smith Tracts, n.d., Crenshaw and Miller Family Papers, folder 5, SHC.

11. Maud Carter Clement, *Frontiers along the Upper Roanoke River, 1740–1776: A Story of First Settlement* (Lynchburg, VA: J. P. Bell, 1964), 7–9, 13–18, 40–50; and William Walter Hoest, "The Plantation in a Regional Economy: Pocket Plantation, 1762–1785" (master's thesis, University of Virginia, 1977), 6–8.

12. William S. Powell, *When the Past Refused to Die: A History of Caswell County North Carolina, 1777–1977* (Durham, NC: Moore, 1977), 37–38; and Smyth, *Tour in the United States*, 243–249. Such a high proportion of early settlers lived along watercourses that Halifax's first tax officials in the 1750s created their lists of tithables by simply surveying up each river and creek; see Wirt Johnson Carrington, *A History of Halifax County, Virginia* (Baltimore: Regional, 1969), 21; and Pocahontas Wight Edmunds, *History of Halifax:* vol. 1, *Narration* (n.p., n.d.), 5–6.

13. Michael B. Barber and Eugene B. Barfield, "Native Americans on the Virginia Frontier in the Seventeenth Century: Archeological Investigations along the Interior Roanoke River Drainage," in *Diversity and Accommodation: Essays on the Cultural Composition of the Virginia Frontier*, ed. Michael J. Puglisi (Knoxville: University of Tennessee Press, 1997), 138, 140; and Frances Hallam Hurt, *An Intimate History of the American Revolution in Pittsylvania County, Virginia* (Danville, VA: Womack, 1976), 44–45.

14. Lorena S. Walsh, "Plantation Management in the Chesapeake, 1620–1820," *Journal of Economic History* 49, 2 (June 1989): 396–401; Paul G. E. Clemens, *The Atlantic Economy and Colonial Maryland's Eastern Shore: From Tobacco to Grain* (Ithaca: Cornell University Press, 1980), 22–23, 168–205; and Kulikoff, *Tobacco and Slaves*, 120.

15. Philip D. Morgan, *Slave Counterpoint: Black Culture in the Eighteenth-Century Chesapeake and Lowcountry* (Chapel Hill: University of North Carolina Press, 1998), 170.

16. Kulikoff, *Tobacco and Slaves*, 120–121. For a similar observation on the prevalence of tobacco culture on a majority of Tidewater farms throughout the colonial period, see Carr, Menard, and Walsh, *Robert Cole's World*, 70–71.

17. Paul S. Sutter, "What Gullies Mean: Georgia's 'Little Grand Canyon' and Southern Environmental History," *Journal of Southern History* 76, 3 (August

2010): 606–610, quote on 606; Laurence C. Walker, *The Southern Forest: A Chronicle* (Austin: University of Texas Press, 1991), 34–36, 56–57; Laurence C. Walker, *The North American Forests: Geography, Ecology, and Silviculture* (Boca Raton, FL: CRC, 1999), 96.

18. N. M. Kirk et al., *Soil Survey of Pittsylvania County, Virginia* (Washington, DC: Government Printing Office, 1922), 6–7, 11; John C. Nicholson, *Soil Survey of Pittsylvania County and the City of Danville, Virginia* (Washington, DC: Government Printing Office, 1994), 13–23; W. Edward Hearn and Frank P. Drane, *Soil Survey of Caswell County, North Carolina* (Washington, DC: Government Printing Office, 1994), inserted soil map; R. C. Jurney et al., *United States Department of Agriculture Soil Survey of Halifax County, Virginia* (Washington, DC: Government Printing Office, 1938), 9, 40; Wrightman W. Garner, *The Production of Tobacco* (New York: Blakiston, 1951), 89; E. M. Rowalt, *Soil Defense in the Piedmont*, USDA Farmers' Bulletin No. 1767 (Washington, DC: Government Printing Office, 1937), 2–4; Walker, *Southern Forest*, 56–57. For descriptions of Southside soil series, see the entries for Appling, Cecil, and Durham sandy loams at the United States Department of Agriculture Natural Resources Conservation Service's website, http://soils.usda.gov/technical/classification/osd/index.html.

19. Bartram in *William Bartram, the Search for Nature's Design: Selected Art, Letters, and Unpublished Writings*, ed. Thomas Hallock and Nancy E. Hoffman, with Joel T. Fry, associate ed. (Athens: University of Georgia Press, 2010), 177–178.

20. Ibid., 178; Byrd, *Westover Manuscripts*, 44–63; Ausband, *Byrd's Line*, chap. 3; Walker, *North American Forests*, 96, 163–164; Timothy Silver, *A New Face on the Countryside: Indians, Colonists, and Slaves in South Atlantic Forests, 1500–1800* (Cambridge: Cambridge University Press, 1990), 19–22.

21. Siegel, *Roots of Southern Distinctiveness*, 12–13; Charles J. Farmer, *In the Absence of Towns, Settlement and Country Trade in Southside Virginia, 1730–1800* (Lanham, MD: Rowman and Littlefield, 1993), 29, 96; Beeman, *Evolution of the Southern Backcountry*, 24–25.

22. Byrd, *History of the Dividing Line*, 90–100, 104–105, 119; Byrd, *Journey*, 288, 293–299.

23. William Byrd, *William Byrd's Natural History of Virginia; or, the Newly Discovered Eden*, ed. Richard Croom Beatty and William Mulloy (Richmond, VA: Dietz, 1940), 51.

24. Byrd, *History of the Dividing Line*, 90–100, 104–105, 119; and Byrd, *Journey*, 288, 293–299.

25. Byrd, *History of the Dividing Line*, 101.

26. Byrd, *William Byrd's Natural History*, quote on 9.

27. John Lederer, *The Discoveries of John Lederer, in Three Several Marches from Virginia to the West of Carolina, and Other Parts of the Continent* (London: Samuel Heyrick, 1672; reprint Rochester, NY, 1902), 15–16.

28. Ibid., 13, 15; Michael B. Barber and Eugene B. Barfield, "Native Americans on the Virginia Frontier in the Seventeenth Century: Archeological Investigations along the Interior Roanoke River Drainage," in *Diversity and Accommodation: Essays on the Cultural Composition of the Virginia Frontier,* ed. Michael J. Puglisi (Knoxville: University of Tennessee Press, 1997), 138–140; and Forest Hazel, "Occaneechi-Saponi Descendants in the North Carolina Piedmont: The Texas Community," *Southern Indian Studies* 40 (October 1991): 6–9. On these buffer zones and their influence on nature and landscape in another portion of the South, see Richard White, *The Roots of Dependency: Subsistence, Environment, and Social Change among the Choctaw, Pawnees, and Navajos* (Lincoln: University of Nebraska Press, 1983).

29. Byrd, *History of the Dividing Line,* 113.

30. Byrd, *Journey,* 277, 288, 297–298; Byrd, *History of the Dividing Line,* 101–102. For the importance of river cane in the Piedmont and the mountains and the plant's adaptation to periodic disturbance, see Mart A. Stewart, "From King Cane to King Cotton: Razing Cane in the Old South," *Environmental History* 12, 1 (January 2007): 59–79; Silver, *New Face on the Countryside,* 22; Donald Edward David, *Where There Are Mountains: An Environmental History of the Southern Appalachians* (Athens: University of Georgia Press, 2000), 71–73; and Albert E. Radford, Harry E. Ahles, and C. Ritchie Bell, *Manual of the Vascular Flora of the Carolinas* (Chapel Hill: University of North Carolina Press, 1968), 60.

31. Byrd, *Journey,* 270, 273, quote on 273.

32. Beeman, *Evolution of the Southern Backcountry,* 33, 67; Siegel, *Roots of Southern Distinctiveness,* 16–21.

33. Compiled from Marian Dodson Chiarito, *Will Book 0, 1752–1773, Halifax County, Virginia* (Nathalie, VA: Clarkton, 1982), 1–60; and Hoest, "Plantation in a Regional Economy," 8, 33. For Edward Booker's will, see Chiarito, page 41, and for Paul Chiles's will and estate inventory, see pages 16 and 37. The percentage of slave ownership among recorded Halifax wills was probably higher than indicated as a number of wills failed to enumerate property, leaving the "whole estate both real & personal" to a single individual.

34. Jacob M. Price, "The Rise of Glasgow in the Chesapeake Tobacco Trade, 1707–1775," *William and Mary Quarterly* 11, 2 (April 1954): 179–199; Gately, *Tobacco,* 80–105; Goodman, *Tobacco in History,* 59–75; Jacob M. Price, *Tobacco Atlantic Trade: The Chesapeake, London, and Glasgow, 1675–1775* (Brookfield, VT: Variorum, 1995); and Eric Burns, *Smoke of the Gods: A Social History of Tobacco* (Philadelphia: Temple University Press, 2007), 28–35, 93.

35. Christopher E. Hendricks, *The Backcountry Towns of Colonial Virginia* (Knoxville: University of Tennessee Press, 2006), 64–69. The proposed town of Eden should not be confused with Byrd's large tract of land by the same name farther upstream in North Carolina. For Byrd's glowing advertisement of Eden, see Byrd, *William Byrd's Natural History.* Byrd's Eden

speculation came as he was planning to develop two other Virginia town sites, on the falls of the James and Appomattox rivers; these settlements would become the cities of Richmond and Petersburg. See Pierre Marambaud, "William Byrd of Westover: Cavalier, Diarist, and Chronicler," *Virginia Magazine of History and Biography* 78, 2 (April 1970): 145.

36. Hendricks, *Backcountry Towns*, 76–81.

37. Carville Earle and Ronald Hoffman, "Urban Development in the Eighteenth-Century South," *Perspectives in American History* 10 (1976): 5–78. Historical treatments of the South's rural nature are numerous, but two additional engaging works that take the connection between Virginia landscapes and the social structure of the staple-producing countryside seriously are Rhys Isaac, *The Transformation of Virginia, 1740–1790* (Chapel Hill: University of North Carolina Press, 1982), and Beeman, *Evolution of the Southern Backcountry*. For an earlier, and differing, account of the rural nature of the plantation South, see Eugene D. Genovese, *The World the Slaveholders Made: Two Essays in Interpretation* (New York: Pantheon, 1969), part 2, chap. 3.

38. Farmer, *In the Absence of Towns*, 107–109; Hendricks, *Backcountry Towns*, 83; Joseph T. Rainer, " 'Commercial Scythians' in the Great Valley of Virginia: Yankee Peddlers' Trade Connections to Antebellum Virginia," in *After the Backcountry: Rural Life in the Great Valley of Virginia, 1800–1900*, ed. Kenneth E. Koons and Warren R. Hofstra (Knoxville: University of Tennessee Press, 2000), 62–73.

39. For examples, see John Noble's Daybook, 1798–1827, entries for 1798, and John Noble's Ledger, 1803–1804, entries throughout, both in Wyllie Family Papers, Virginia Historical Society, Richmond, VA (hereafter VHS); and Diary of unidentified peddler, October 30, 1807–January 22, 1808, VHS. For a full treatment of merchants and early backcountry consumers, see Ann Smart Martin, *Buying into the World of Goods: Early Consumers in Backcountry Virginia* (Baltimore: Johns Hopkins University Press, 2008).

40. Archibald Glen to D. Buchanan, August 30, 1798; James A. Glenn to Isabella Glenn, October 28, 1798; and James A. Glenn to Isabella Glenn, June 17, 1799, all in Caswell County Historical Association Collection, folder 8, SHC. Glenn operated a country store in Halifax County and carried hogsheads from the Southside to Petersburg, where he bought trade goods for the return wagon trip.

41. *Fifth Census; Enumeration of the Inhabitants of the United States—1830* (Washington, DC: Duff Green, 1832), 10–13, 84–85, 90–91; *Compendium of the Enumeration of the Inhabitants and Statistics of the United States—Sixth Census* (Washington, DC: Thomas Allen, 1841), 34–35, 42–43; *The Seventh Census of the United States: 1850* (Washington, DC: Robert Armstrong, 1853), 259–260, 307; *Population of the United States in 1860; Compiled from the Original Returns of the Eighth Census* (Washington, DC: Government Printing Office, 1864), 358, 516–517; and Siegel, *Roots of Southern Distinctiveness*, 16.

42. John Goodall Bruce, *The Bruce Family: Descending from George Bruce (1650–1715)* (Parsons, WV: McClain, 1977), 87–89, 92; "Bruce Family," *Virginia Magazine of History and Biography* 11, 3 (January 1904): 330; Edmunds, *History of Halifax*, 247–248; Carrington, *History of Halifax*, 120–127; and Kathleen Bruce, "Materials for Southern Virginia Agricultural History," *Agricultural History* 4, 1 (January 1930): 12–13. In-depth material on the Bruce family's wealth and activities can be found in the Bruce Family Papers, Small Special Collections Library at the University of Virginia (hereafter UVA), or on microfilm in *RASP*, series E, part 3, reels 7–30.

43. "The Bruce Family (Continued)," *Virginia Magazine of History and Biography* 11, 4 (April 1904): 441.

44. Henry Wiencek, *The Hairstons: An American Family in Black and White* (New York: St. Martin's, 1999), 8–9, 49.

45. Cosmopolite, "Richest Man in Virginia," *Richmond Whig and Public Advertiser*, February 7, 1851, p. 1. This article was reprinted for a wider circulation in the New Orleans journal *De Bow's Review* in 1853 under the same title. See Cosmopolite, "The Richest Man in Virginia," *De Bow's Review* 14, 1 (January 1853): 70–71.

46. Ulrich B. Phillips, *American Negro Slavery: A Survey of the Supply, Employment and Control of Negro Labor as Determined by the Plantation Regime* (New York: D. Appleton, 1918), 232–233, quote on 232. Cosmopolite credited Hairston with ownership of between 1,600 and 1,700 slaves and with management of 1,000 more. Estimating Hairston's slaveholding proves extremely difficult as his plantations were spread across a number of counties in at least two states, and for most of the antebellum period, Samuel managed hundreds of slaves technically owned by his elderly mother, Ruth. Based on several searches of the census slave schedules, it seems safe to state that Hairston owned at least 1,000 slaves in Virginia and North Carolina during the 1840s and 1850s, and perhaps as many as twice that number, and managed several hundred more. Even the lower end of this range would make Hairston perhaps the largest slave owner in the South at the outbreak of the war. In his study of Deep South planters on the eve of the Civil War, Joseph Menn found John Burnside of Louisiana to be the largest sugar planter, with 940 slaves; Levin Marshall of Louisiana and Mississippi the largest cotton planter, with 932 slaves; and Stephen King of Georgia the largest rice planter, with 582 slaves. Menn's study did not include South Carolina. See Joseph Karl Menn, "The Large Slaveholders of the Deep South, 1860," 2 vols. (Ph.D. diss., University of Texas, 1964), 1:233–234.

47. Hairston Family Plantation Record Books 1 and 3, Peter Wilson Hairston Papers, box 8, folders 104 and 106, SHC. Ruth's holdings, managed primarily by Samuel, included twelve plantations in Henry, Patrick, and Pittsylvania counties in Virginia and in Stokes and Davie counties North Carolina.

48. Compiled from the U.S. Census Bureau Slave Schedule, Henry County, Virginia, 1860; and the Historical Census Data Browser (hereafter cited as

HCDB), available at http://fisher.lib.virginia.edu/collections/stats/histcen-sus/index.html.

49. Compiled from the U.S. Census Bureau Slave Schedules for the northern and southern districts of Pittsylvania and Halifax counties, and for Caswell County, 1860, and the HCDB.

50. Wiencek, *Hairstons*, 60–61, 91–92.

51. Edmunds, *History of Halifax*, 247–248; Carrington, *History of Halifax County*, 120–126; and Bruce, "Materials for Southern Virginia," 12. Extensive evidence of Bruce's milling business at Meadsville, Virginia, survive in a number of mill account books and ledgers found in *RASP*, series E, part 3, reels 24 and 25.

52. When a farmer began his plant beds seemed a matter of personal preference to some extent. Charles Coleman of Clover, in Halifax County, began a plant bed as early as mid-December, whereas his neighbor William Sims prepared his beds in late January. In his influential treatise on raising tobacco, Peter Minor advised that plant beds could be created as late as March in a pinch. Charles Coleman Diaries, December 21, 1842, *RASP*, series D, reel 12 (hereafter cited as Coleman Diaries); William H. Sims Farm Book, January 18, 27, 1855, *RASP*, series E, part 1, reel 2 (hereafter cited as Sims Farm Book); Peter Minor, *The Cultivation and Management of Tobacco, from Plant Bed to the Prize* (Baltimore: J. Robinson, 1822), 3–4.

53. Minor, *Cultivation and Management of Tobacco*, 3–4. Although most plant beds were small, planters often made multiple beds, and sometimes they raised many more plants than necessary to provide insurance against the failure of the first planting or to share with neighbors. In 1845, George Jeffreys of Caswell sowed four plant beds that covered 2,400 square yards. See Diary of George Washington Jeffreys, February 8, 1845, William Bethell Williamson Papers, box 1, folder 7, SHC (hereafter cited as Jeffreys Diary).

54. Samuel C. Shelton, "The Culture and Management of Tobacco," *Southern Planter* 21, 4 (April 1861): 210–211; William Tatham, *An Historical and Practical Essay on the Culture and Commerce of Tobacco* (London: Vernor and Hood, 1800), 7–9; Coleman Diaries, December 23, 1842; and Sims Farm Book, January 27, 1855.

55. W. W. W. Bowie, "An Essay of the Culture and Management of Tobacco," *American Farmer* 10, 2 (August 1854): 33; Shelton, "Culture and Management of Tobacco," 211; Minor, *Cultivation and Management of Tobacco*, 3–4; and Jeffreys Diary, December 20, 1844, and January 15, 1845. For the small size of tobacco seeds, see Anne Radford Phillips, "Farm Women of Stokes County, North Carolina, and the Production of Flue-Cured Tobacco, 1925 to 1955: Continuity and Change" (Ph.D. diss., University of Maryland, 1990), 55, n. 39.

56. "Cultivation of Tobacco," *American Farmer, and Spirit of the Agricultural Journals of the Day* 2, 51 (May 12, 1841): 406; Bowie, "Essay on the Culture and Management of Tobacco," 34; Shelton, "Culture and Management of

Tobacco," 211; and Coleman Diaries, January 21, 1843. Most of these steps in plant bed preparation and maintenance went back to at least the mid-1600s—see Carr, Menard, and Walsh, *Robert Cole's World*, 55–56.

57. Like making plant beds, transplanting time was flexible within certain limits. Charles Coleman tended to start his planting during the last week of May, William Sims began his 1855 planting on June 1, and Vincent Shelton, Jr., of Pittsylvania did not finish his 1850 planting until July 16. The limiting factor was days remaining until frost—farmers wanted to ensure their crop had time to ripen fully before cold weather set in. Tobacco was tremendously susceptible to cold; even a light frost could result in "the destruction of every plant" (Tatham, *Historical and Practical Essay*, 23). Coleman Diaries, May 27, 1843, May 24, 1844, May 24, 1845, May 23, 1847, May 15, 1848; Sims Farm Book, June 1, 1855; and R. D. Ricketts, ed., *A Diary Kept by William C. Shelton for the Year 1850: The Daily Journal of a School Master and Farmer of Pittsylvania County, Virginia* (Danville, VA: Virginia–North Carolina Piedmont Genealogical Society, 1995), 50.

58. "Culture and Curing of Tobacco in Virginia: A Short History of the Types of Tobacco Produced in Virginia and Maryland," in Francis Walker and Charles Seaton, *Report of the Productions of Agriculture as Returned at the Tenth Census* (Washington, DC: Government Printing Office, 1883), 203; Shelton, "Culture and Management of Tobacco," 212; and Tatham, *Historical and Practical Essay*, quote on 6–7.

59. Bowie, "Essay on the Culture and Management of Tobacco," 34; Minor, *Cultivation and Management of Tobacco*, 5–6; "Culture and Curing of Tobacco," 203–204; and Jeffreys Diary, May 5, 1845.

60. "Culture and Curing of Tobacco," 204; and Bowie, "Essay on the Culture and Management of Tobacco," 34. Despite the stereotype of southern farms as bastions of mule agriculture, in the Southside during the antebellum period horses seemed much more common as work animals. For example, in Pittsylvania, Halifax, and Caswell in 1860, horses outnumbered mules and asses an astonishing 10,479 to 2,458. Even oxen were a more common sight than mules, with 4,336 working the counties' fields in 1860. In the example of just one Halifax County plantation, in 1852 James C. Bruce's Berry Hill had 10 oxen and 29 horses (16 of which Bruce classified as working animals) but no mules. Aggregate figures compiled from Joseph C. G. Kennedy, *Agriculture of the United States in 1860; Compiled from the Original Returns of the Eighth Census* (Washington, DC: Government Printing Office, 1864), 104, 154, 158. Information on Berry Hill comes from James C. Bruce's Plantation Inventory, 1852, Bruce Family Papers, *RASP*, series E, part 3, reel 16.

61. Sims Farm Book, July 2, 1855.

62. This term had old roots, going back to the seventeenth-century Chesapeake. See Carr, Menard, and Walsh, *Robert Cole's World*, 58.

63. "Farmers' Meeting at the Exchange Hotel," *Southern Planter* 17, 4 (April 1857): 226; "Tobacco," *American Farmer and Rural Register* 1, 7 (July 1872): 226;

Minor, *Cultivation and Management of Tobacco*, 6; Shelton, "Culture and Management of Tobacco," 213, and "Cultivation of Tobacco," 406.

64. Coleman Diaries, August 14, 1843. Recording the ubiquity of tobacco cultivation in Virginia, William Tatham noted in regard to topping and suckering: "Many of the Virginians let the thumb nail grow long, and harden it *in the candle*, for this purpose: not for the use of *gouging* out people's eyes, as some have thought fit to insinuate." Tatham, *Historical and Practical Essay*, 18.

65. Coleman Diaries, August 10, 1848. Both of these alternative suggestions come from Minor, *Cultivation and Management of Tobacco*, 7.

66. Charles L. Perdue, Jr., Thomas E. Barden, and Robert K. Phillips, *Weevils in the Wheat: Interviews with Virginia Ex-Slaves* (Charlottesville: University of Virginia Press, 1976), 322. Charles Coleman also recorded putting his "small hands" to work worming (Coleman Diaries, August 20, 1843).

67. Perdue, Barden, and Phillips, *Weevils in the Wheat*, 148.

68. Robert D. Key, "On Firing Tobacco through Flues, the Benefits of Cutting When Fully Ripe," *Farmer's Register* 4, 1 (May 1836): 42; A Plaster, "On Curing Tobacco," *Farmer's Register* 4, 5 (September 1836): 313; Caswell, "On Curing Yellow Tobacco," *Southern Planter* 19, 8 (August 1859): 492; and Tatham, *Historical and Practical Essay*, quote on 23.

69. Ricketts, *Diary Kept by William C. Shelton*, 72.

70. For the practice of "sunning" tobacco for a time before moving the crop to a barn, see Jeffreys Diary, October 3, 1844.

71. Minor, *Cultivation and Management of Tobacco*, 9–10; Bowie, "Essay on the Culture and Management of Tobacco," 35; A Plaster, "On Curing Tobacco," 313; Tatham, *Historical and Practical Essay*, 37–38; and John Fraser Hart and Eugene Cotton Mather, "The Character of Tobacco Barns and Their Role in the Tobacco Economy of the United States," *Annals of the Association of American Geographers* 51, 3 (September 1961): 275.

72. Perdue, Barden, and Phillips, *Weevils in the Wheat*, 148; Ruth Little-Stokes, *An Inventory of Historic Architecture, Caswell County, North Carolina: The Built Environment of a Burley and Bright-Leaf Tobacco Economy* (Yanceyville, NC: Caswell County Historical Association, 1979), 20, 106; Catherine W. Bishir and Michael T. Southern, *A Guide to the Historic Architecture of Piedmont North Carolina*, Richard Hampton Jenrette Series in Architecture and the Decorative Arts (Chapel Hill: University of North Carolina Press, 2003), 182–183. Billings, *Tobacco*, 408–409; Tatham, *Historical and Practical Essay*, 29–34; and Hart and Mather, "Character of Tobacco Barns," 274, 288–293. Hart and Mather provide a particularly good examination of log barns. Their study focuses on flue-cured barns, but the basic form remained the same as during the fire-cured era.

73. "Letter from the Courthouse," *Danville Register*, October 5, 1870, p. 2.

74. Minor, *Cultivation and Management of Tobacco*, 12; Coleman Diaries, January 24, 1844; and Jeffreys Diary, January 6, May 24, 1845.

75. Tatham, *Historical and Practical Essay*, 47–48; Minor, *Cultivation and Management of Tobacco*, 12–13. In 1800, legislation had restricted hogsheads to fifty-four inches in height.

76. In just a few examples of the increasing size of hogsheads, in 1841 Elisha Barksdale of Halifax sold a lot of hogsheads that averaged 1,700 pounds; in 1847 George Clement of Pittsylvania was selling hogsheads weighing as much as 1,820 pounds, and in 1841 James C. Bruce of Halifax sold several hogsheads weighing 1,960 pounds apiece. Carrington, Gibson, and Thomson to Elisha Barksdale, September 14, 1841, in the Peter Barksdale Papers, *RASP*, series F, part 3, reel 33; Receipt of George Clement, August 5, 1847, Pocket Plantation Papers, *RASP*, series E, part 1, reel 14; and Receipt of James C. Bruce, June 16, 1841, Bruce Family Papers, *RASP*, series E, part 3, reel 12.

77. Robert, *Tobacco Kingdom*, 48.

78. Danville Tobacco Association, *100 Years of Progress, 1869–1969* (Danville, VA: Womack, 1969), 16; John Richard Dennett, *The South as It Is, 1865–1866*, ed. Henry M. Christman (Athens: University of Georgia Press, 1986), 88–89, 94; Tatham, *Historical and Practical Essay*, 55–66; Hoest, "Plantation in a Regional Economy," 12–15; Robert, *Tobacco Kingdom*, 53–54; and Beeman, *Evolution of the Southern Backcountry*, 35–37.

79. Danville Tobacco Association, *100 Years of Progress*, 17–18; Clement, *Frontiers along the Upper Roanoke River*, 34; Hagan, *Story of Danville*, 28, 52–52; Tatham, *Historical and Practical Essay*, 63–66; Beeman, *Evolution of the Southern Backcountry*, 37; Robert, *Tobacco Kingdom*, 55–57, 63; Seigel, *Roots of Southern Distinctiveness*, 12–13; Jeffreys Diary, September 5, 1844; Receipts of James C. Bruce, January 5, 1843, November 16, 1848, Bruce Family Papers, *RASP*, series E, part 3, reels 12 and 14; Cecelski, *Waterman's Song*, 43–46; Account Book of Moses E. Gilliam, 1842–1843, 20–90, Spragins Family Papers, VHS; and Tobacco Shipping Ledger, 1817–1825, Bruce Family Papers, *RASP*, series E, part 3, reel 24. Gilliam's warehouse also stored lesser quantities of manufactured tobacco, flour, and cotton.

80. Robert, *Tobacco Kingdom*, 64–65; and Aaron W. Marrs, *Railroads in the Old South: Pursuing Progress in a Slave Society* (Baltimore: Johns Hopkins University Press, 2009), xv, map 1.

81. Henry Howe, *Historical Collections of Virginia; Containing a Collection of the Most Interesting Facts, Traditions, Biographical Sketches, Anecdotes, &c. Relating to Its History and Antiquities, Together with Geographical and Statistical Descriptions* (Charleston, SC: Babcock, 1845), 290, 429.

82. Daily Journal of James Bruce, 1838–1841, entries throughout, Bruce Family Papers, *RASP*, series E, part 3, reel 22. For similar accounts of small game hunting and the composition of game in the region, see Daniel Tatum Merritt's Diary, December 17, 1860, February 6, 1863, VHS (hereafter cited as Merritt Diary); Ricketts, *Diary Kept by William C. Shelton*, 1, 14; Robert Enoch Withers, *Autobiography of an Octogenarian* (Roanoke, VA: Stone, 1907), 92–93; and Jeffreys Diary, 1844–1845, entries throughout.

83. Jeffreys Diary, October 2, 1844.
84. Withers, *Autobiography of an Octogenarian*, 89–91; and Henry Goings, *Rambles of a Runaway from Southern Slavery* (Stratford, Canada: J. M. Robb, 1869), 9.
85. Merritt Diary, March 8, 1858.
86. Siegel, *Roots of Southern Distinctiveness*, 12–13, 165; Jack Temple Kirby, *Poquosin: A Study of Rural Landscape and Society* (Chapel Hill: University of North Carolina Press, 1995), 3–4; Farmer, *In the Absence of Towns*, 29, 96; and Beeman, *Evolution of the Southern Backcountry*, 24–25.
87. Warren R. Hofstra, *The Planting of New Virginia: Settlement and Landscape in the Shenandoah Valley* (Baltimore: John Hopkins University Press, 2004), 277–278; and Warren R. Hofstra, "Land Policy and Settlement in the Northern Shenandoah Valley," in *Appalachian Frontiers: Settlement, Society, and Development in the Preindustrial Era*, ed. Robert D. Mitchell (Lexington: University Press of Kentucky, 1991), 121–123.

CHAPTER 2. LET THERE BE BRIGHT

1. Tilley, *Bright-Tobacco Industry*, 24–26.
2. In addition to Tilley's *Bright-Tobacco Industry*, works with a version of the Slade story (more or less) include: Robert, *Tobacco Kingdom;* Jane Gray Hagan, *The Story of Danville* (New York: Stratford House, 1950), 51–52; Cornelius Oliver Cathey, *Agricultural Developments in North Carolina, 1783–1860* (Chapel Hill: University of North Carolina Press, 1956), 125; S. Huntington Hobbs, Jr., *North Carolina: An Economic and Social Profile* (Chapel Hill: University of North Carolina Press, 1958), 108; Clement Eaton, *A History of the Old South* (New York: Macmillan, 1966), 214; Joseph Clarke Robert, *The Story of Tobacco in America*, 2nd ed. (Chapel Hill: University of North Carolina Press, 1967), 183; Danville Tobacco Association, *100 Years of Progress*, 15, 55; William S. Powell, *When the Past Refused to Die: A History of Caswell County North Carolina, 1777–1977* (Durham, NC: Moore, 1977), 469–470; S. N. Hawks, Jr., *Principles of Flue-Cured Tobacco Production*, 2nd ed. (Raleigh: North Carolina State University, 1978), 5; Little-Stokes, *Inventory of Historic Architecture*, 3; Siegel, *Roots of Southern Distinctiveness*, 102; William S. Powell, *North Carolina through Four Centuries* (Chapel Hill: University of North Carolina Press, 1989), 310–311; Robert C. McAdams, "The Tobacco Culture of Wilson County, North Carolina" (Ph.D. diss., University of Tennessee, 1996), 43; Eldred E. Prince, Jr., and Robert R. Simpson, *Long Green: The Rise and Fall of Tobacco in South Carolina* (Athens: University of Georgia Press, 2000), 47; and Adrienne M. Petty, "Standing Their Ground: Small Farm Owners in North Carolina's Tobacco Belt, 1920–1982" (Ph.D. diss., Columbia University, 2004), 19. Drawing on one of Tilley's early oral presentations of her research, Robert included a synopsis of the Slade story in a footnote in *The Tobacco Kingdom* prior to Tilley's publication. A notable exception is Cathey, *Agricultural Developments*, 125. Cathey mentions the

Slade brothers as the progenitors of bright tobacco and makes no mention of Stephen Slade, though he does date their activities to 1839. Cathey, in fact, does not include Tilley's work in his bibliography.

3. Examples include Robert L. Ragland, "A Scrap from the History of Yellow Tobacco," *Southern Planter and Farmer* 37, 5 (May 1876): 344; J. B. Cameron, *A Sketch of the Tobacco Interests in North Carolina* (Oxford, NC: W. A. Davis, 1881), 9–10; Arnold, *History of the Tobacco Industry*, 32; J. B. Killebrew and Herbert Myrick, *Tobacco Leaf: Its Culture and Cure, Marketing and Manufacture, a Practical Handbook on the Most Approved Methods in Growing, Harvesting, Curing, Packing and Selling Tobacco, Also of Tobacco Manufacture*, 2nd ed. (New York: Orange Judd, 1909), 10–11; Carl Avery Werner, *Tobaccoland: A Book about Tobacco; Its History, Legends, Literature, Cultivation, Social and Hygienic Influences, Commercial Development, Industrial Processes and Governmental Regulation* (New York: Tobacco Leaf, 1922), 42; Lewis Cecil Gray, *History of Agriculture in the Southern United States to 1860*, 2 vols. (Washington, DC: Carnegie Institution, 1933), 2:757; and Franklin C. Erickson, "The Tobacco Belt of North Carolina," *Economic Geography* 21, 1 (January 1945): 58. For Abisha's brothers, see Tilley, *Bright-Tobacco Industry*, 23.

4. Hahn, *Making Tobacco Bright*, 13–14, 138.

5. "Bright Tobacco, An Old Negro the First to Cure It," *Progressive Farmer* 1, 10 (April 14, 1886): 4. Ironically enough, "Captain John Lee" was almost certainly John G. Lea, a prominent tobacconist who spearheaded a campaign of racial violence in Reconstruction Caswell as head of the local Ku Klux Klan. See chapter 5.

6. Ibid. Replacement of the name Abisha with Elisha was a common error in a number of subsequent histories. According to Tilley, there was no one by the name of Elisha Slade in Caswell County at the time. She speculates that Elisha is a hybrid of Abisha's name and that of his brother Elias, though it seems just as likely that contemporaries mistook the rather unique name Abisha for the more common given name (Tilley, *Bright-Tobacco Industry*, 23).

7. The classic case here is Lowcountry tidal rice culture, which geographer Judith Carney argues was based to a certain extent on black contributions rooted in their environmental knowledge and past agricultural experience. See Carney, *Black Rice*. Daniel Littlefield's *Rice and Slaves: Ethnicity and the Slave Trade in Colonial South Carolina* (Baton Rouge: Louisiana State University Press, 1981) and Peter Wood's *Black Majority: Negroes in Colonial South Carolina from 1670 through the Stono Rebellion* (New York: Alfred A. Knopf, 1974) both make a similar argument. My research into a coastal Georgia plantation suggests that a similar process may have taken place in sea island cotton culture as well. See Swanson, *Remaking Wormsloe Plantation: The Environmental History of a Lowcountry Landscape* (Athens: University of Georgia Press, 2012), chap. 2.

8. Hahn has also made this last point in *Making Tobacco Bright*, 138.

9. David E. Nye, *America as Second Creation: Technology and Narratives of New Beginnings* (Cambridge: MIT Press, 2003), 10–11. I would like to thank Shane Hamilton for bringing Nye's analysis of creation myths to my attention.

10. Ibid., 12.

11. John Ott in Robert L. Ragland, *Major Ragland's Instructions How to Grow and Cure Tobacco, Especially Fine Yellow* (Richmond, VA: Southern Fertilizing Company, 1885), 2.

12. Goodman, *Tobacco in History*, 3–4.

13. Sidney W. Mintz, *Sweetness and Power: The Place of Sugar in Modern History* (New York: Viking, 1985).

14. Lorena S. Walsh, "Plantation Management in the Chesapeake, 1620–1820," *Journal of Economic History* 49, 2 (June 1989): 396; Carr, Menard, and Walsh, *Robert Cole's World*, 42; Morgan, *American Slavery*, 302. Other than a varietal name and accounts of its value, "no adequate description of it [sweet-scented tobacco] remains." See Charles E. Gage, "Historical Factors Affecting Tobacco Types and Uses and the Evolution of the Auction Market," *Agricultural History* 11, 1 (January 1937): 44. For a general description of sweet-scented, see George Watterston, *A Memoir on the History, Culture, Manufactures, Uses, &c. of the Tobacco Plant* (Washington City [DC]: Jonathan Elliot, 1817), 7.

15. Thomas Emort, "Agriculture," *American Farmer* 4, 20 (August 9, 1822): 157.

16. P. W. Harper, "Desultory Remarks on the Making of Tobacco," *Farmer's Register* 3, 12 (April 1836): quotes on 712; Robert L. Ragland, "The Yellow Tobacco of North Carolina and Virginia," *American Farmer* 3, 12 (June 15, 1884): 178; and Robert, *Tobacco Kingdom*, 41–43.

17. Robert L. Ragland, *Cultivation and Curing of Fine Yellow and Shipping Tobacco from the Plant-Bed to Market* (Richmond, VA: J. W. Fergusson, 1878?), 4. Ragland's comments concerned antebellum seed selection practices. For similar descriptions, see Yellow Pryor, "Management of Tobacco," *Farmer's Register* 1, 10 (March 1834): 632–633; Bowie, "Essay on the Culture and Management," 33; and Shelton, "Culture and Management of Tobacco," 210.

18. Jeffreys Diary, February 26, 1845.

19. N. A. Venable, "Tobacco," *Southern Planter* 3, 5 (May 1843): 102.

20. Robert L. Ragland, *Wholesale Catalogue of Reliable Tobacco Seeds* (Hyco, VA: R. L. Ragland Seed, 1894), 9–10.

21. Shelton, "Culture and Management of Tobacco," 209.

22. Robert L. Ragland, *Tobacco, from the Seed to the Salesroom* (Richmond, VA: William Ellis Jones, 1880), 5.

23. For the genetic similarity of various tobacco types, as well as some skepticism about the nature of tobacco varieties, see Hahn, *Making Tobacco Bright*, appendix. Hahn interprets the relative lack of seed advertisements and specificity as evidence that seed selection mattered little, though it seems equally plausible that the lack of published advertisements reflected

farmers' belief that seed selection was an intensely local, site-specific activity. The writings of Jeffreys, Shelton, and others seem to reflect the latter view.

24. E. H. Mathewson, *The Culture of Flue-Cured Tobacco*, USDA Bulletin No. 16 (Washington, DC: Government Printing Office, 1913), 4; A. P. Brodell, *Cost of Producing Virginia Dark and Bright Tobacco and Incomes from Farming, 1922–1925*, Virginia Polytechnic Institute Agricultural Experiment Station, Bulletin No. 255 (Blacksburg: Virginia Polytechnic Institute and State University, 1927), 4; Rowalt, *Soil Defense in the Piedmont*, 2–4; Wrightman W. Garner, *The Production of Tobacco* (New York: Blakiston, 1951), 88–89, 362; Bowie, "Essay on the Culture and Management," 34.

25. Tilley, *Bright-Tobacco Industry*, 13–14. For the localized nature of quality bright tobacco soils, see "Farmers' Meeting at the Exchange Hotel," *Southern Planter* 17, 4 (April 1857): 225.

26. Warner, *Tobaccoland*, 174. For an early antebellum identification of the quality of these sandy, or "grey" lands, see "Cultivation of Tobacco," *American Farmer* 2, 51 (May 12, 1841): 405–407.

27. Shelton, "Culture and Management of Tobacco," 212.

28. Garner, *Production of Tobacco*, 108–109; and Mathewson, *Culture of Flue-Cured Tobacco*, 4.

29. Killebrew and Myrick, *Tobacco Leaf*, 80–84; and Augustus Voeleker in John Ott, *Tobacco in Virginia and North Carolina* (Richmond, VA: Southern Fertilizer, 1877), quote on 15.

30. "Chemical Properties of Tobacco," *Southern Planter* 20, 4 (April 1860): 216.

31. Garner, *Production of Tobacco*, 331–332; "Farmers' Meeting," 226; Robert L. De Coin, *History and Cultivation of Cotton and Tobacco* (London: Chapman and Hall, 1864), 258; and D. Layton Davis and Mark T. Nielsen, eds., *Tobacco: Production, Chemistry and Technology* (Malden, MA: Blackwell Science, 1999), 128.

32. Cameron, *Sketch of the Tobacco Interests*, 44; and Shelton, "Culture and Management of Tobacco," 212.

33. Mathewson, *Culture of Flue-Cured Tobacco*, 5.

34. W. Edward Hearn and Frank P. Drane, *Soil Survey of Caswell County, North Carolina* (Washington, DC: Government Printing Office, 1910), 9; Killebrew and Myrick, *Tobacco Leaf*, 11.

35. Killebrew and Myrick, *Tobacco Leaf*, 11.

36. Thomas Whitehead, ed., *Virginia: A Synopsis of the Geology, Geography, Climate and Soil of the State, Together with Its Resources of Mines, Forests and Fields, Its Flocks and Its Herds* (Richmond: Virginia State Board of Agriculture, 1889), 46–47.

37. Robert D. Key, "On Firing Tobacco through Flues. The Benefit of Cutting When Fully Ripe," *Farmer's Register* 4, 1 (May 1836): 41–42; Minor, *Cultivation and Management of Tobacco*, 16; Tilley, *Bright-Tobacco Industry*, 18–19; and Robert, *Tobacco Kingdom*, 44–45. Robert gives the date of Pendleton's invention as 1827.

38. Robert L. Ragland, "A Scrap from the History of Yellow Tobacco," *Southern Planter* 37, 5 (May 1876): 343–344; Robert L. Ragland, "The Yellow Tobacco of North Carolina and Virginia," *American Farmer* 3, 12 (June 15, 1884): 178; Whitehead, ed., *Virginia*, 45–46; Tilley, *Bright-Tobacco Industry*, 20–21; and Robert, *Tobacco Kingdom*, 45, quote in the latter.

39. Shelton, "Culture and Management of Tobacco," 215.

40. R. J. S., "The Curing of Tobacco with Charcoal," *Southern Planter* 18, 10 (October 1858): 596. According to Tilley, these instructions were likely from one of the Slade brothers or a close neighbor.

41. Ibid., 596.

42. Davis and Nielsen, eds., *Tobacco*, 128, 131; Hawks, *Principles of Flue-Cured Tobacco*, 191–195; and T. C. Tso, *Physiology and Biochemistry of Tobacco Plants* (Stroudsburg, PA: Dowden, Hutchinson, and Ross, 1972), 171–172.

43. Hahn, *Making Tobacco Bright*.

44. Davis and Nielsen, eds., *Tobacco*, 131.

45. R. J. S., "Curing of Tobacco," 595–596; Caswell, "On Curing Tobacco Yellow," *Southern Planter* 19, 8 (August 1859): 492; and Shelton, "Culture and Management of Tobacco," 212–214.

46. Shelton, "Culture and Management of Tobacco," 209.

47. D. Garland to John T. Garland, May 22, 1846, Caswell County Historical Association Collection, SHC.

48. Frederick Law Olmsted, *The Cotton Kingdom: A Traveler's Observations on Cotton and Slavery in the American Slave States, 1853–1861* (New York: De Capo, 1996), 70.

49. Perry in Perdue, Barden, and Phillips, *Weevils in the Wheat*, 223–224.

50. Johnson in ibid., 160.

51. On African American labor, agriculture, and environmental knowledge, see Kimberly N. Ruffin, *Black on Earth: African American Ecoliterary Traditions* (Athens: University of Georgia Press, 2010), chap. 1.

52. Hunt in Perdue, Barden, and Phillips, *Weevils in the Wheat*, 148. Hunt's owner apparently harvested the tobacco leaves in the field rather than cutting the entire stalk to cure, a method that became increasingly popular with regional cultivators after the Civil War.

53. Robert L. Ragland, *Major Ragland's Instructions How to Grow and Cure Tobacco, Especially Fine Yellow* (Richmond, VA: Southern Fertilizing, 1885), 25.

54. The following paragraphs are drawn from the Jeffreys Diary, 1845 entries. Dates for direct quotations are cited below.

55. Ibid., January 20, 1845.

56. Ibid., January 9, 1845.

57. Ibid., January 20, 1845.

58. Ibid., January 10, 1845. Broomsedge, also commonly known as broomstraw or poverty grass, is an early colonizer of old fields.

59. Ibid., January 20, 1845.

60. Ibid., October 3, 1844.

61. Ibid., January 20, 1845.

62. Robert, *Tobacco Kingdom,* 132–157, esp. 143.

63. For postwar prices, see Arthur G. Peterson, *Historical Study of Prices Received by Producers of Farm Products in Virginia, 1801–1927* (Blacksburg: Virginia Agricultural Experiment Station, 1929), 172, table 79.

64. The receipts were for 913 hogsheads and loose lots of tobacco. The average hogshead was in excess of 1,200 pounds during the antebellum period, and the loose lots varied in size from a few hundred to a few thousand pounds. For a complete list of the sources examined to compile these figures, see the appendix.

65. "To the Planters of the Tobacco Growing Sections of Virginia and North Carolina," *Milton Chronicle,* February 12, 1857, p. 2; and "Tobacco," *Milton Chronicle,* May 21, 1858, p. 2.

66. "The Dan River Valley—Its Superior Products," *Danville Republican,* July 24, 1856, p. 2. The article quoted prices as high as $45.75 per hundredweight on the Richmond market.

67. Advertisement, *Virginia Echo,* August 19, 1859, p. 3.

68. Williams and Carrington to William Bailey, June 17, 1859, William Bailey Papers, *RASP,* series E, part 1, reel 1.

69. Receipt of Elisha Barksdale, September 10, 1852, Southside Virginia Family Papers, *RASP,* series E, part 3, reel 2.

70. Receipt of John T. Garland, September 3, 1847, Glenn Family Papers, Caswell County Historical Association Collection, SHC.

71. Advertisement, *Virginia Echo,* August 19, 1859, p. 3.

72. "Visit to Mount Vernon," *Virginia Echo,* September 9, 1859, quote on 2; and Tilley, *Bright-Tobacco Industry,* 21–22.

73. As Barbara Hahn has noted, demand for a new consumer good did not often emerge fully formed, or from any intrinsic qualities of a commodity per se, but often came from a combination of consumer desires and producer marketing, to varying degrees. Hahn, *Making Tobacco Bright,* 222.

74. Arnold, *History of the Tobacco Industry,* 58.

75. Gregg D. Kimball, *American City, Southern Place: A Cultural History of Antebellum Richmond* (Athens: University of Georgia Press, 2006), 23.

76. Clara Fountain, *Danville: A Pictorial History* (Virginia Beach, VA: Donning, 1979), 17; and Hagan, *Story of Danville,* 9.

77. Robert, *Tobacco Kingdom,* 166–168; and Powell, *When the Past,* 116.

78. Suzanne G. Schnittman, "Slavery in Virginia's Urban Tobacco Industry, 1840–1860" (Ph.D. diss., University of Rochester, 1987), 27.

79. Robert, *Tobacco Kingdom,* 178.

80. List of tobacco purchased and manufactured, November 1, 1854, William Thomas Sutherlin Papers, SHC. William Thomas Sutherlin, who was the primary investor in Sutherlin and Ferrell, began manufacturing tobacco in Danville in 1844. See "Maj. William T. Sutherlin," *Southern Planter and Farmer* 10, 10 (October 1875): 596.

81. Schnittman, "Slavery in Virginia's Urban Tobacco Industry," 28; and Diane Barnes, *Artisan Workers in the Upper South: Petersburg, Virginia, 1820–1865* (Baton Rouge: Louisiana State University Press, 2008), 160–161.

82. Schnittman, "Slavery in Virginia's Urban Tobacco Industry," 13; Robert, *Tobacco Kingdom*, 197–201; Barbara Hahn, "Making Tobacco Bright: Institutions, Information, and Industrialization in the Creation of an Agricultural Commodity, 1617–1937" (Ph.D. diss., University of North Carolina at Chapel Hill, 2006), 101–102; Perdue, Barden, and Phillips, *Weevils in the Wheat*, 153; Herbert R. Northrup, *The Negro in the Tobacco Industry* (Philadelphia: University of Pennsylvania Press, 1970), 2; and Morgan, *Emancipation in Virginia's Tobacco Belt*, 17–18; The records of the Wyllies, a Danville tobacco manufacturing family, indicate that at least some tobacco factories hired white boys for some jobs. See Receipt of Henry Boman, February 4, 1860, Wyllie Family Papers, VHS.

83. Hagan, *Story of Danville*, 9; and John E. Edwards, "Danville, Virginia," *Christian Advocate* 60, 52 (December 24, 1885): 832, quote in the latter.

84. Receipts of William T. Sutherlin, December 20, 1847, January 1, July 1, 1848 (two), William Thomas Sutherlin Papers, SHC; and Schnittman, "Slavery in Virginia's Urban Tobacco Industry," 141. Schnittman has calculated that the average monthly lease for a male slave in Danville's factories was $12.40, and for a female slave, $9.00.

85. U.S. Census Bureau Slave Schedules, southern district of Pittsylvania and southern district of Halifax counties, Virginia, 1860. For the Wyllie company hiring workers, see note 1, above.

86. This description of factory operations draws heavily on Schnittman, "Slavery in Virginia's Urban Tobacco Industry," 22–25, 161; and Barnes, *Artisan Workers*, 163–164. For the prevalence of chewing tobacco in local manufacturing, see Tilley, *Bright-Tobacco Industry*, 490–491. For the preservative role of flavorings in plug manufacturing, see *The American Tobacco Story* (Richmond, VA: American Tobacco, 1964), 17; and Hahn, *Making Tobacco Bright*, 54–57.

87. Gail Cooper, "Custom Design, Engineering Guarantees, and Unpatentable Data: The Air Conditioning Industry, 1902–1935," *Technology and Culture* 35, 3 (July 1994): esp. 524–531.

88. David Nye has argued that these understandings transmitted from the countryside to early industrial enterprises that relied primarily on organic power were commonplace. Nye, *Consuming Power: A Social History of American Energies* (Cambridge: MIT Press, 1998), 35–36.

89. Charles B. Dew, *Bond of Iron: Master and Slave at Buffalo Forge* (New York: W. W. Norton, 1994), 262.

90. Robert, *Story of Tobacco*, 88.

91. For an example, see Advertisement, *Milton Chronicle*, November 1, 1849, p. 3.

92. Robert, *Story of Tobacco*, 89.

93. Lorenzo Ivy in Perdue, Barden, and Phillips, *Weevils in the Wheat*, 153.

94. Johnson H. Owen to William T. Sutherlin, March 5, 1861, William T. Sutherlin Papers, SHC.

95. Jackson, *My Father's Name*, 70.

96. Powell, *When the Past*, 116, 129; and HCDB. For additional references to Caswell facilities, see Robert, *Tobacco Kingdom*, 166–168, 177; Katherine Kerr Kendall, *Caswell County, 1777–1877: Historical Abstracts of Minutes of Caswell County, North Carolina* (Raleigh, NC: Multiple Images, 1976), 86; and Little-Stokes, *Inventory of Historic Architecture*, 194. Little-Stokes claims that the five facilities in Yanceyville in 1860 averaged 144 workers each, though other sources strongly suggest that 144 total workers between the five factories was the more likely figure.

97. Robert, *Tobacco Kingdom*, 166–168; and Schnittman, "Slavery in Virginia's Urban Tobacco Industry," 47.

98. Receipt of William Finney, September 22, 1863, Southside Virginia Family Papers, *RASP*, series E, part 3, reel 4; and Charles B. Motley, *Yes There Is a Dry Fork Virginia* (Danville, VA: Bassett, 1977), 92.

99. Receipts of Philip H. Howerton, November 11, 1850, May 29, 1851, Charles H. Cabaniss Papers, Duke; and Edmunds, *History of Halifax:* vol. 1, *Narration*, 120, 171, 189. John Hatchett also operated a factory somewhere in Halifax in 1858; see Receipt of John J. Hatchett, September 29, 1858, Hatchett Family Papers, Duke.

100. U.S. Census Bureau Slave Schedule, Caswell County, North Carolina, 1860; and U.S. Census Bureau Slave Schedule, northern district of Pittsylvania County, Virginia, 1860.

101. The diary of Pittsylvania's William Shelton documents just such production: Ricketts, ed., *Diary Kept by William C. Shelton*, 88–89.

102. Schnittman, "Slavery in Virginia's Urban Tobacco Industry," 103.

103. William H. Edwards to William A. J. Finney, February 16, 1859, Philip Thomas to William A. J. Finney, January 12, 1859, both in William A. J. Finney Papers, *RASP*, series F, part 3, roll 34; and Jason Dowell to William A. J. Finney, July 29, 1854, Southside Virginia Family Papers, *RASP*, series E, part 3, reel 4.

104. Robert, *Tobacco Kingdom*, 197–201.

105. Hahn, "Making Tobacco Bright," 60.

106. Schnittman, "Slavery in Virginia's Urban Tobacco Industry," 47.

107. For the use of a tree and beam by the Reynolds family in antebellum Patrick County, Virginia, which eventually led to the R. J. Reynolds tobacco empire, see Michele Gillespie, *Katherine and R. J. Reynolds: Partners of Fortune in the Making of the New South* (Athens: University of Georgia Press, 2012), 16.

108. "Public Sale Notice," *Milton Chronicle*, September 16, 1869, p. 3. There was no indication of how long the company had been defunct, though its operations may have been halted by the war. For the similarity of antebellum and early postwar manufacturing facilities, see Tilley, *Bright-Tobacco Industry*, 489–490.

109. For examples of local merchants documenting tobacco product sales, see Claiborne Store Ledger, 1855–1858, throughout, John F. Claiborne Papers, Duke; Account Book of William Clark Grasty, 1858, pp. 138, 249, and 1859–1860, William Clark Grasty Papers, Duke (hereafter cited as Grasty Account Books).

110. Receipts of Philip Howerton, November 11, 1850, May 29, 1851, Philip H. Howerton Papers, Duke; Receipt of John J. Hatchett, September 29, 1858, Hatchett Family Papers, Duke; Receipts of William T. Sutherlin, January 21, 31, March 27, 1860, William Thomas Sutherlin Papers, SHC; and Patton, Smith, and Putman to John Sutherlin, May 4, 1858, William Thomas Sutherlin Papers, Duke.

111. Robert, *Story of Tobacco*, 93.

112. Sutherlin and Ferrell, List of tobacco purchased and manufactured, November 1, 1854, William Thomas Sutherlin Papers, SHC; Blow and March to John Sutherlin, July 23, 1858, and C. C. Anderson to William Sutherlin, August 11, 1863, both in William Thomas Sutherlin Papers, Duke; and Receipt of John J. Hatchett, September 29, 1858, Hatchett Family Papers, Duke.

113. M. Shilton to William T. Sutherlin, May 19, 1860, William Thomas Sutherlin Papers, SHC.

114. Receipt of William T. Sutherlin, August 20, 1859, William Thomas Sutherlin Papers, SHC; Receipt of John J. Hatchett, September 29, 1858, Hatchett Family Papers, Duke; and B. H. and O. H. P. Thomas to John Sutherlin, June 11, 1858, William Thomas Sutherlin Papers, Duke.

115. Receipt of William T. Sutherlin, August 20, 1859, and Samuel Ayers to William T. Sutherlin, January 4, 1862, both in William Thomas Sutherlin Papers, SHC; Williams and Cheatam to John Sutherlin, July 15, 1858, William Thomas Sutherlin Papers, Duke; Goodman, *Tobacco in History*, 100; Hahn, *Making Tobacco Bright*, 58.

116. For an analysis of California's agricultural marketing, see Douglas Cazaux Sackman, *Orange Empire: California and the Fruits of Eden* (Berkeley: University of California Press, 2005), 84–116; and Steven Stoll, *The Fruits of Natural Advantage: Making the Industrial Countryside in California* (Berkeley: University of California Press, 1998).

117. Sackman, *Orange Empire*, 87.

118. For sales to merchants in these locations, see B. H. and O. H. P. Thomas to John Sutherlin, June 11, 1858, William Thomas Sutherlin Papers, Duke; and Receipt of William T. Sutherlin, June 26, 1860, William Thomas Sutherlin Papers, SHC.

119. For the dominance of Danville-area bright leaf as plug wrappers, see Hawks, *Principles of Flue-Cured Tobacco*, 5. For the argument that brand names primarily served as advertisements from manufacturers to merchants and agents, see Hahn, *Making Tobacco Bright*, 58–59.

120. G. Melvin Herndon, *William Tatham and the Culture of Tobacco* (Coral Gables, FL: University of Miami Press, 1969), 414; Bennett, "King Bacca's

Throne," 25; Schnittman, "Slavery in Virginia's Urban Tobacco Industry,"
24–25; and Blow and March to John Sutherlin, May 12, 1858, William
Thomas Sutherlin Papers, Duke.

121. Jason Dowell to William A. J. Finney, July 29, 1854, Southside Virginia
Family Papers, *RASP*, series E, part 3, reel 4.

122. S. Wyatt and Company's Tobacco Circular, May 24, 1860, William Thomas
Sutherlin Papers, SHC.

123. Compton and Hughes to John Sutherlin, April 19, 1858, William Thomas
Sutherlin Papers, Duke.

124. Patton, Smith, and Putman to John Sutherlin, May 4, 1858, William
Thomas Sutherlin Papers, Duke.

125. Prince and Simpson, *Long Green*, 15; Charles E. Gage, "Historical Factors
Affecting American Tobacco Types and Uses and the Evolution of the Auc-
tion Market," *Agricultural History* 11, 1 (January 1937): 48; John Van Willigen
and Susan C. Eastwood, *Tobacco Culture: Farming Kentucky's Burley Belt*
(Lexington: University Press of Kentucky, 1998), 11–12; and William C.
Rense, "The Perique Tobacco Industry of St. James Parish, Louisiana: A
World Monopoly," *Economic Botany* 24, 2 (April–June 1970): 123–124.

126. *American Tobacco Story*, 17.

127. Bird L. Ferrell to John Sutherlin, October 11, 1858, William Thomas Sutherlin Papers, Duke; Armistead T. Moore to William T. Sutherlin, April 26,
1860, and John T. Garland to William T. Sutherlin, March 1857, both at William Thomas Sutherlin Papers, SHC.

128. Thomas Barksdale to John Sutherlin, May 10, 1858, William Thomas
Sutherlin Papers, Duke; and John T. Garland to William T. Sutherlin,
March, 1857, William Thomas Sutherlin Papers, SHC. Barksdale com-
manded twenty-five dollars per hundredweight for his bright leaf.

129. John T. Garland to William T. Sutherlin, March 1857, B. Brown to William
T. Sutherlin, May 35, 1860, and Armistead T. Moore to William T. Sutherlin, April 26, 1860, all in William Thomas Sutherlin Papers, SHC; Thomas
Lindsey to John Sutherlin, April 26, 1858, and Thomas E. Barksdale to John
Sutherlin, May 10, 1858, both in William Thomas Sutherlin Papers, Duke;
and D. Garland to John T. Garland, May 22, 1846, Caswell County Historical Association Collection, SHC, quote in the latter.

130. "To the Planters of the Tobacco Growing Sections of Virginia and North
Carolina," *Milton Chronicle*, February 12, 1857, p. 2.

131. Virginia Agricultural Census, Halifax and Pittsylvania Counties, 1860, National Archives and Records Administration (hereafter cited as NARA),
T1132, rolls 6 and 7.

132. "High Prices for Tobacco," *Virginia Echo*, October 21, 1859, p. 2.

CHAPTER 3. BRIGHT LEAF, BRIGHT PROSPECTS

1. Sean Wilentz, *The Rise of American Democracy: Jefferson to Lincoln* (New
York: W. W. Norton, 2005), 456–457, 463–465, quote from *New Era*, on 456.

For a general theory of the connection between economic depressions and agricultural innovation, see Earle, "Myth of the Southern Soil Miner," 175–210. Steven Stoll makes similar connections between changing agricultural practice and the earlier depression of the 1820s. See Stoll, *Larding the Lean Earth: Soil and Society in Nineteenth-Century America* (New York: Hill and Wang, 2002), 40–54.

2. Patrick Henry in Walker, *Southern Forest,* 171.

3. John Taylor, *Arator; Being a Series of Agricultural Essays, Practical and Political* (Georgetown, DC: n.p., 1813), quote on 82; Jack Temple Kirby, *Poquosin: A Study of Rural Landscape and Society* (Chapel Hill: University of North Carolina Press, 1995), 65–67; David R. Montgomery, *Dirt: The Erosion of Civilizations* (Berkeley: University of California Press, 2007), 126; Garret Ward Shelton and C. William Hill, Jr., *The Liberal Republicanism of John Taylor of Caroline* (Madison, NJ: Fairleigh Dickinson University Press, 2008), 69–88; and Benjamin Cohen, *Notes from the Ground: Science, Soil, and Society in the American Countryside* (New Haven: Yale University Press, 2009), 42–47.

4. Edmund Ruffin, *Nature's Management: Writings on Landscape and Reform, 1822–1859,* ed. Jack Temple Kirby (Athens: University of Georgia Press, 2000); Stoll, *Larding the Lean Earth,* 150–160; William Mathew, *Edmund Ruffin and the Crisis of Slavery in the Old South: The Failure of Agricultural Reform* (Athens: University of Georgia Press, 1988); Kirby, *Poquosin,* 61–85; Drew A. Swanson, "Fighting over Fencing: Agricultural Reform and Antebellum Efforts to Close the Virginia Open Range," *Virginia Magazine of History and Biography* 117, 2 (2009): 106–108; Cohen, *Notes from the Ground,* 85–94; and Montgomery, *Dirt,* 129–130. For a good general biography of Ruffin as a reformer, see David F. Allmendinger, Jr., *Ruffin: Family and Reform in the Old South* (New York: Oxford University Press, 1990), and for Ruffin's political career, see Betty L. Mitchell, *Edmund Ruffin: A Biography* (Bloomington: Indiana University Press, 1981).

5. Ruffin, *Nature's Management,* 323, emphasis in the original.

6. Avery Craven, *Edmund Ruffin, Southerner: A Study in Secession,* reprint (Baton Rouge: Louisiana State University Press, 1991), 61; and Edmund Ruffin, *An Essay on Calcareous Manures* (Petersburg, VA: J. W. Campbell, 1832).

7. Ruffin, *Nature's Management,* 325; Robert, *Tobacco Kingdom,* 24–31; and John Hartwell Cocke, "Tobacco, the Bane of Virginia Husbandry," *Southern Planter* 18 (1858): 716–719.

8. Address of James C. Bruce to the Mecklenberg and Granville Agricultural Clubs, July 14, 1847, pp. 7–15, Bruce Family Papers, *RASP,* series E, part 3, reel 14.

9. Compiled from *Fifth Census; Enumeration of the Inhabitants of the United States—1830* (Washington, DC: Duff Green, 1832), 10–13, 84–85, 90–91; *Compendium of the Enumeration of the Inhabitants and Statistics of the United States—Sixth Census* (Washington, DC: Thomas Allen, 1841), 34–35, 42–43;

The Seventh Census of the United States: 1850 (Washington, DC: Robert Armstrong, 1853), 259–260, 307; and *Population of the United States in 1860; Compiled from the Original Returns of the Eighth Census* (Washington, DC: Government Printing Office, 1864), 358, 516–517.

10. Advertisement, *Milton Chronicle*, May 22, 1844, p. 4. For similar, see advertisement in *Richmond Enquirer*, November 11, 1832, p. 1.

11. "Going to Texas," *Milton Chronicle*, August 3, 1842, p. 2. On the booming appeal of Texas to Virginians during the 1840s in particular, see Fischer and Kelly, *Bound Away*, 181.

12. Jeffreys Diary, January 30, February 17, 1845.

13. Charles J. Clement to George Clement, April 24, 1850, Pocket Plantation Papers, *RASP*, series E, part 1, reel 15.

14. Peter Hairston to George Hairston, March 22, 1859, Hairston Family Papers, VHS.

15. Mathew, *Edmund Ruffin and the Crisis of Slavery*, 196–197.

16. For an eloquent phrasing of this question in extended form, see Lynn A. Nelson, *Pharsalia: An Environmental Biography of a Southern Plantation, 1780–1880* (Athens: University of Georgia Press, 2007), 1–12.

17. Avery Craven, *Soil Exhaustion as a Factor in the Agricultural History of Virginia and Maryland, 1606–1860* (Urbana: University of Illinois Press, 1926).

18. Arthur R. Hall, *Early Erosion-Control Practices in Virginia*, USDA Miscellaneous Publication No. 256 (Washington, DC: Government Printing Office, 1937).

19. Earle, "Myth of the Southern Soil Miner," 175–210.

20. Mathew, *Edmund Ruffin and the Crisis of Slavery*.

21. Stoll, *Larding the Lean Earth*.

22. Nelson, *Pharsalia*.

23. Andrew Nelson Lytle, "The Hind Tit," in *I'll Take My Stand: The South and the Agrarian Tradition*, rev. ed. (Baton Rouge: Louisiana State University Press, 1977), 205.

24. "List of Subscribers," *Farmer's Register* 1, 12 (May 1834): 771, 773, 775; Account Book of Robert Wilson, June 30, 1854, Robert Wilson Account Books, oversize volume SV-1896/75, SHC; Receipt of William R. Hatchett, January 1, 1855, Hatchett Family Papers, Duke; "Payments to the *Southern Planter*, from April 1st to May 1st, 1852," *Southern Planter* 12, 5 (May 1852): 154; and "The Cry Is Still They Come!," *American Farmer* 5, 3 (September 1849): 83.

25. For selected examples, see E., "Remarks on the Soil and Cultivation of a Part of the County of Halifax," *Farmer's Register* 5, 2 (June 1837): 117–120; R. N., "Which Is the Best Route for a Railway to the South-West?" *Farmer's Register* 4, 6 (October 1836): 369–374; "Halifax, VA, August 12th, 1838," *Farmer's Register* 6, 7 (October 1, 1838): 442–443; Edmund Ruffin, "Extracts from Private Correspondence," *Farmer's Register* 1, 4 (September 1833): 216–218; George Read, "Agricultural Intelligence—Report of Seasons, Crops, &c.," *Farmer's Register* 1, 10 (March 1834): 636; A Young Planter,

"To the Editors of the Southern Planter," *Southern Planter* 3, 5 (May 1843): 107; "The Coming Crop," *Southern Planter* 16, 1 (January 1856): 5; Thomas Walton, "Magnum Bonum Plums," *Southern Planter,* 17, 9 (September 1857): 548; G. T. N., "Great Yield of Wheat," *Southern Planter* 11, 11 (November 1851): 340–341; and "The Silk Culture," *American Farmer* 1, 4 (June 1839): 29.

26. William S. Powell, *When the Past Refused to Die: A History of Caswell County North Carolina, 1777–1977* (Durham, NC: Moore, 1977), 477; Coleman Diaries, August 1, 1844; Receipt of James C. Bruce, October 1, 1853, Bruce Family Papers, *RASP,* series E, part 3, reel 16; and Account Book of Robert Wilson, December 31, 1857, Robert Wilson Account Books, oversize volume SV-1896/75, SHC.

27. *Address of James C. Bruce, Esq., President of the Union Agricultural Society of Virginia and North Carolina,* pamphlet, 1854, in Bruce Family Papers, *RASP,* series E, part 3, reel 1; "Progress of Agriculture," *Southern Planter* 11, 2 (February 1851): 37; and "John R. Edmunds of Halifax," *Virginia Echo,* October 21, 1859, p. 2. Three farmers from Halifax and Pittsylvania, including Bruce and Edmunds, served as judges at the 1853 VSAS annual farm exhibition in Richmond ("Virginia State Agricultural Society," *Southern Planter* 13, 8 [August 1853]: 227–230).

28. John R. Edmunds, "Address of John R. Edmunds, Esq. before the Virginia State Agricultural Society, November 3, 1853," *Southern Planter* 13, 12 (December 1853): 11.

29. Ibid., 1–10. See John P. Norton, *Elements of Scientific Agriculture; or The Connection between Science and Art of Practical Farming* (New York: C. M. Saxton, Barker, 1850); Mary Somerville, *Physical Geography* (Philadelphia: Blanchard and Lea, 1853); Justus von Liebig, *Organic Chemistry in Its Applications to Agriculture and Physiology,* ed. Lyon Playfair (London: Taylor and Walton, 1840); Cuthbert W. Johnson, *The Farmer's and Planter's Encyclopaedia of Rural Affairs: Embracing All the Most Recent Discoveries in Agricultural Chemistry* (Philadelphia: Lippincott, Grambo, 1851); J. B. Boussingault, *Rural Economy, in Its Relations with Chemistry and Physiology to the Details of Practical Farming* (London: H. Bailliere, 1845); Henry Coleman, *European Agriculture and Rural Economy: From Personal Observation,* 2 vols. (Boston: Phillips, Sampson, 1850); and David Low, *Elements of Practical Agriculture: Comprehending the Cultivation of Plants, the Husbandry of the Domestic Animals, and the Economy of the Farm* (London: Longman, Brown, Green, and Longmans, 1847).

30. Address of James C. Bruce to the Mecklenberg and Granville Agricultural Clubs, July 14, 1847, pp. 7–15, quote on 8, Bruce Family Papers, *RASP,* series E, part 3, reel 14.

31. *Address of James C. Bruce, Esq., President of the Union Agricultural Society of Virginia and North Carolina,* pamphlet, 1854, in Bruce Family Papers, *RASP,* series E, part 3, reel 16.

32. Leo Rogin, *The Introduction of Farm Machinery in Its Relation to the Productivity of Labor in the Agriculture of the United States during the Nineteenth*

Century (Berkeley: University of California Press, 1931), 52–57; Gray, *History of Agriculture*, 795; Earl W. Hayter, *The Troubled Farmer, 1850–1900: Rural Adjustment to Industrialism* (Dekalb: Northern Illinois University Press, 1968), 4–5; Eugene D. Genovese, *The Political Economy of Slavery: Studies in the Economy and Society of the Slave South*, 2nd ed. (Middletown, CT: Wesleyan University Press, 1989), 56–57; and Earle, "Myth of the Southern Soil Miner," 199. Avery Craven pointed to the diversification of plows in the Chesapeake in 1926, in *Soil Exhaustion as a Factor*, 91.

33. At least one historian has argued that antebellum Virginia farms exhibited a greater diversification of plows than the South in general, though this statement was made in regard to the Tidewater only. See Rogin, *Introduction of Farm Machinery*, 54–55.

34. For selected examples, see Hall, *Early Erosion-Control*, 13–15; Accounts of George Clements with Joab Oaks, 1836–1850, in Pocket Plantation Records, *RASP*, series E, part 1, reel 11; Account of Melchesidick Spraggins with B. Blankenship, 1823, and Account of Melchesidick Spraggins with Joseph Crews, 1830, both in Southside Virginia Family Papers, *RASP*, series E, part 3, reel 1; Grasty Account Books, 1836–1846; Account of Elisha Barksdale with Isaac Kirk, July 2, 1841, Southside Virginia Family Papers, *RASP*, series E, part 3, reel 2; E., "Remarks on the Soil," 119; Jeffreys Diary, January 9, 1845; Hairston Family Plantation Record Books 1 and 3, 1852, Peter Wilson Hairston Papers, box 8, folders 104 and 106, SHC.

35. Hall, *Early Erosion-Control*, 13–15; "Deep Land Culture," *Arator* 1, 11 (February 1856): 336; "Agricultural Warehouse," *People's Advocate*, June 27, 1856, p. 1.

36. Hairston Family Plantation Record Books 1 and 3, 1852, Peter Wilson Hairston Papers, box 8, folders 104 and 106, SHC.

37. Coleman Diaries, October 10, December 15, 1842, April 16, 1843; Sims Farm Book, January 6, February 19, 1855, William Bailey Papers, *RASP*, series E, part 1, reel 2; Account Book of E. A. Coleman, 1840s, entries throughout, Papers of Ethelbert Algernon Coleman, UVA; Grasty Account Books, 1849–1851, pp. 1, 248; and G. Melvin Herndon, "The Significance of the Forest to the Tobacco Plantation Economy in Antebellum Virginia," *Plantation Society in the Americas* 1, 3 (October 1981): 430–431.

38. Rowalt, *Soil Defense in the Piedmont*, 17–18.

39. Hall, *Early Erosion-Control*, 21; and E., "Remarks in the Soil," 118. For similar efforts in the South Carolina Piedmont during the antebellum period, see Arthur R. Hall, *The Story of Soil Conservation in the South Carolina Piedmont 1800–1860*, USDA Miscellaneous Publication No. 407 (Washington, DC: Government Printing Office), esp. 23–27.

40. Stanley W. Trimble, *Man-Induced Soil Erosion on the Southern Piedmont, 1700–1970* (Ankeny, IA: Soil Conservation Society of America, 1974), 26.

41. Rental Agreement between George Clement and Wyatt Wallace, November 20, 1845, Pocket Plantation Papers, *RASP*, series E, part 1, reel 14.

42. Contract between Elizabeth Spraggins and William Terry, October 12, 1841, Southside Virginia Family Papers, *RASP*, series E, part 3, reel 1.

43. Contract between William H. Armistead and H. C. Moon and Parham Moon, January 1, 1846, Philip H. Howerton Papers, Duke. Armistead traveled extensively out of state, but he made arrangements with a relative to ensure that the Moons followed the conservation terms of the contract. See William H. Armistead to Philip Howerton, May 12, 1848, Philip H. Howerton Papers, Duke. For a similar agreement from the same Caswell community, see Contract between Thomas Bruce and A. B. Spooner, July 25, 1853, Philip H. Howerton Papers, Duke.

44. Robert, *Tobacco Kingdom*, 30.

45. E., "Remarks on the Soil," 119.

46. Jeffreys Diary, February 21, 1845.

47. James C. Bruce to Josiah Wills, November 25, 1842, Bruce Family Papers, VHS; Daily Journal of James Bruce, entries for 1840, Bruce Family Papers, *RASP*, series E, part 3, reel 22; Sims Farm Book, March 3, 1855; Coleman Diaries, February 17, 1843; Smith and Maddux to Nathaniel Ragsdale, March 24, 1855, Ragsdale Family Papers, VHS; E., "Remarks on the Soil," 118; and Receipt of Rufus H. Owen, September 14, 1861, Owen Family Papers, VHS. See also, Hall, *Early Erosion-Control*, 10–11. Bruce ordered clover seed in lots as large as eighty bushels, indicating that he was reselling the seed to neighboring landowners. For regular purchases of small amounts of clover, timothy, and orchard grass seed by area farmers, see Grasty Account Books, 1859–1860, pp. 15, 49, 51, 64, 312, 389, and throughout.

48. Shawn William Miller, *An Environmental History of Latin America* (New York: Cambridge University Press, 2007), 147–154; Stoll, *Larding the Lean Earth*, 188–190; Jack Temple Kirby, *Mockingbird Song: Ecological Landscapes of the South* (Chapel Hill: University of North Carolina Press, 2006), 89; Montgomery, *Dirt*, 185–187; and Richard A. Wines, *Fertilizer in America: From Waste Recycling to Resource Exploitation* (Philadelphia: Temple University Press, 1985), 33–53.

49. Aaron W. Marrs, *Railroads in the Old South: Pursuing Progress in a Slave Society* (Baltimore: Johns Hopkins University Press, 2009), 114, table 3; Jim Ronlett to Philip Howerton, June 1, 1851, Philip H. Howerton Papers, Duke; Advertisement, *Milton Chronicle*, July 7, 1853, p. 4; Advertisement, *Milton Chronicle*, April 12, 1861, p. 3; and "Agricultural Warehouse," 1.

50. W. H. Wesson to William Bailey, March 19, 1857, William Bailey Papers, *RASP*, series E, part 1, reel 1; Receipt of F. R. Cousins, October 14, 1859, F. R. Cousins Papers, Duke; Ledger, 1858–1860, p. 478, Account Books of J. H. Hargrave, Duke; Receipts of Philip H. Howerton, April 28, September 29, 1855, May 1, 1857, Philip H. Howerton Papers, Duke; Grasty Account Books, 1858, entries throughout, 1859–1860, entries throughout; and Maddux and Co. to Nathaniel Ragsdale, March 24, 1859, Ragsdale Family Papers, VHS.

51. Mathew, *Edmund Ruffin and the Crisis of Slavery*, 120–121. Lime and plaster seem to have been common stock in regional country stores, especially during the 1850s. For examples of the purchase and use of lime and plaster by farmers and planters in the three counties, see Powell, *When the Past*, 476–477; Advertisement, *Milton Chronicle*, April 4, 1850, p. 3; Advertisement, *Milton Chronicle*, July 7, 1853, p. 4; Coleman Diaries, April 2, 1844; Johnson, Clark and Co. to Samuel P. Wilson, July 6, 1860, Samuel Pannill Wilson Papers, UVA; Samuel Carter to Philip Howerton, November 19, 1841, Philip H. Howerton Papers, Duke; Receipts of Philip Howerton, April 28, May 17, 22, 1855, May 1, 1857, in Philip H. Howerton Papers, Duke; and Grasty Account Books, 1849–1851, p. 58, 1858, pp. 133, 179, 186, and 1859–1860, entries throughout.

52. Kathleen Bruce, "Materials for Southern Virginia Agricultural History," *Agricultural History* 4, 1 (January 1930): 12; E., "Remarks on the Soil," 118; Edmunds, *History of Halifax:* vol. 1, *Narration*, 247–248; and N. J. Whitfield to James C. Bruce, February 21, 1840, M. V. Campbell to James C. Bruce, March 12, 1840, and N. J. Whitfield to James C. Bruce, March 30, 1840, all in Bruce Family Papers, *RASP*, series E, part 3, reel 11.

53. Toll List of the Roanoke Navigation Company, October 19, 1853, Bruce Family Papers, *RASP*, series E, part 3, reel 16. These three were the only items exempt from a toll. The fees on other goods ranged from five cents for a barrel of flour or pork to thirty cents per hogshead of tobacco.

54. Stoll, *Larding the Lean Earth*, 123–134.

55. Garner, *Production of Tobacco*, 331–332; and Killebrew and Myrick, *Tobacco Leaf*, 80–84.

56. Minor, *Cultivation and Management*, 4; Sims Farm Book, March 3, 1855; Jeffreys Diary, January 18, 1845; and Merritt Diary, March 15, 25, 1842.

57. Minor, *Cultivation and Management of Tobacco*, 4–5.

58. Herndon, "Significance of the Forest," 430–431; "Value of Ashes," *Milton Chronicle*, August 18, 1841, p. 4; and Coleman Diaries, January 10, 1848.

59. Merritt Diary, March 15, 25, 1842; Sims Farm Book, January 10, 1855; Coleman Diaries, January 10, 1845; and Jeffreys Diary, February 20, 1845. According to Arthur Hall, these practices were relatively widespread across the antebellum plantation South: Hall, *Story of Soil Conservation*, 14–15.

60. "Wool Growing in Virginia," *Milton Chronicle*, July 7, 1853, p. 1.

61. Virginia Agricultural Census, Halifax and Pittsylvania Counties, 1850, NARA, T1132, rolls 2 and 3.

62. Powell, *When the Past*, 111–112; "Oregon Pea," *Milton Chronicle*, May 16, 1854, p. 4; William A. Gilliam to William Bailey, March 13, 1860, William Bailey Papers, *RASP*, series E, part 1, reel 1; "Silk Culture," 29; and "To Land Owners," *Southern Planter* 14, 10 (October 1854): 312.

63. Cookbooks of Eliza Ann Clark, ca. 1840 and ca. 1852–1862, in Spraggins Family Papers, VHS.

64. Jeffreys Diary, February 1, April 19, 1845.

65. Sam Bowers Hilliard, *Hog Meat and Hoecake: Food Supply in the Old South, 1840–1860* (Carbondale: Southern Illinois University Press, 1972), 92–102; Kirby, *Poquosin*, 98–103; and Kirby, *Mockingbird Song*, 113–124. For a local reference to these practices, see Merritt Diary, December 5–7, 1842, March 8, 1858.

66. Daily Journal of James Bruce, October 29, 1838, Bruce Family Papers, *RASP*, series E, part 3, reel 22.

67. Ibid., December 20, 1838, December 21, 1840; and Weather Notebook of James Bruce, December 16, 1858, February 1, 1859, Bruce Family Papers, *RASP*, series E, part 3, reel 24.

68. Account Book of William Mitchell, 1859–1862, entries throughout, Southside Virginia Family Papers, *RASP*, series E, part 3, reel 3; Coleman Diaries, December 13, 1843, December 14, 1844, January 10, 11, 1845, December 18, 1846, December 20, 1847, and January 1, 1849; Ricketts, ed., *Diary Kept by William C. Shelton*, 92; and Maria B. Owens, Pork Memo, 1840, Owen Family Papers, VHS.

69. Hilliard, *Hog Meat and Hoecake*, 102.

70. "Agricultural Warehouse," 1.

71. William Daniel to Charles H. Cabaniss, August 16, 1833, Charles H. Cabaniss Papers, Duke; and Advertisement Draft of Philip Howerton and Thomas Easley, November 12, 1851, Diagram of Threshing Machine, n.d., and List of Timber for "Thrashing" Machine, n.d., all in Philip H. Howerton Papers, Duke.

72. Coleman Diaries, July 25, 1843.

73. For examples of Southside orchard and vineyard culture, see Rebecca F. Barksdale to ?, June 1, 1855, Peter Barksdale Papers, *RASP*, series F, part 3, reel 33; Coleman Diaries, January 20, February 20, April 11, 14, 1843; Sims Farm Book, 1855, entries throughout; and Daily Journal of James Bruce, 1838 entries throughout, Bruce Family Papers, *RASP*, series E, part 3, reel 22.

74. Receipt of William T. Sutherlin, December 27, 1856, William Thomas Sutherlin Papers, SHC.

75. Elijah Hundley, Journal No. 1, 1860 entries, Plantation Records of Elijah Hundley, Small Special Collections, UVA.

76. Merritt Diary, entries throughout.

77. Virginia Agricultural Census, Halifax and Pittsylvania Counties, 1850, NARA, T1132, rolls 2 and 3.

78. The following composite sketch of Coleman's farming activities is drawn from the Coleman Diaries, 1842–1849. Entry dates are provided for direct quotations. In many ways, Coleman's agricultural intensification program mirrored that of planter William Massie of Nelson County, as described in Nelson, *Pharsalia*.

79. November 25, 1842. Coleman most likely refers to an English translation of Jean Antoine Chaptal, *Chimie appliquée à l'agriculture* (Paris: Huzard, 1823).

80. March 6, 1843.

81. December 21, 1842, December 13, 1843.

82. Ruffin actually envisioned agricultural reform as the key to ensuring slavery survival in the older section of the Atlantic South. He argued that it was general agricultural malaise that forced eastern planters to sell surplus slaves to western cotton and sugar lands and declared, "Nothing can check this forced emigration of blacks . . . except increased production of food, obtained by enriching our lands, and the consequent increase of farming profits." See Ruffin, *Essay,* appendix C, 164.

83. Stoll, *Larding the Lean Earth,* 124–125, 133–134; and Mathew, *Edmund Ruffin and the Crisis of Slavery,* 204–213.

84. E L. and E. P. Love, *The Art of Curing Fancy Yellow Tobacco,* reprint (Culler, NC: W. C. Phillips, 1892). This edition is a reprint of a pamphlet that first appeared in the late 1860s. While the Loves began publishing their pamphlets following the Civil War, they or their neighbors experimented with bright tobacco before the conflict, and their advice seems to have reflected prewar practices in many cases.

85. Jeffreys Diary, January 29, February 12, 1845, quote on the latter date.

86. Agricola, *Series of Essays,* 23–24.

87. Jeffreys Diary, May 6, 1845.

88. Garner, *Production of Tobacco,* 108–109, 331–332; J. B. Killebrew and Herbert Myrick, *Tobacco Leaf: Its Culture and Cure, Marketing and Manufacturing, A Practical Handbook on the Most Approved Methods in Growing, Harvesting, Curing, Packing and Selling Tobacco, Also of Tobacco Manufacture,* reprint (New York: Orange Judd, 1909), 80–84; Mathewson, *Culture of Flue-Cured Tobacco,* 5; and Tilley, *Bright-Tobacco Industry,* 181–182.

89. For Virginia farmers' concerns about deforestation, see Swanson, "Fighting over Fencing," 110–117; and G. Melvin Herndon, "The Significance of the Forest to the Tobacco Plantation Economy in Antebellum Virginia," *Plantation Society in the Americas* 1, 3 (October 1981): 430–431.

90. This calculation assumes a five-tier barn sixteen- or twenty-feet square, with twenty-foot walls, a fifth of which were composed of mud chinking between the logs, and does not include the wood needed for rafters, shingles, eaves, or the door. It is based on descriptions in Minor, *Cultivation and Management of Tobacco,* 9–10; Shelton, "Culture and Management of Tobacco," 215–216; Robert L. Ragland, *Tobacco, from Seed to the Salesroom* (Richmond, VA: William Ellis Jones, 1880), 3–4; Ruth Little-Stokes, *An Inventory of Historic Architecture, Caswell County, North Carolina: The Built Environment of a Burley and Bright-Leaf Tobacco Economy* (Yanceyville, NC: Caswell County Historical Association, 1979), 106; Catherine W. Bishir and Michael T. Southern, *A Guide to the Historic Architecture of Piedmont North Carolina* (Chapel Hill: University of North Carolina Press, 2003), 182–183; and the author's study of surviving barns.

91. For various estimates of the wood needed to cure one barn of yellow tobacco, see J. D. Cameron, *A Sketch of the Tobacco Interests in North Carolina*

(Oxford, NC: W. A. Davis, 1881), 99; W. H. Snow, *Snow's Modern Barn System of Raising and Curing Tobacco,* 3rd ed. (Baltimore: Isaac Friedenwald, 1890), 13; Anne Radford Phillips, "Farm Women of Stokes County, North Carolina and the Production of Flue-Cured Tobacco, 1925–1955: Continuity and Change" (Ph.D. diss., University of Maryland, 1990), 35; Robert C. McAdams, "The Tobacco Culture of Wilson County, North Carolina" (Ph.D. diss., University of Tennessee, 1996), 72; and Bennett, "King Bacca's Throne," 78. For average estimated wood consumption during the mid-1800s, see Michael Williams, *Americans and Their Forests: A Historical Geography* (New York: Cambridge University Press, 1989), 334. Tobacco production for the three counties comes from the 1860 agricultural census. Average production per barn comes from Samuel Shelton, "Culture and Management of Tobacco," 216.

92. Williams, *Americans and Their Forests,* 341–344; and Donald E. Davis, *Where There are Mountains: An Environmental History of the Southern Appalachians* (Athens: University of Georgia Press, 2000), 148–153.

93. Anonymous farmer's notes on how "To Fire with Coal," n.d., Southside Virginia Family Papers, *RASP,* series E, part 3, reel 5.

94. Jeffreys Diary, January 10, 1845.

95. Drane, *Soil Survey of Caswell County,* 21, 26; Jurney et al., *Soil Survey of Halifax County,* 17, 21, 33–34, 40–42; Kirk et al., *Soil Survey of Pittsylvania,* 28, 33–34; "Farmers' State Club," *Southern Planter* 11, 3 (March 1851): 76. The early twentieth-century soil surveys of Caswell, Halifax, and Pittsylvania list productivity figures for unfertilized soil series. New corn and wheat strains had slightly improved productivity, but erosion and agricultural use had also lessened soil fertility by the early 1900s, and thus the surveys should provide relatively accurate statistics for determining antebellum productivity.

96. Ruffin, in Mathew, *Edmund Ruffin and the Crisis of Slavery,* 57.

97. For examples of the detailed instructions provided to the region's antebellum bright leaf cultivators, see R. J. S., "The Curing of Tobacco with Charcoal," *Southern Planter* 18, 10 (October 1858): 595–596; "Caswell," "On Curing Tobacco Yellow," *Southern Planter* 19, 8 (August 1859): 492; and Shelton, "Culture and Management of Tobacco," 209–217. The first two articles were written by Caswell planters, and the last by a Pittsylvania grower.

98. Love and Love, *Art of Curing,* 3, 6–8.

99. Cohen, *Notes from the Ground,* esp. 17–48.

100. Ellen Glasgow, *The Deliverance: A Romance of the Virginia Tobacco Fields* (New York: Doubleday Page, 1904), 8.

101. Tilley, *Bright-Tobacco Industry,* 25–30.; and Seigel, *Roots of Southern Distinctiveness,* 102.

102. Jeffreys Diary, January 20, 1845.

103. Robert L. Ragland, "A Scrap from the History of Yellow Tobacco," *Southern Planter and Farmer* 37, 5 (May 1876): 344; and Tilley, *Bright-Tobacco Industry,* 26.

104. For examples of these interactions, see List of tobacco purchased and manufactured, November 1, 1854; John T. Garland to William T. Sutherlin, March 1857; Armistead T. Moore to William T. Sutherlin, April 26, 1860; Bedford Brown to William T. Sutherlin, May 35, 1860, all in William Thomas Sutherlin Papers, SHC; Thomas Lindsey to John Sutherlin, April 26, 1858, William Thomas Sutherlin Papers, Duke; and Account Book of Robert Wilson, 1848–1861, Robert Wilson Account Books, oversize volume SV-1896/77, SHC.

105. Lawrence P. Jackson, *My Father's Name: A Black Virginia Family after the Civil War* (Chicago: University of Chicago Press, 2012), 62–63.

106. Stoll, *Larding the Lean Earth,* 108–119; and Address of James C. Bruce to Mecklenburg and Granville Agricultural Clubs, pp. 7–15, Bruce Family Papers, *RASP,* series E, part 3, reel 14.

107. Anthony M. Tang, *Economic Development in the Southern Piedmont, 1860–1950* (Chapel Hill: University of North Carolina Press, 1958), 46.

108. Robert, *Tobacco Kingdom,* 63; Kathleen Bruce, "Materials for Southern Virginia Agricultural History," *Agricultural History* 4, 1 (January 1930): 12; Charles B. Motley, *Yes There Is a Dry Fork Virginia* (Danville, VA: Bassett, 1977), 9; L. Beatrice Hairston, *A Brief History of Danville, Virginia, 1728–1954* (Richmond, VA: Dietz, 1955), 15–16; and Clement, *History of Pittsylvania,* 240–241.

109. Motley, *Yes There Is a Dry Fork,* 30–31; A. W. Barksdale to R. L. Barksdale, September 23, 1857, Peter Barksdale Papers, *RASP,* series F, part 3, reel 33, quote in the latter.

110. Vincent Witcher, "Richmond and Danville Railway," *American Railway Times* 8, 3 (January 1856): 1; and Robert, *Tobacco Kingdom,* 67–71.

111. "Remarks of Mr. Cunningham of Person on the Dan River and Yadkin Railroad," *Milton Chronicle,* February 22, 1855, p. 1; Edward Pollock, *Illustrated Sketch Book of Danville, Virginia, Its Manufactures and Commerce* (Danville, VA: John M. Hutchings and Son, 1885), 71; Durwood T. Stokes, "Charles Napoleon Bonaparte Evans and the Milton Chronicle," *North Carolina Historical Review* 44, 3 (July 1969): 253; Siegel, *Roots of Southern Distinctiveness,* 110–111, 153; and Robert, *Tobacco Kingdom,* 70–71.

112. Marrs, *Railroads in the Old South,* 113–114.

113. Killebrew and Myrick, *Tobacco Leaf,* 11. See also, Hearn and Drane, *Soil Survey of Caswell,* 9.

114. H. P. Womack to Pleasant H. Womack, April 19, 1847, and H. P. Womack to Pleasant H. Womack, April 24, 1853, both in Hatchett Family Papers, box 2, Duke.

115. "Extraordinary Prices Paid for Tobacco," *Milton Chronicle,* February 12, 1857, p. 3. Some of the details of this article were reprinted in the *Southern Planter:* "Farmers' Meeting at the Exchange Hotel," *Southern Planter* 17, 4 (April 1857): 225.

116. Evans, quoted in Stokes, "Charles Napoleon Bonaparte Evans," 254.

117. Peter Hairston to George Hairston, March 22, 1859, Hairston Family Papers, VHS.
118. Receipts of William Bailey, September 29, 1858–May 19, 1859, William Bailey Papers, *RASP*, series E, part 1, reel 1; Receipts of Elisha Barksdale, September 1, 1841–September 10, 1852, Receipts of Joel Hubbard, April 16, 1840–June 16, 1859, and John Tyree to Joel Hubbard, May 9, 1857, all at Southside Virginia Family Papers, UVA; Account Ledger of John Sims and William Sims, 1828–1867, William Bailey Papers, *RASP*, series E, part 1, reel2; Receipts of William T. Sutherlin, June 16, 1849–December 30, 1859, William Thomas Sutherlin Papers, SHC; Receipts of Sutherlin and Ferrell, October 4, 1858–October 18, 1858, William Thomas Sutherlin Papers, Duke; and Receipt of Philip H. Howerton, May 29, 1851, Charles H. Cabaniss Papers, Duke.
119. Jeffreys Diary, January 20, February 26, 1845; Tilley, *Bright-Tobacco Industry*, 22–25; Coleman Diaries, December 16, 1846; and Ricketts, ed., *Diary Kept by William C. Shelton*.
120. Receipt of Philip H. Howerton, September 29, 1855, Philip H. Howerton Papers, Duke; Receipt of F. R. Cousins, October 14, 1859, F. R. Cousins Papers, Duke; and Receipt of Rufus H. Owen, September 14, 1861, Owen Family Papers, VHS.
121. George Fitzhugh, *Cannibals All! or, Slaves without Masters* (Richmond, VA: A. Morris, 1857), xiii.
122. Cohen, *Notes from the Ground*, 85–99; Mark Fiege, *The Republic of Nature: An Environmental History of the United States* (Seattle: University of Washington Press, 2012), 176.

CHAPTER 4. TOBACCO GOES TO WAR

1. For a summary of the evacuation from Richmond to Danville, see Nelson D. Lankford, *Richmond Burning: The Last Days of the Confederate Capital* (New York: Viking, 2002), 71, 104–105; Cassye Averett Young, *The Last Capitol of the Confederacy, Danville, Virginia, as the President of the Confederacy Saw It on His Stay Here from April 3 to April 10, 1865* (Danville, VA: Danville Printing, 1955), 4–9; and Myrta Lockett Avary, *Dixie after the War: An Exposition of Social Conditions Existing in the South, during the Twelve Years Succeeding the Fall of Richmond* (New York: Doubleday, Page, 1906), 47–52.
2. Lisa Brady, "The Wilderness of War: Nature and Strategy in the American Civil War," *Environmental History* 10, 3 (July 2005): 422. See also Lisa Brady, *War upon the Land: Military Strategy and the Transformation of Southern Landscapes during the American Civil War* (Athens: University of Georgia Press, 2012); Megan Kate Nelson, *Ruin Nation: Destruction and the American Civil War* (Athens: University of Georgia Press, 2012).
3. John Inscoe has called for more studies of slavery and economy in portions of the South behind the battle lines. See Inscoe, *Race, War, and Remem-*

brance in the Appalachian South (Lexington: University Press of Kentucky, 2008), 81.

4. Siegel, *Roots of Southern Distinctiveness,* 153.

5. Tilley, *Bright-Tobacco Industry,* 358.

6. Siegel, *Roots of Southern Distinctiveness,* 149–152; and "Maj. William T. Sutherlin," *Southern Planter and Farmer* 10, 10 (October 1875): 596.

7. Jeffrey W. McClurken, *Take Care of the Living: Reconstructing Confederate Veteran Families in Virginia* (Charlottesville: University of Virginia Press, 2009), 12–13; Jack P. Maddex, Jr., *The Virginia Conservatives, 1867–1879: A Study in Reconstruction Politics* (Chapel Hill: University of North Carolina Press, 1970), 22; Richard Lowe, *Republicans and Reconstruction in Virginia, 1856–1870* (Charlottesville: University Press of Virginia, 1991), 10–11; and James M. McPherson, *Battle Cry of Freedom: The Civil War Era,* reprint (New York: Ballantine, 1989), 273–274, 277–278.

8. United Daughters of the Confederacy, *War Recollections of the Confederate Veterans of Pittsylvania County, Virginia, 1861–1865* (Chatham, VA: Randall D. Reynolds, 1961), 5–6; and Clement, *History of Pittsylvania County,* 246–247.

9. Jeffrey McClurken, "After the Battle: Reconstructing the Confederate Veteran Family in Pittsylvania County and Danville, Virginia, 1860–1900" (Ph.D. diss., Johns Hopkins University, 2002), 5.

10. Edmunds, *History of Halifax,* 497–498; Carrington, *History of Halifax,* 516–518; Alfred J. Morrison, *Halifax County, Virginia, a Handbook: Prepared under the Direction of the Board of Supervisors* (Richmond, VA: Everett Waddey, 1907), 86.

11. Robert Enoch Withers, *Autobiography of an Octogenarian* (Roanoke, VA: Stone, 1907), 125; and United Daughters of the Confederacy, *War Recollections,* 12–13, quote on 13.

12. Hahn, "Making Tobacco Bright," 113–114; McPherson, *Battle Cry of Freedom,* 382–383; and J. G. Randall and David Donald, *The Civil War and Reconstruction,* 2nd ed. (Lexington, MA: D. C. Heath, 1969), 440–442.

13. Ludwell Johnson, "Contraband Trade during the Last Year of the Civil War," *Mississippi Valley Historical Review* 49, 4 (March 1963): 647.

14. William Sours to John Sours, February 12, 1861, Sours Family Papers, folder 21, SHC.

15. Receipts of James C. Bruce, March 11, April 11, 1861, both in Bruce Family Papers, *RASP,* series E, part 3, reel 19.

16. Williams and Carrington to William Bailey, April 19, 1861, William Bailey Papers, *RASP,* series E, part 1, reel 1.

17. Van Bienthussen and Crafton to William T. Sutherlin, September 7, 1861; Pemberton and Carter to William T. Sutherlin, September 9, 1861; and Hill and Warren to William T. Sutherlin, September 9, 1861, all in William Thomas Sutherlin Papers, SHC. Quote in the latter.

18. B. F. Grasty to William T. Sutherlin, August 20, 1861, William Thomas Sutherlin Papers, SHC.

19. Thomas Samuel to Sutherlin and Ferrell, February 19, 1863, William Thomas Sutherlin Papers, Duke.

20. For examples of communications between Sutherlin and his agents in the field, see Johnson H. Owen to William T. Sutherlin, September 20, 1861; Johnson H. Owen to William T. Sutherlin, William Thomas Sutherlin Papers, SHC; W. W. Bunch to William T. Sutherlin, January 4, 1861, William Thomas Sutherlin Papers, SHC; and S. R. Neal to William Sutherlin, November 26, 1862, William Thomas Sutherlin Papers, Duke.

21. Advertisement, *Danville Appeal,* March 26, 1864, p. 4.

22. For this practice during the antebellum era, see Robert, *Story of Tobacco in America,* 94; Gillespie, *Katherine and R. J. Reynolds,* 26–27.

23. Samuel Ayers to William T. Sutherlin, January 4, 1862, William Thomas Sutherlin Papers, SHC; and Elijah Torrian to Sutherlin and Ferrell, February 19, 1863, William Thomas Sutherlin Papers, Duke.

24. Johnson H. Owen to William T. Sutherlin, August 27, 1861, 71, SHC.

25. Johnson H. Owen to William T. Sutherlin, January 18, 1862, and John Sutherlin to William T. Sutherlin, January 21, 1862, William T. Sutherlin Papers, folder 15, SHC; and James M. Norman to William T. Sutherlin, June 1, 1862, William Thomas Sutherlin Papers, SHC. John Sutherlin advised William that tobacco was selling two to three dollars higher than it had twelve months previous.

26. E. M. Gardient to William T. Sutherlin, August 19, 1861, William Thomas Sutherlin Papers, SHC; and Middleton and Son to William T. Sutherlin, June 20, 1862, William Thomas Sutherlin Papers, SHC.

27. James R. Millner, "To the Public," March 20, 1863, pp. 3–5, 16, 23–24, in *Confederate Imprints, 1861–1865* (New Haven: Research Publications, 1974) (hereafter cited as *Confederate Imprints*), reel 94; and John Sutherlin to William T. Sutherlin, January 19, 1863, William Thomas Sutherlin Papers, Duke.

28. William T. Sutherlin, "A Reply to the Publication of James R. Millner, Dated March 20th, 1863," July 28, 1863, 3, 20, 27, 32, in *Confederate Imprints,* reel 95, italics in the original.

29. Though hard figures are impossible to obtain, there is substantial anecdotal evidence of widespread wartime tobacco cultivation. For examples, see Hahn, "Making Tobacco Bright," 119–120; Probate inventory of John T. Muse, 1864, Southside Virginia Family Papers, *RASP,* series E, part 3, reel 3; Receipt of William A. J. Finney, September 2, 1863, Southside Virginia Family Papers, *RASP,* series E, part 3, reel 4; Receipt of Joel Hubbard, October 14, 1865, Southside Virginia Family Papers, *RASP,* series E, part 3, reel 3; Receipt of W. C. Tate, April 2, 1866, Southside Virginia Family Papers, *RASP,* series E, part 3, reel 6; and Merritt Diary, September 12, 1863. Though after the end of the war, tobacco receipts from late 1865 and early 1866 record the sale of 1865 crops started in plant beds in the winter of 1864/1865.

30. Anderson Willis to William T. Sutherlin, August 28, 1861, William Thomas Sutherlin Papers, SHC. Anderson's letter also illustrated the wartime demand for bright leaf. He offered his crop to Sutherlin but informed the manufacturer that there was another Danville buyer waiting in the wings.

31. James H. Brewer, *The Confederate Negro: Virginia's Craftsmen and Military Laborers, 1861–1865* (Durham, NC: Duke University Press, 1969), 125.

32. George Jones, in McClurken, *Take Care of the Living*, 30.

33. William R. Hatchett to Allen Hatchett, October 1, 1862, Hatchett Family Papers, Duke. Continued tobacco production could pay in other ways as well. When Caswell soldier and tobacco farmer Bartlett Yancey Malone was captured and imprisoned at Point Lookout, Maryland, his father sent him a box of bright chewing tobacco, which he parceled out and sold to his fellow prisoners for the healthy sum of $55.70. Bartlett Yancey Malone, *The Diary of Bartlett Yancey Malone*, ed. J. G. De Roulhac Hamilton, Henry M. Wagstaff, and William W. Pierson, Jr. (Chapel Hill: University of North Carolina, 1919), 53–54.

34. "To Planters," *Milton Chronicle*, June 28, 1861, p. 2.

35. Account Book of Robert Wilson, 1861–1865, 112, Robert Wilson Account Books, oversize volume SV-1896/77, SHC.

36. A. Wellis to William T. Sutherlin, February 22, 1863, William Thomas Sutherlin Papers, Duke.

37. Duval Porter, ed., *Men, Places and Things, as Noted by Benjamin Simpson* (Danville, VA: Dance Brothers, 1891), 206.

38. Receipt of William A. J. Finney, September 22, 1863, Southside Virginia Family Papers, *RASP*, series E, part 3, reel 4.

39. John Booker to William Sims, January 22, 1862, Bailey Family Papers, VHS.

40. John Sutherlin to William T. Sutherlin, February 13, 1863, William Thomas Sutherlin Papers, Duke.

41. Brandt, *Cigarette Century*, 25.

42. This per capita estimate is based on a national population of 31,443,008 and $21,820,535 worth of manufactured tobacco products. It assumes an average value of 30 to 40 cents per pound for tobacco products, a range drawn from wide reading. Price fluctuations outside of these estimates would vary the per capita figure, but the point that the nation consumed a great deal of tobacco would remain valid. See *Manufactures of the United States in 1860; Compiled from the Original Returns of the Eighth Census* (Washington, DC: Government Printing Office, 1865), 732; and Kennedy, *Population of the United States*, vii.

43. Dickens, in Gately, *Tobacco*, 174.

44. Bell Irvin Wiley, *The Life of Billy Yank: The Common Soldier of the Union* (New York: Bobbs-Merrill, 1952), 101–102, quote on 101.

45. Bell Irvin Wiley, *The Life of Johnny Reb: The Common Soldier of the Confederacy* (New York: Charter, 1962), both quotes on 172; and Malone, *The Diary of*

Barlett Yancey Malone, 53–54. The letters of a Wisconsin soldier suggested that bored troops even chewed tobacco (and spit) when in local churches. See John O. Holzhueter, "William Wallace's Civil War Letters: The Virginia Campaign," *Wisconsin Magazine of History* 57, 1 (Autumn 1973): 39.

46. Editorial, *Milton Chronicle,* November 9, 1863, p. 2. See also Durwood T. Stokes, "Charles Napoleon Bonaparte Evans and the *Milton Chronicle,*" *North Carolina Historical Review* 44, 3 (July 1969): 258.

47. For selected examples of farmers and planters who made efforts to plant grain in preference to tobacco over the course of the war, see Elijah Hundley, Journal No. 1, Plantation Records of Elijah Hundley, Small Special Collections, UVA; Williams and Carrington to William Bailey, March 28, 1862, Bailey Family Papers, VHS; Merritt Diary; Account Ledger of John Sims and William Sims, 1829–1881, William Bailey Papers, *RASP,* series E, part 1, reel 2. William Sims made perhaps the most dramatic shift in production: his 1861 records record forty-one hogsheads of tobacco produced on his Halifax plantation, yet his 1862 accounts listed none (though Sims did produce at least some tobacco throughout the course of the war).

48. William Ayers to William T. Sutherlin, January 22, 1862, William Thomas Sutherlin Papers, SHC.

49. E. S. Yarbrough, "Yarbrough's Foundry," unpublished typescript, 1960, p. 2.

50. Joseph C. G. Kennedy, *Agriculture of the United States in 1860, Compiled from the Original Returns of the Eighth Census* (Washington, DC: Government Printing Office, 1864), 104–106, 154–156, 158–160.

51. Drane, *Soil Survey of Caswell County,* 21, 26; Jurney et al., *Soil Survey of Halifax County,* 17, 21, 33–34, 40–42; Kirk et al., *Soil Survey of Pittsylvania,* 28, 33–34; "Farmers' State Club," *Southern Planter* 11, 3 (March 1851): 76. The early twentieth-century soil surveys of Caswell, Halifax, and Pittsylvania list productivity figures for unfertilized soil series. New corn and wheat strains had slightly improved productivity, but erosion and agricultural use had also lessened soil fertility by the early 1900s, and thus the surveys should provide relatively accurate statistics for determining antebellum productivity. This claim is also supported by antebellum anecdotal evidence.

52. Sam Bowers Hilliard, *Hog Meat and Hoecake: Food Supply in the Old South, 1840–1860* (Carbondale: Southern Illinois University Press, 1972), 106–109. In 1860, Caswell, Halifax, and Pittsylvania were all below the 2.2 swine per person ratio that Hilliard defined as the pork sufficiency line.

53. James W. Richmond to William Bailey, June 27, 1851, Bailey Family Papers, VHS. Richmond was a drover from Scott County, Virginia, who offered to drive a herd of hogs to meet the needs of northeastern Halifax County.

54. John Booker to William Sims, January 22, 1862, and August 15, 1863, both in Bailey Family Papers, VHS; Merritt Diary, October 24, 1863; and Receipt of Samuel P. Wilson, October 3, 1864, Samuel Pannill Wilson Papers, Small Special Collections, UVA.

55. Tax Receipt of Samuel P. Wilson, November 6, 1865, Samuel Pannill Wilson Papers, Small Special Collection, UVA.

56. Account Ledger of John Sims and William Sims, 1829–1881, William Bailey Papers, *RASP*, series E, part 1, reel 2.

57. Hairston Family Plantation Record Book 2, Peter Wilson Hairston Papers, box 8, folder 105, SHC.

58. Probate inventory of John T. Muse, 1864, Southside Virginia Family Papers, *RASP*, series E, part 3, reel 3.

59. Colonel Larkin Smith, "Notice to Farmers and Quartermasters," June 13, 1863, Samuel Pannill Wilson Papers, Small Special Collections, UVA.

60. Commissary Receipts of Samuel P. Wilson, 1863–1865, Samuel Pannill Wilson Papers, Small Special Collections, UVA.

61. Tax in Kind Receipt of Pleasant H. Womack, July 7, 1864, Hatchett Family Papers, Duke.

62. For an example of an impressment notice, see Impressment Notice, March 16, 186_, in George Hairston Papers, folder 18, SHC.

63. Major and Quartermaster Jason G. Paxton, "To the Farmers of Campbell, Franklin, Henry, Patrick, Grayson, Carroll, Floyd, the Western Part of Pittsylvania and Halifax, and the Southern Part of Bedford Counties," Broadside, November 13, 1863, in *Confederate Imprints*, reel 83; McClurken, "After the Battle," 64.

64. Commissary Receipts of Samuel P. Wilson, 1863–1865, Samuel Pannill Wilson Papers, Small Special Collection, UVA. The tax in kind continued under the new regulations as well, with quartermasters writing off 10 percent of seizures as tax, and issuing small payments of IOUs for the remainder of the seized goods.

65. For examples, see Sallie J. Sims to William Sims, n.d. (probably 1864), and James Young to Phoebe Bailey, February 29, 1864, both in Bailey Family Papers, VHS.

66. Benjamin Lines Farinholt to William Sims, July 31, 1864, and William Sims to Phoebe Bailey, August 17, 1864, both in Bailey Family Papers, VHS. Quote in the latter. Seizures and impressment of agricultural produce appears to have been carried out in a fairly uniform manner—that is to say, in the region privilege did not protect a well-to-do planter from impressment—as evinced by the fact that the Sims family had personal ties with General Lee yet still faced food and slave impressments. For the family's familiarity with Lee, see Robert E. Lee to Maria C. Sims, May 16, 1862, Robert E. Lee Papers, folder 1, SHC.

67. Robert Enoch Withers, *Autobiography of an Octogenarian* (Roanoke, VA; Stone, 1907), 209.

68. The body of literature on slavery, paternalism, and slave resistance is far too vast to explore here, but one of the best places to start remains Eugene Genovese's *Roll, Jordan, Roll: The World the Slaves Made* (New York: Random House, 1974).

69. Jackson, *My Father's Name*, 107–109.
70. Brewer, *Confederate Negro*, 8; Blair, *Virginia's Private War*, 122–124; Jackson, *My Father's Name*, 152–153. For an example of slave rentals to the Confederate government, see Hairston Family Plantation Record Book 1, 46, Peter Wilson Hairston Papers, box 8, folder 104, SHC.
71. Brewer, *Confederate Negro*, 41–42, 87–88, 141–142. This railroad spur allowed easier transportation of Caswell tobacco to Danville by the end of the war and was the idea of a number of wealthy Southside planters and tobacco manufacturers, including William Sutherlin. See Subscribers to the Piedmont Railroad Company, February 8, 1862, William Thomas Sutherlin Papers, Duke.
72. William Sims to Phoebe Bailey, August 17, 1864, Bailey Family Papers, VHS; and Commissary Receipts of Samuel P. Wilson, 1863–1865, Samuel Pannill Wilson Papers, Small Special Collection, UVA.
73. Blair, *Virginia's Private War*, 100–101.
74. In Brewer, *Confederate Negro*, 125.
75. William Sims to William Bailey, March 15, 1861, Bailey Family Papers, VHS.
76. Phoebe Bailey to William Sims, March 1, 1864, Bailey Family Papers, VHS.
77. James I. Robertson, Jr., "Houses of Horror: Danville's Civil War Prisons," *Virginia Magazine of History and Biography* 69, 3 (July 1961): 329–331; W. H. Newlin, *An Account of the Escape of Six Federal Soldiers from Prison at Danville, VA: Their Travels by Night through the Enemy's Country to the Union Pickets at Gauley Bridge, West Virginia, in the Winter of 1863–64*, reprint of 1870 ed. (Cincinnati: Western Methodist, 1886), 9–10; Edward Pollock, *Illustrated Sketch Book of Danville, Virginia, Its Manufactures and Commerce* (Danville, VA: John M. Hutchings and Son, 1885), 46; and Perdue, Barden, and Phillips, *Weevils in the Wheat*, 152. The various accounts of Danville's prisons alternate between referring to the facilities as converted factories and warehouses. Based on prisoner accounts of the prisons, especially references to multi-floor structures, they were the former.
78. Jeffrey McClurken, "After the Battle: Reconstructing the Confederate Veteran Family in Pittsylvania County and Danville, Virginia, 1860–1900" (Ph.D. diss., Johns Hopkins University, 2002), 54–57; Tilley, *Bright-Tobacco Industry*, 36; and Siegel, *Roots of Southern Distinctiveness*, 153. Alternately, farmers and planters with enough barns may have stored cured tobacco on their farms for almost a year before having to clear their barns for the following year's crop, a practice that could have temporarily reduced the need for storage space in town.
79. Withers, *Autobiography of an Octogenarian*, 195. Hesseltine incorrectly gives 2,400 as the maximum prisoner population in the town. See William B. Hesseltine, *Civil War Prisons: A Study in War Psychology*, reprint of 1930 ed. (New York: Frederick Ungar, 1971), 169.
80. For refugees in Danville and the surrounding countryside, see ibid., 191; Siegel, *Roots of Southern Distinctiveness*, 153; and United Daughters of the Confederacy, *War Recollections*, 81–84.

81. Robertson, "Houses of Horror," 345. For accounts of the unsanitary conditions and disease, see ibid., 331–335; George Haven Putnam, *A Prisoner of War in Virginia, 1864–5* (New York: G. P. Putnam's Sons, 1912), 33–35; Alfred S. Roe, *In a Rebel Prison; or, Experiences in Danville, VA* (Providence: Rhode Island Soldiers and Sailors Historical Society, 1891), 6; Newlin, *Account of the Escape*, 9–10; and Withers, *Autobiography of an Octogenarian*, 198–200.

82. Roe, *In a Rebel Prison*, 10, 32–33. Roe is the only surviving prisoner to mention this practice.

83. Putnam, *Prisoner of War*, 36–37.

84. Roe, *In a Rebel Prison*, 12, 15; and Robertson, "Houses of Horror," 336–338, quote on 337.

85. Newlin, *Account of the Escape*, 10–11.

86. Merritt Diary, September 9, November 17, 1862, October 24, 1863; and A. B. Cabarriss to William Sims, March 31, 1863, in Bailey Family Papers, VHS.

87. Bird L. Ferrell to J. W. Ferrell, February 15, 1864, William Thomas Sutherlin Papers, Duke.

88. William Sours to John Sours, October 9, 1865, Sours Family Papers, folder 21, SHC.

89. Avary, *Dixie after the War*, 150.

90. Booker, Kerr and Lyon (of Richmond) to William Sims, January 21, 1865, Bailey Family Papers, VHS.

91. Advertisement, *Milton Chronicle*, April 10, 1863, p. 1; and "Game," *Danville Daily New Era*, June 1, 1865, p. 1. At the outbreak of the war, prime Peruvian guano sold for fifty dollars per ton in Milton (Advertisement, *Milton Chronicle*, April 12, 1861, p. 3).

92. McClurken, *Take Care of the Living*, 32–33.

93. Thomas Patrick to ——— Patrick, January 31, 1865, and Neal and Whittock to William T. Sutherlin, January 5, 1865, both in William Thomas Sutherlin Papers, SHC.

94. Receipt of Joel Hubbard, October 14, 1865, Southside Virginia Family Papers, *RASP*, series E, part 3, reel 3; and Receipt of W. C. Tate, April 2, 1866, Southside Virginia Family Papers, *RASP*, series E, part 3, reel 6.

95. Lankford, *Richmond Burning*, 86, 90, 104; and Young, *Last Capitol*, 5.

96. Lankford, *Richmond Burning*, 104–105; Siegel, *Roots of Southern Distinctiveness*, 158–159; and Withers, *Autobiography of an Octogenarian*, 218–225.

97. Siegel, *Roots of Southern Distinctiveness*, 159; "Warning to Marauders," *Sixth Corps*, May 3, 1865, p. 2. The issues of the *Sixth Corps* newspaper contained messages emphasizing the importance of order and loyalty and sought to prevent local opposition to the occupation.

98. Newlin, *Account of the Escape*, 22–24.

99. Alanson A. Haines, *History of the Fifteenth Regiment New Jersey Volunteers* (New York: Jenkins and Thomas, 1883), 310–311, quotes on 311.

100. Charles M. Snyder, "A Teen-Age G.I. in the Civil War," *New York History* 35 (1954): 29.

101. Dennett, *South as It Is*, 93.

102. Ibid., 103.

103. Ibid., 94–95.

104. Edward A. Pollard, *The Virginia Tourist: Sketches of the Springs and Mountains of Virginia* (Philadelphia: J. B. Lippincott, 1871), 18. Pollard's description was likely of the countryside closer to Richmond.

105. John Townsend Trowbridge, *The South: A Tour of Its Battlefields and Ruined Cities, a Journey through the Desolated States, and Talks with the People, 1867*, ed. J. H. Segars (Macon, GA: Mercer University Press, 2006), 225. Trowbridge named no particular commonwealth locale in his description; rather he applied his remarks to all of the state's tobacco district.

106. Bagby, in James Tice Moore, *Two Paths to the New South: The Virginia Debt Controversy* (Lexington: University Press of Kentucky, 1974), 9.

107. Frederick Law Olmsted, *The Cotton Kingdom: A Traveler's Observations on Cotton and Slavery in the American Slave States, Based upon Three Former Volumes of Journeys and Investigations by the Same Author* (New York: Mason Bros., 1861).

108. William Sours to John Sours, November 19, 1865, Sours Family Papers, folder 21, SHC.

109. Avary, *Dixie after the War*, 52–56.

110. John Gilliam to William T. Sutherlin, June 14, 1865, William Thomas Sutherlin Papers, SHC; and U.S. Census Bureau Slave Schedule, Pittsylvania County, 1860. One of Sutherlin's first personal orders following the end of the war reflected his wealth. In October he sent off to New York for a number of cases of champagne, claret, and sherry, a cask of bourbon, a walnut bed, and Javanese coffee. See Receipt of William T. Sutherlin, October 31, 1865, William Thomas Sutherlin Papers, SHC.

111. Compiled from HCDB. The exact figure from the 1860 census was 38,592 slaves out of a total population of 74,839.

CHAPTER 5. FIRE IN THE FIELDS

1. Clement, *History of Pittsylvania County*, 255–256. On Clement's background, see Jackson, *My Father's Name*, 82–84.

2. The Virginia branch of the BRFAL was established May 31, 1865, to mediate black and white relations during Reconstruction (Richard Lowe, *Republicans and Reconstruction in Virginia, 1856–1870* [Charlottesville: University Press of Virginia, 1991], 29).

3. Complaint of Griffin Cobb vs. John Blackwell, September 12, 1867, Greensboro Field Office, Records of the Bureau of Refugees, Freedmen, and Abandoned Lands, National Archives and Records Administration, Washington, DC (hereafter BRFAL), RG 105, M1909, roll 21. National Archives BRFAL documents are also available digitally through Ancestry.com.

4. Mark Fiege has beautifully described these connections for the antebellum cotton cycle in *Republic of Nature*, 106–117.

5. Ibid., 411.

6. Ibid., 413.

7. On the eco-social changes wrought by the war on southern landscapes, see Megan Kate Nelson, *Ruin Nation: Destruction and the American Civil War* (Athens: University of Georgia Press, 2012); Lisa M. Brady, *War upon the Land: Military Strategy and the Transformation of Southern Landscapes during the American Civil War* (Athens: University of Georgia Press, 2012); G. Terry Sharrer, *A Kind of Fate: Agricultural Changes in Virginia, 1861–1920* (West Lafayette, IN: Purdue University Press, 2002).

8. Kirby, *Mockingbird Song*, 128–129; Jackson, *My Father's Name*, 147, 149–150. A full 25 percent of all military-age men in Pittsylvania died in service over the course of the war. Almost half of the survivors were wounded or suffered from a serious illness. McClurken, *Take Care of the Living*, 4.

9. Merritt Diary, May 29, 1866.

10. Duval Porter, ed., *Men, Places and Things, as Noted by Benjamin Simpson* (Danville, VA: Dance Brothers, 1891), 9–10. Taxes on land suddenly bereft of the labor necessary to work it created an added burden. See Tax Receipt of Samuel Hairston, March 7, 1866 (two), Peter Wilson Hairston Papers, folder 54, SHC.

11. The necessity of expensive equipment coupled with the lack of capital in the postwar South greatly hampered more capital-intensive crop cultures. In just one example, Louisiana sugar producers needed $25 million to cover operating expenses in 1867, money impossible to find during southern Reconstruction. See Rodrigue, *Reconstruction in the Cane Fields*, 59. See also J. Carlyle Sitterson, *Sugar Country: The Cane Sugar Industry in the South, 1753–1950* (Lexington: University of Kentucky Press, 1953), chap. 11.

12. Receipts of Joel Hubbard, August 2, 13, 4, 1867, and Receipt of W. C. Tate, April 2, 1866, all in Southside Virginia Family Papers, *RASP*, series E, part 3, reels 3 and 6; Receipt of Philip Howerton, February 6, 1866, Philip H. Howerton Papers, Duke; and William R. Hatchett to Allen Hatchett, April 9, 1866, Hatchett Family Papers, Duke.

13. Merritt Diary, May 14, 1865.

14. Bird L. Ferrell to J. W. Ferrell, November 23, 1865, William Thomas Sutherlin Papers, Duke.

15. William H. Sims to Phoebe Bailey, May 25, 1865, in Bailey Family Papers, VHS. For Sims's 1863 slaveholding, see Account Ledger of John Sims and William Sims, 1829–1881, in William Bailey Papers, *RASP*, series E, part 1, reel 2.

16. Jackon, *My Father's Name*, 79.

17. Withers, *Autobiography of an Octogenarian*, 229.

18. Despite his early affiliation with the BRFAL, Withers was an ex-Confederate colonel and staunch conservative who ran unsuccessfully for governor in 1868 and served as a Democratic U.S. senator from 1875 to 1881. In his office and on the campaign trail, Withers used appeals to racism to combat

Republican and Readjuster calls for reform, equating a vote for either party with a vote for black political domination. See James Tice Moore, *Two Paths to the New South: The Virginia Debt Controversy, 1870–1883* (Lexington: University Press of Kentucky, 1974), 137; and Lowe, *Republicans and Reconstruction*, 151–152.

19. Withers, *Autobiography of an Octogenarian*, 229–230. The Pittsylvania bureau was at first located in one of William Sutherlin's Danville houses (not the same mansion used by Confederate President Davis), though it is unclear if Sutherlin rented the building to the agency or if federal officials commandeered the structure. Either way, Sutherlin acquiesced to the BRFAL's actions in a desire to see the town return to business as usual. See William B. Payne to William T. Sutherlin, December 26, 1865, William Thomas Sutherlin Papers, SHC.

20. Kerr-Ritchie, *Freedpeople in the Tobacco South*, 40.

21. Jackson, *My Father's Name*, 74, 76–77 (quote on 74).

22. Captain J. F. Wilcox to R. S. Lacey, January 20, 1866, Danville Letterbook, BRFAL, RG 105, M1913, roll 71 (hereafter Danville Letterbook); and Bird L. Ferrell to J. W. Ferrell, November 23, 1865, William Thomas Sutherlin Papers, Duke.

23. C. Thurston Chase to Orlando Brown, June 29, 1868, Danville Letters Sent and Received, BRFAL, RG 105, M1913, roll 72; quote in Captain J. F. Wilcox to R. S. Lacey, December 20, 1865, Danville Letterbook.

24. John Flintoff in Powell, *History of Caswell County*, 228. See also Edward King, *The Southern States of North America: A Record of Journeys in Louisiana, Texas, the Indian Territory, Missouri, Arkansas, Mississippi, Alabama, Georgia, Florida, South Carolina, North Carolina, Kentucky, Tennessee, Virginia, West Virginia and Maryland*, vol. 3 (London: Blackie and Son, 1875), 554.

25. Kerr-Ritchie, *Freedpeople in the Tobacco South*, 41.

26. Steven Hahn, *A Nation under Our Feet: Black Political Struggles in the Rural South from Slavery to the Great Migration* (Cambridge: Belknap/Harvard University Press, 2003), 135.

27. Eric Foner, *Reconstruction: America's Unfinished Revolution, 1863–1877* (New York: Harper and Row, 1988), 70–71.

28. Contract between Robert Wilson and David Wilson et al., February 20, 1869, Hairston and Wilson Family Papers, box 2, folder 32, SHC. For similar regional contracts, see Contract between William T. Sutherlin and John Kelly, January 13, 1870, and Contract between William T. Sutherlin, John Edwards, and Nat Jones, January 1, 1871, both in William Thomas Sutherlin Papers, Duke.

29. Slave narrative of William Williams, in George P. Rawick, ed., *The American Slave: A Composite Autobiography*, vol. 12 (Westport, CT: Greenwood, 1972), 115–116; Stewart, *"What Nature Suffers to Groe,"* 195–196; and Kirby, *Mockingbird Song*, 216–217; Scott E. Giltner, *Hunting and Fishing in the New*

South: Black Labor and White Leisure after the Civil War (Baltimore: Johns
Hopkins University Press, 2008), esp. 10–44.

30. Perdue, Barden, and Phillips, *Weevils in the Wheat*, 310; and Rawick, ed.,
American Slave, 12:116.

31. Jackson, *My Father's Name*, 104.

32. For an analysis of these economies of scale, see James R. Irwin, "Slave Agriculture and Staple Crops in the Virginia Piedmont" (Ph.D. diss., University of Rochester, 1986).

33. Dennett, *South as It Is*, 95.

34. Maddex, *Virginia Conservatives*, 186.

35. Circular Letter by Virginia BRFAL Assistant Commissioner Orlando
Brown, June 15, 1865, in Steven Hahn et al., eds., *Freedom: A Documentary
History of Emancipation, 1861–1867: Selected from the Holdings of the National Archives of the United States*, series 3, vol. 1, *Land and Labor, 1865* (Chapel Hill: University of North Carolina Press, 2008), 223–224. For BRFAL
forcing freedpeople to sign contracts, see the above volume, 27–28. In his
tour of the South during 1865 and 1866, Whitelaw Reid reported that freedpeople in and around Lynchburg were extremely reluctant to sign contracts
for more than a few weeks or a month at a time in late 1865 as most were
convinced the federal government would redistribute land with the coming
of the new year. Freedpeople in Pittsylvania, Halifax, and Caswell seemed
much less resistant to annual contracts. Reid, *After the War: A Southern
Tour: May 1, 1865, to May 1, 1866* (Cincinnati: Moore, Wilstach, and Baldwin, 1866), 335–337.

36. Contracts, Indentures, and Papers Regarding Cases, Danville Field Office,
BRFAL, RG 105, M1913, roll 72, nos. 2–246.

37. Calculated from ibid. The 108 contracts include a few from Caswell and
Halifax, but all arrangements were made in Danville. See the discussion of
contract violations below for evidence of contract nature in Halifax.

38. William B. Payne to William T. Sutherlin, December 26, 1865; Hairston
Family Plantation Record Book 1, 59–60, Peter Wilson Hairston Papers, box
8, folder 104, SHC; and Contract between Robert Wilson and Joe Wilson
et al., July 3, 1865, Contract between W. F. Walters and Robin Dickinson,
November 11, 1865, Contract between Robert Wilson and Doctor Douglas,
December 26, 1865, Contract between Robert Wilson and Ellick and Phibby
Boswell, January 4, 1866, and Contract between Robert Wilson and Peyton
and Margaret Freeman, January 6, 1866, all in Hairston and Wilson Family
Papers, folder 32, SHC.

39. As Leon Litwack stresses, there were a multiplicity of postwar contract
terms, but share contracting seemed popular in most southern regions
from the crop season of 1866 on due to "nothing more than economic necessity," Litwack, *Been in the Storm So Long: The Aftermath of Slavery* (New
York: Alfred A. Knopf, 1979), 412. For some notable regional and national
studies that reach this conclusion, see Roger L. Ransom and Richard Sutch,

One Kind of Freedom: The Economic Consequences of Emancipation (New York: Cambridge University Press, 1977), 87–103; Ronald Davis, *Good and Faithful Labor: From Slavery to Sharecropping in the Natchez District, 1860–1890* (Westport, CT: Greenwood Press, 1982), 99–102; Julie Saville, *The Work of Reconstruction: From Slave to Wage Laborer in South Carolina, 1860–1870* (New York: Cambridge University Press, 1994), 110–111; Ralph Shlomowitz, "The Origins of Southern Sharecropping," *Agricultural History* 53, 3 (July 1979): 563–565; Joseph P. Reidy, *From Slavery to Agrarian Capitalism in the Cotton Plantation South: Central Georgia, 1800–1880* (Chapel Hill: University of North Carolina Press, 1992), 148; Thavolia Glymph, "Freedpeople and Ex-Masters: Shaping a New Order in the Postbellum South, 1865–1868," in Thavolia Glymph, Harold D. Woodman, Barbara J. Fields, and Armstead L. Robinson, *Essays on the Postbellum Southern Economy* (College Station: Texas A&M University Press, 1985), 48–72; Barbara J. Fields, *Slavery and Freedom on the Middle Ground: Maryland during the Nineteenth Century* (New Haven: Yale University Press, 1985). Lynda Morgan likewise concluded that some form of shares was the most common initial agreement in the Virginia Piedmont, and Kerr-Ritchie's sampling of the same region found slightly over half of forty-one early contracts were for wages. See Lynda Morgan, *Emancipation in Virginia's Tobacco Belt*, 188; and Kerr-Ritchie, *Freedpeople in the Tobacco South*, 50–51. Stephen Hahn has stressed the importance of extended family group, or "squad," labor agreements throughout much of the South in the first years after Appomattox. Although the Pittsylvania contracts are often between a landowner and several freedpeople, they are rarely larger than immediate family units. See Hahn, *Nation under Our Feet*, 169–170.

40. Saville, *Work of Reconstruction*, 111. Indeed, the predominate form of early wage contracts in Pittsylvania was very similar to that of the sugar districts of Louisiana during Reconstruction, as recorded by Rodrigue, *Reconstruction in the Cane Fields*, chap. 3; Sitterson, *Sugar Country*, 219–225; Howard Ashley White, "The Freedman's Bureau in Louisiana" (Ph.D. diss., Tulane University, 1955), 129–130. Even in regions dominated by large plantations, such as the Georgia Lowcountry, sharecropping remained a popular alternative to wage labor. See William S. McFeely, *Sapelo's People: A Long Walk into Freedom* (New York: W. W. Norton, 1994), 134–135.

41. In this regard tobacco planters were like Louisiana sugar planters. See Rodrigue, *Reconstruction in the Cane Fields*, 2.

42. See the discussion of slave leases in chapter 2.

43. Compiled from Contracts, Indentures, and Papers Regarding Cases, Danville Field Office, BRFAL, RG 105, M1913, roll 72. The lone example of a sharecropper receiving three-fourths of the crop was an unusual case. Landowner Robert Wilson promised Frank Saw the unusually generous terms in exchange for Saw's labor clearing new tobacco land. See Contract between Robert Wilson and Frank Saw, January 4, 1870, Hairston and Wil-

son Family Papers, folder 32, SHC. For Sims's contract arrangements, see "List of Tobacco Produced," n.d. (probably the late 1860s or early 1870s), William Bailey Papers, *RASP,* series E, part 1, roll 1.

44. "List of wages paid, May 18, 1867," in Samuel Pannill Wilson Papers, Small Special Collections, UVA. The description of Wilson's postwar contracts was drawn from twenty-five labor receipts, ranging from February 8, 1868, to April 7, 1880, in Samuel Pannill Wilson Papers. Wilson was a substantial planter before the war; he owned sixty-two slaves in 1860 (the manuscript census mistakenly listed Wilson as just "Samuel Pannill"). (U.S. Census Bureau Slave Schedule, northern district of Halifax County, Virginia, 1860.) The records of Ruth Hairston show similar variability in contracts, wages, and credit on a single plantation. See Hairston Family Plantation Record Books 1 and 2, Peter Wilson Hairston Papers, box 8, folders 104 and 105, SHC.

45. Contract between N. C. Miller and Laborers, August 9, 1865, Crenshaw and Miller Family Papers, folder 4, SHC.

46. Contract between Robert Wilson and David Wilson et al., February 20, 1869, Hairston and Wilson Family Papers, folder 32, SHC.

47. For the prevalence of verbal contracts in the region, see Deposition of James Allen, January 4, 1866, in Contracts, Indentures, and Papers Regarding Cases, Danville Field Office, BRFAL, RG 105, M1913, roll 72.

48. Kerr-Ritchie, *Freedpeople in the Tobacco South,* 56–57, quote on 56.

49. McFeely, *Sapelo's People,* 82; Jackson, *My Father's Name,* 74–75.

50. For particularly lucid and concise summaries of these arguments, see Shlomowitz, "Origins of Southern Sharecropping," 570–575; and Reidy, *From Slavery to Agrarian Capitalism,* 148–150.

51. John C. Ott, *Tobacco: The Outlook for 1875; with an Account of the Production, Consumption and Movement of This Staple in the United States . . . and Some Observations on Farm Labor in the South* (Richmond, VA: Southern Fertilizer, 1875), 27–31.

52. These delayed-wage contracts were not unique. For example, they bear a strong resemblance to the system Henry Middleton developed on his rice plantation near Georgetown, South Carolina. See Eric Foner, *Nothing but Freedom: Emancipation and Its Legacy* (Baton Rouge: Louisiana State University Press, 1983), 87–90.

53. Halifax Courthouse Complaint Book, 1868, complaints nos. 152–162, 200–206, 234–240, BRFAL, RG 105, M1913, roll 97.

54. These two courts also heard a limited number of cases from Caswell County, where there was no field office. Wilcox's replacement claimed the courts were needed because "Freedmen are not paid promptly and . . . advantage is taken of their ignorance." From notes in the back of Danville Letterbook, dated February 1, 1867.

55. It is highly likely that an even higher percentage of the total cases involved contract disputes as a number of cases listed only the names and races of the litigants and no case information.

56. Captain J. F. Wilcox to Colonel Orlando Brown, January 26, 1866, Danville Letterbook, BRFAL, RG 105, M1913, roll 71.

57. These figures are calculated from Halifax Courthouse Complaint Book, 1868, BRFAL, RG 105, M1913, roll 97; and Cases Reported June 1868, Danville Letterbook. The Danville book records complaints for only a portion of the year.

58. Halifax Courthouse Complaint Book, 1868, complaints nos. 21, 30, BRFAL, RG 105, M1913, roll 97.

59. For examples of this process on a regional farm, see Receipts of W. T. Clark, May 9, June 6, 21, 1867, Peter Wilson Hairston Papers, folder 54, SHC. Clark served as a farm manager for Samuel Hairston and collected small lots of tobacco from a number of tenants or sharecroppers and shipped the leaf to a broker in Richmond.

60. Halifax Courthouse Complaint Book, 1868, complaints nos. 149, 150, 243.

61. See, for example, ibid., complaint no. 106; and Bird L. Ferrell to J. W. Ferrell, September 14, 1868, William Thomas Sutherlin Papers, Duke.

62. F. W. Liedtke to William W. Holden, September 22, 1868, in William Woods Holden, *The Papers of William Woods Holden*, 2 vols., ed. Horace W. Raper and Thornton W. Mitchell. (Raleigh: North Carolina Division of Archives and History, 2000), 1:374–375.

63. Captain J. F. Wilcox to R. S. Lacey, January 22, 1866, Danville Letterbook; for quotation, see Bettie L. Clark to Phoebe Bailey, September 4, no year (1860s), in Bailey Family Papers, VHS.

64. Halifax Courthouse Letterbook, 1868, November 30, 1868, BRFAL, RG 105, M1913, roll 97. For the Barksdale family's antebellum prominence in Clover's tobacco agriculture, see the Peter Barksdale Papers, *RASP*, series F, part 3, roll 33.

65. Halifax Courthouse Letterbook, 1868, October 5, 1868, BRFAL, RG 105, M1913, roll 97.

66. Ibid., November 28, 1868.

67. Halifax Courthouse Complaint Book, 1868, complaint no. 21, BRFAL, RG 105, M1913, roll 97.

68. Compiled from indentures in Contracts, Indentures, and Papers Regarding Cases, Danville Field Office, BRFAL, RG 105, M1913, roll 72.

69. Russell R. Menard, "Transitions to African Slavery in British America, 1630–1730: Barbados, Virginia and South Carolina," in *Slavery in America: A Reader and Guide*, ed. Kenneth Morgan (Athens: University of Georgia Press, 2005), 31.

70. Captain J. F. Wilcox to R. S. Lacey, December 13, 1865, Danville Letterbook; Jackson, *My Father's Name*, 41.

71. Captain J. F. Wilcox to R. S. Lacey, December 28, 1865, Danville Letterbook.

72. Stetson Kennedy, *After Appomattox: How the South Won the War* (Gainesville: University Press of Florida, 1995), 38.

73. Karin L. Zipf, *Labor of Innocents: Forced Apprenticeship in North Carolina, 1715–1919* (Baton Rouge: Louisiana State University Press, 2005), 74–76.

74. Advertisement, *Danville Register,* June 25, 1867, p. 2. For a similar advertisement using the same image, see Jackson, *My Father's Name,* 41.

75. J. F. Wilcox to R. S. Lacey, January 7, 1866, Danville Letterbook.

76. See, for example, ibid.; J. F. Wilcox to R. S. Lacey, February 5, 1866, Danville Letterbook; Complaint of Griffin Cobb vs. John Blackwell, September 12, 1867, Greensboro Field Office, BRFAL, RG 105, M1909, roll 21.

77. Jackson, *My Father's Name,* 86–87.

78. Depositions in the Case of *U.S. Government vs. James Hunt,* February 1, 1866, Contracts, Indentures, and Papers Regarding Cases, Danville Field Office, BRFAL, RG 105, M1913, roll 72.

79. C. Thurston Chase to Orlando Brown, June 29, 1868, Danville Letters Sent and Received, BRFAL, RG 105, M1913, roll 72.

80. Katherine Kerr Kendall, *Caswell County, 1777–1877: Historical Abstracts of Minutes of Caswell County, North Carolina* (Raleigh, NC: Multiple Images, 1976), 102.

81. "John G. Lea's Confession to the Ku Klux Klan Murder of John W. Stephens," given July 2, 1919, to the North Carolina Historical Commission (hereafter cited as "Lea's Confession"). In this astonishing document, Lea testified to his participation in Stephens's assassination, describing the event in detail. Lea's confession was sealed by the Historical Commission and opened only following his death in 1935. Full text of the document was reprinted in "65-Year-Old Homicide Is Solved," October 2, 1935, clipping in Caswell County Historical Association Collection, folder 24, SHC, and is available online at http://www.rootsweb.ancestry.com/~nccaswel/misc/confession.htm (last accessed May 7, 2013).

82. Hahn, *Nation under Our Feet,* 183. For the composition and motivation of Union Leagues, see Hahn, *Nation under Our Feet,* 177–186; and Jane Dailey, *Before Jim Crow: The Politics of Race in Postemancipation Virginia* (Chapel Hill: University of North Carolina Press, 2000), 49–50.

83. S. M. Lotten to William North, July 22, 1867, Letters Received, Greensboro Field Office, BRFAL, RG105, M1909, roll 19.

84. J. E. Cook to William W. Holden, August 22, 1868; John W. Stephens to William W. Holden, August 29, 1868; W. J. Dawes to Jacob F. Chur, September 8, 1868, all in Holden, *Papers of William Woods Holden,* 1:358, 1:364–365, 1:369, quote in Cook to Holden.

85. William Woods Holden. *Memoirs of W. W. Holden,* vol. 2 (Durham, NC: Seeman, 1911), 2:154–155; Allen W. Trelease, *White Terror: The Ku Klux Klan Conspiracy and Southern Reconstruction* (Westport, CT: Greenwood, 1979), 192.

86. G. T. F. Gannaway to William W. Holden, December 28, 1868, and John W. Stephens to William W. Holden, August 29, 1868, both in Holden, *Papers of William Woods Holden,* 1:442–443, 1:364–365; Albion W. Tourgée to Jos.

Abbott, May 24, 1870, available at http://www.digitalhistory.uh.edu/recon-struction/section4_tourgeekkk.htm (last accessed February 13, 2009); Myrta Lockett Avary, *Dixie after the War: An Exposition of Social Conditions in the South, during the Twelve Years Succeeding the Fall of Richmond* (New York: Doubleday, Page, 1906), 277; and "Lea's Confession," quote in the latter.

87. Federal Writers' Project, *Born in Slavery: Slave Narratives from the Federal Writers' Project, 1936–1938, North Carolina Narratives*, vol. 11, part 2, p. 11. Available at http://memory.loc.gov/mss/mesn/112/015011.tif (last accessed March 4, 2011).

88. Edgar E. Folk and Bynum Shaw, *W. W. Holden: A Political Biography* (Winston-Salem, NC: John F. Blair, 1982), 214.

89. Forrest did visit nearby Greensboro in February of 1870 to recruit black labor for two Alabama railroad projects. If anything, as Allen Trelease has pointed out, Forrest's visit probably angered local Klansmen with his attempt to appropriate black laborers. Trelease, *White Terror*, 211–212.

90. Holden, *Memoirs*, 141.

91. Stephens was nicknamed "Chicken" Stephens due to a conviction in Rockingham County for poultry theft, and he served time for a shooting related to the incident. Stephens's mother, who lived with her son at the time, died in 1869. The coroner's report ruled that she fell on a wash basin and cut her throat, but town gossip implicated Stephens in the death. See A. J. Stedman, *Murder and Mystery: History of the Life and Death of John W. Stephens, State Senator of North Carolina, from Caswell County* (Greensboro, NC: Patriot, 1870), 7–18, Diary of Jacob Doll, June 30, 1869; Trelease, *White Terror*, 479, n. 24; Luther M. Carlton, "Assassination of John Walter Stephens," in *An Annual Publication of Historical Papers: Legal and Biographical Studies*, series 2 (Durham, NC: Historical Society of Trinity College, 1898), 2–4.

92. Stedman, *Murder and Mystery*, 7–18; "Lea's Confession"; Hamilton, *Reconstruction in North Carolina*, 351, 376. Albion Tourgée's famous Reconstruction novel, *A Fool's Errand*, was based in part on events in Caswell and Alamance counties during the Klan uprisings. On Tourgée's career and his brief relationship with Stephens, see Mark Elliott, *Color-Blind Justice: Albion Tourgée and the Quest for Racial Equality from the Civil War to* Plessy v. Ferguson (New York: Oxford University Press, 2006).

93. First quote in C. E. C., "The Kuklux: The Klan at Its Old Work in North Carolina," *New York Times*, 29 June, 1872, p. 1; second quote in Stedman, *Murder and Mystery*, 39.

94. George Anderson to Edna Watkins, January 22, 1914, Watkins Family Papers, folder 1, SHC.

95. "Lea's Confession."

96. Ibid.; Stedman, *Murder and Mystery*, 36–37.

97. J. G. De Roulhac Hamilton, *Reconstruction in North Carolina* (New York: Columbia University Press, 1914), 473. See also Trelease, *White Terror*, 213. A

proponent of the Dunning school of Reconstruction historiography, Hamilton portrayed Stephens as a completely evil character and celebrated his assassination as mere justice. For an alternate view of Stephens's character, see Carlton, "Assassination of John Walter Stephens," 1–12.

98. William Faulkner, *The Hamlet* (New York: Random House, 1940); Trelease, *White Terror*, 192.

99. "Run, Nigger, Run," *Milton Chronicle*, 28 October, 1869, p. 1; Avary, *Dixie after the War*, 274–275; Stedman, *Murder and Mystery*, 35; "John G. Lea's Confession," first quote in Stedman, second quote in "Lea's Confession." The *Chronicle*'s claims about a black army did reflect one of Stephens's earlier statements. Following the Yanceyville riot in 1868, he asked Holden to "send (Col) [troops] in preference to white, if you can," John W. Stephens to William W. Holden, August 29, 1868, in Holden, *Papers of William Woods Holden*, 1:364–365.

100. "Lea's Confession"; Stedman, *Murder and Mystery*, 19–27; Diary of Jacob Doll, May 21, 22, 1870; George Anderson to Edna Watkins, January 22, 1914; "Life in North Carolina," *New York Times*, February 26, 1873; "Murder Will Out," *Raleigh Signal*, March 24, 1892. See also, Vanessa S. Walker, *Their Highest Potential: An African American School Community in the Segregated South* (Chapel Hill: University of North Carolina Press, 1996), 13–14.

101. Dairy of Jacob Doll, May 22, 1870.

102. Holden, *Memoirs*, 1:122–123; William C. Harris, *William Woods Holden: Firebrand of North Carolina Politics* (Baton Rouge: Louisiana State University Press, 1987), 287–290; Folk and Shaw, *W. W. Holden*, 216–217; Allen W. Trelease, "Republican Reconstruction in North Carolina: A Roll-Call Analysis of the State House of Representatives, 1868–1870," *Journal of Southern History* 42, 3 (August 1976): 331–332. The Klan would force Shoffner to flee the state shortly after passage of the bill (Trelease, *White Terror*, 203).

103. "Lea's Confession"; Powell, *When the Past Refused to Die*, 246–247; Harris, *William Woods Holden*, 291; and Archibald Henderson to Susan M. Fain, August 18, 1870, Archibald E. Henderson Papers, Duke. Events in Caswell lived on long after the end of Reconstruction, merging into the mythical corpus of literature on the violent aftermath of the Civil War. Lightly fictionalized accounts of Stephens's murder and the Kirk-Holden War can be found in Albion Tourgée, *A Fool's Errand, by One of the Fools* (New York: Fords, Howard, and Hulbert, 1879), and Charles Oscar Beasley, *Those American R's: Rule, Ruin, Restoration, by One Who Has Been R'd* (Philadelphia: E. E. Wensley, 1882).

104. "North Carolina," *Halifax Record*, 4 September, 1870, p. 2; Harris, *William Woods Holden*, 291–297; Holden, *Memoirs*, 1:175; *State v. F. A. Wiley and others*, 64 NC 821, 1870 WL 1889 (NC); "An Act for Amnesty and Pardon," *Public Laws and Resolutions, Together with the Private Laws, of the State of North Carolina, Passed by the General Assembly at Its Session 1872–73* (Raleigh:

Stone and Uzzell, 1873), 298–300. The amnesty act left open prosecution for serious felonies, including murder, but this clause was often overlooked.

105. Harris, *William Woods Holden*, 299–308; Trelease, *White Terror*, 223–225; and Hahn, *Nation under Our Feet*, 285.

106. "Lea's Confession."

107. Trelease, *White Terror*, 379, 408–409. Lea would go on to a career as a successful tobacconist in Danville following Reconstruction, and general white sentiment in the county supported the assassination of Stephens until well into the mid-twentieth century, when one local newspaper could still describe the Caswell Union League as "a sort of post–Civil War CIA which was, in its way, as devious an [*sic*] the Ku Klux Klan and even more menacing by the fact that it enjoyed the favor and backing of the Federal Government." See "Caswell Night Riders . . . Lea Family Played Large Part in Ku Klux Klan Affairs," *Greensboro Daily News*, April 6, 1941, and "Drama of 1870: 'Chicken' Stephens Executed by Klan," n.d. [1960s], both clippings in Caswell County Historical Association Collection, folder 24, SHC, quote in the latter.

108. Dennett, *South as It Is*, 96–98.

109. William R. Hatchett to Allen Hatchett, April 9, 1866, Hatchett Family Papers, Duke.

110. Indenture between A. E. Hairston and John Owen, January 1, 1871, George Hairston Papers, *RASP*, series J, part 11, roll 4.

111. Elijah Hundley, Journal No. 2, 1872–1875 entries, Plantation Records of Elijah Hundley, Small Special Collections, UVA.

112. Data compiled from HCDB, 2004, retrieved December 19, 2008, from the University of Virginia, Geospatial and Statistical Data Center: http://fisher .lib.virginia.edu/collections/stats/histcensus/index.html.

113. Susannah H. Jones, "Labor and Landownership in Halifax County, Virginia, 1865–1880" (master's thesis, University of Virginia, 1980), 2–3, 13; and Charles S. Aiken, *The Cotton Plantation South since the Civil War* (Baltimore: Johns Hopkins University Press, 1998), 9–10.

114. *Acts and Joint Resolutions Passed by the General Assembly of the State of Virginia, at Its Session of 1872–73* (Richmond, VA: Shepperson, 1873), 357–358.

115. *Public Laws of the State of North Carolina, Passed by the General Assembly at Its Session 1876–77* (Raleigh, NC: M. S. Littlefield, 1877), chap. 283. This North Carolina statute was at odds with an 1869 law guaranteeing a tenant's right to a portion of the crop. See *Public Laws*, 1868–1869, 308–309.

CHAPTER 6. A BARREN AND FRUITFUL LAND

1. This description of the Brumfields comes from the unpublished manuscript "Bright Leaf: An Account of a Virginia Farm," written by descendent Mary Brumfield Garnett in 1971, drawing on family records. This paper is available in the Small Special Collections at UVA.

2. William B. Payne to William Sutherlin, December 26, 1865, William Thomas Sutherlin Papers, SHC; Agreement between William T. Sutherlin

and John Kelly, January 13, 1870, and Agreement between William T. Sutherlin, John Edwards, and Nat Jones, January 1, 1871, both in William Thomas Sutherlin Papers, Duke; Ledger, January 1887–1890 and Ledger, 1880, 1891–1892, both in William Thomas Sutherlin Papers, SHC; and "Maj. William T. Sutherlin," *Southern Planter and Farmer* 10, 10 (October 1875): quote on 597.

3. Michael Williams, *Americans and Their Forests: A Historical Geography* (New York: Cambridge University Press, 1989), 238–285, 358–360; Earle, "Myth of the Southern Soil Miner," 205–208; Trimble, *Man-Induced Soil Erosion;* Mikko Saikku, *This Delta, This Land: An Environmental History of the Yazoo-Mississippi Floodplain* (Athens: University of Georgia Press, 2005), 110–137; and Kirby, *Mockingbird Song,* 132–137.

4. The best summary of the growth of the loose-leaf auction sales system remains Tilley, *Bright-Tobacco Industry,* 197–250.

5. Robert, *Tobacco Kingdom,* 102–103, 108; Hahn, "Making Tobacco Bright," 84; Hagan, *Story of Danville,* 55; and Joseph Clarke Robert, "Rise of the Tobacco Warehouse Auction System in Virginia, 1800–1860," *Agricultural History* 7, 4 (October 1933): 177.

6. Pollock, *Illustrated Sketch Book of Danville,* 32–35, 127; Bennett, "King Bacca's Throne," 43; Hagan, *Story of Danville,* 54; Danville Tobacco Association, *100 Years of Progress,* 54–56; Pete Daniel, "Reasons to Talk about Tobacco," *Journal of American History* 96, 3 (December 2009): 664. Various dates are given for the opening of Neal's, but the consensus dates the warehouse's operation to the late 1850s, with 1858 the most likely opening date. For at least a few years, the Virginia General Assembly granted Neal's legal sanction to sell loose lots weighing less than 1,000 pounds.

7. Arnold, *History of the Tobacco Industry,* 44–45; Maddex, *Virginia Conservatives,* 172–173.

8. Cameron, *Sketch of the Tobacco Interests,* 82–83; Kirk Munroe, "Danville, Virginia," *Harper's Weekly* 31 (January 29, 1887): 75; Advertisement for Graves Warehouse, *Milton Chronicle,* May 27, 1869, p. 1; Advertisement for Graves Warehouse, *Halifax Record,* December 6, 1871, p. 4; Advertisement for the Liberty Warehouse, *Milton Chronicle,* January 20, 1875, p. 3; "Star Warehouse," *Danville Daily Post,* December 25, 1879, p. 4; Advertisement for the Banner Warehouse, *Milton Chronicle,* December 9, 1880, p. 3; and Danville Tobacco Association, *100 Years of Progress,* quote on 66. For daily warehouse transactions, see various receipts from Pace's, Graves's and Hollands's warehouses, 1872–1878, in Southside Virginia Family Papers, *RASP,* series E, part 3, reel 1. For the records of a Danville tobacco buyer who patronized a number of these auction warehouses, see Daybook of Captain Young, 1871–1872, Duke.

9. Newspaper boosterism continued in the postwar period as well, with local papers often running articles on high auction prices. For examples, see "Tobacco," *Milton Chronicle,* September 2, 1869, p. 3; "That Fine Tobacco,"

Milton Chronicle, September 16, 1869, p. 2; "Danville Tobacco Market," *Milton Chronicle,* March 10, 1881, p. 3; no title, *Caswell News,* January 20, 1888, p. 3; "Just as I Told You," *Halifax Advertiser,* September 12, 1885, p. 1.

10. William Cronon, *Nature's Metropolis: Chicago and the Great West* (New York: W. W. Norton, 1992), 97–147.

11. USDA Bureau of Agricultural Economics, *Standard Grades for Flue-Cured Tobacco* (Washington, DC: Government Printing Office, 1930), 8–10; and USDA Bureau of Agricultural Economics, *Proposed Classification of Leaf Tobacco Covering Forms, Classes, Types, and Groups of Tobacco* (Washington, DC: Government Printing Office, 1929), 4–9. Barbara Hahn argues that this typing and classification gave institutional approval to the idea of bright leaf, thus making amorphous tobacco variations into a rigidly defined agricultural crop. It seems equally likely that the USDA grading campaign merely lent structure and definition to ongoing auction system practices that openly admitted that bright tobacco spanned a wide spectrum. See Hahn, "Making Tobacco Bright," 177–200.

12. While moving against the general trend, bright tobacco's increasing emphasis on linking crop variations to producer expertise was not entirely unique. For the similar example of sea island cotton, see Stewart, *"What Nature Suffers to Groe,"* 121.

13. By the mid-1870s, Danville already claimed to be "the largest loose leaf tobacco market in the world." See *Dibrell Brothers, Incorporated, 1873–1973* (Danville, VA: Dibrell Brothers, 1973), 5.

14. "Planters Warehouse," *Danville Daily Post,* December 25, 1879, p. 4; Cameron, *Sketch of the Tobacco Interests,* 67–69; Munroe, "Danville," 75; Receipt of Charles Coles, January 20, 1875, Southside Virginia Family Papers, *RASP,* series E, part 3, reel 5; and Danville Tobacco Association, *100 Years of Progress,* 49, 51.

15. The distinction between farm units and farmland under one owner is an important one, as farm management and land ownership were often divorced.

16. Figures compiled from the HCDB.

17. In one 1848 shipment alone, Bruce shipped fifty-four hogsheads of tobacco, the produce of more than one hundred acres under typical productivity. See Receipt of James C. Bruce, November 16, 1848, Bruce Family Papers, *RASP,* series E, part 3, reel 14.

18. Figures drawn from Virginia Agricultural Census, Halifax and Pittsylvania Counties, 1880, NARA, T1132, rolls 24 and 27.

19. Population figures compiled from HCDB.

20. The preceding crop and livestock figures were calculated from Joseph C. G. Kennedy, *Agriculture of the United States in 1860; Compiled from the Original Returns of the Eighth Census* (Washington, DC: Government Printing Office, 1864), 104–105, 154–155, 158–159; and Francis A. Walker and Charles W. Seaton, *Report on the Productions of Agriculture, as Returned at the Tenth*

Census (Washington, DC: Government Printing Office, 1883), 165, 174, 200, 209, 301, 321. Sam Bowers Hilliard gives 2.2 swine per capita as the pork self-sufficiency mark for the antebellum South. Even assuming an increase in hog weights through selective breeding and improved husbandry— assumptions that are far from given in the bright belt—the 1880 figure of 0.5 swine per capita meant the Southside must have imported a portion of its pork. See Hilliard, *Hog Meat and Hoecake*, 106–111.

21. Tatham, *Historical and Practical Essay*, 29–34.

22. Survey map of George C. Venable farm, Halifax County Deed Book 68, 1880–1881, p. 302, Halifax County Courthouse, Halifax, VA. Surveys of Pittsylvania and Caswell deed books for 1880 revealed a number of similar maps. Clerks drew maps of land deeds in only a portion of cases, but the surviving maps from the three counties, along with other written accounts, recorded similar land arrangements as those on Venable's farm. See Deed Books 75 and 76, 1879–1880, Pittsylvania County Courthouse, Chatham, VA; and Deed Book NN, 1879–1881, Caswell County Courthouse, Yanceyville, NC. For the dispersed pattern of postwar tobacco barns, see Catherine W. Bishir and Michael T. Southern, *A Guide to the Historic Architecture of Piedmont North Carolina* (Chapel Hill: University of North Carolina Press, 2003), 182–183; Bennett, "King Bacca's Throne," chap. 1; and John Fraser Hart and Eugene Cotton Mather, "The Character of Tobacco Barns and Their Role in the Tobacco Economy of the United States," *Annals of the Association of American Geographers* 51, 3 (September 1961): 290. For the prevalence of this agricultural geography in Caswell in particular, see W. C. Kerr, *Report on the Cotton Production of the State of North Carolina, with Discussion of the General Agricultural Features of the State* (Washington, DC: Government Printing Office, 1880), 61.

23. Walker and Seaton, *Report on the Productions of Agriculture*, 301, 321.

24. G. Melvin Herndon, "The Significance of the Forest to the Tobacco Plantation Economy in Antebellum Virginia," *Plantation Society in the Americas* 1, 3 (October 1981): 439.

25. "Forest Trees," *Southern Workman* 4, 9 (September 1875): 64.

26. "Preservation of Timber," *Southern Planter* 42, 10 (October 1881): 590.

27. Whitehead, ed., *Virginia*, 10, 20.

28. Garnett, "Bright Leaf."

29. Stanley W. Trimble, *Man-Induced Soil Erosion on the Southern Piedmont, 1700–1970* (Ankeny, IA: Soil Conservation Society of America, 1974), 72.

30. Compiled from 1860–1900 census figures found at HCDB.

31. J. L. M., "Letter from Pittsylvania," *Southern Planter* 48, 8 (August 1, 1887): 412.

32. For a similar use of line markers in speculation on forest composition, see Robert D. Mitchell, Warren R. Hofstra, and Edward F. Conner, "Reconstructing the Colonial Environment of the Upper Chesapeake Watershed," in Philip E. Curtin, Grace S. Brush, and George W. Fisher, eds., *Discovering*

the Chesapeake: The History of an Ecosystem (Baltimore: Johns Hopkins University Press, 2001), 179–183. The author owes thanks to an anonymous reader for Yale University Press for noting additional details concerning the complexity of surveyor tree selection.

33. The following figures were compiled by the author from Deed Book 75 and Deed Book 76, Pittsylvania County Courthouse, Chatham, VA; Land Survey Book No. 1, part 2, Deed Book 67, and Deed Book 68, Halifax County Courthouse, Halifax, VA; Land Survey Book A and Deed Book NN, Caswell County Courthouse, Yanceyville, NC; and Marian Dodson Chiarito, *Old Survey Book 1, 1746–1782, Pittsylvania County, Virginia* (Nathalie, VA: Clarkton, 1988), 323–355. Halifax and Pittsylvania had reached their current bounds by 1780, but Caswell still included modern-day Person County, and a portion of the Caswell surveys likely came from the latter district.

34. The frequency of pines in the land surveys proved puzzling. I had assumed that the use of pines as markers would increase with deforestation, as several species of pines are old field colonizers, but the opposite proved to be the case. Pines decreased from 18 percent of total markers in 1780 to just 9.1 percent of markers in 1880 surveys. Perhaps the surveyors were reluctant to use any but mature pines as they wanted to ensure durable survey lines, and thus a decrease in pine markers might represent a decline in mature pine stands rather than a decrease in total pine numbers. This percentage of pines as markers should in no way be taken to represent the percentage of pines in regional woodlands as a whole but rather as a reflection of the preferences and prejudices of surveyors. For certain oak species as phytometers of weak soil, see Walker, *Southern Forest*, 54.

35. Love and Love, *Art of Curing*, 19–20; and Ragland, *Major Ragland's Instructions*, 35.

36. Love and Love, *Art of Curing*, 16–17.

37. Ibid., 6.

38. Farmer, "Farm Accounts—Again," *Southern Planter and Farmer* 2, 1 (January 1868): 35. See chapter 7 for a discussion of the postwar expansion of bright tobacco culture.

39. Viator, "A Letter from a Virginian," *Southern Planter and Farmer* 40, 4 (April 1879): 185.

40. Ragland in "Report on the Culture and Curing of Tobacco," in Walker and Seaton, *Report of the Productions of Agriculture*, 202. Also quoted in Tilley, *Bright-Tobacco Industry*, 110. Tilley argues that soil erosion declined in the late nineteenth century as farmers employed better soil conservation practices, but there seems little evidence to support her contention (Tilley, *Bright-Tobacco Industry*, 110). Indeed, the Soil Conservation Service's work in the Southside during the Great Depression indicated that erosion accelerated in the late 1800s. For more on this, see the epilogue.

41. J. L. M., "Letter from Pittsylvania," 412.

42. Trimble, *Man-Induced Soil Erosion*, 16, 91, n. 6.

43. H., "Hints on the Labor Question," *Southern Planter and Farmer* 1, 9 (October 1867): 574.

44. Charles Bruce, "Southside Virginia and the Agricultural Interest," *Southern Planter* 48, 5 (May 1, 1887): 254.

45. "Evening Session," *Southern Planter and Farmer* 1, 1 (February 1867): 44.

46. This notion that tenants were particularly destructive land stewards is a persistent one. See Trimble, *Man-Induced Soil Erosion*, 69, for a typical discussion.

47. Alfred J. Morrison, *Halifax County, Virginia, a Handbook: Prepared under the Direction of the Board of Supervisors* (Richmond, VA: Everett Waddey, 1907), 10; and Tilley, *Bright-Tobacco Industry*, 110–111.

48. Whitehead, ed., *Virginia*, 42.

49. Kerr-Ritchie, *Freedpeople in the Tobacco South*, 178. In 1900, the first year that the census enumerated tenants by race, the three counties actually had more white than black tenants, by a count of 3,456 to 2,439 (compiled from HCDB). This in part reflected a slight majority white population and white landowners' continued desires to employ blacks as wage laborers rather than tenants.

50. Killebrew, *Report of the Culture and Curing*, 202.

51. Compiled from HCDB.

52. Tilley, *Bright-Tobacco Industry*, 156–157; Mathewson, *Culture of Flue-Cured Tobacco*, 11, 14; Cameron, *Sketch of the Tobacco Interests*, 79; Ragland, *Tobacco, from the Seed to the Salesroom*, 10; Arnold, *History of the Tobacco Industry*, 24; Werner, *Tobaccoland*, 174; Love and Love, *Art of Curing*, 14–15; Kerr-Ritchie, *Freedpeople in the Tobacco South*, 111–112; W. F. Massey, "Fertilizer for Bright Tobacco," *Southern Planter* 57, 9 (September 1896): 393; and W. F. Massey, "Tobacco Fertilizers," *Southern Planter* 58, 4 (April 1897): 153–154.

53. *Halifax Record*, April 23, 1886, throughout.

54. John Solomon Otto, *Southern Agriculture during the Civil War Era, 1860–1880* (Westport, CT: Greenwood, 1994), 89–90; Robert L. Ragland, *On the Cultivation and Curing of "Bright Wrappers"* (Richmond: VA: Clemmitt and Jones, n.d.), 3; and Milton Whitney, *Tobacco Soils of the United States: A Preliminary Reoprt upon the Soils of the Principal Tobacco Districts* (Washington, DC: Government Printing Office, 1898), 15. On the rise of southern phosphates in regional agricultural practice, as well as their little-acknowledged importance in cotton culture, see Earle, "Myth of the Southern Soil Miner," 205–208.

55. By the early 1890s, Robert Ragland alone marketed forty-one varieties of tobacco, twenty-nine of which he classified as bright leaf. These included the descriptively named "Gold Finder," "Bullion," and "Oak Hill Yellow." See Robert L. Ragland, *Wholesale Catalogue of Reliable Tobacco Seeds* (Hyco, VA: R. Ragland Seed, 1894), available at Duke.

56. Brianna Rego, "The Polonium Brief: A Hidden History of Cancer, Radiation, and the Tobacco Industry," *Isis* 100 (2009): 453–484, quote on 455; Monique E. Muggli, Jon O. Ebbert, Channing Robertson, and Richard D. Hurt, "Waking a Sleeping Giant: The Tobacco Industry's Response to the Polonium-210 Issue," *American Journal of Public Health* 98, 9 (September 2008): 1643–1650; R. T. Ravenholt, "Tobacco's Global Death March," *Population and Development Review* 16, 2 (June 1990): 228–229. Thanks to Maria Bujenovic for bringing polonium-210 to my attention.

57. Tilley, *Bright-Tobacco Industry*, 167.

58. Ibid., 167–168.

59. For examples with typical language, see Contract of Lacy Lea and Gabe Bigalow, January 1, 1881, Archibald E. Henderson Papers, Duke; Contract of Wood Covington and R. H. Enoch, December 7, 1879, and Contract of Patassco Guano Company and J. L. Kersey and J. L. Law, April 21, 1880, both in Caswell County Deed Book NN, 1879–1880, Caswell Courthouse; and Ledger, 1876–1891, and Promissory Notes of A. G. Pritchett and Son, 1886–1890s, Pritchett Family Papers, UVA.

60. Haley Brothers to Katherine Moses, May 6, 1880, Katherine Spiller (Graves) Moses Papers, VHS. For similar examples of regional landowners securing fertilizer loans for their tenants, see Farm Ledger of Nathanial R. Coleman, 1885 entries, Ethelbert Algernon Coleman Papers, UVA; and List of Tenants' Tobacco, n.d. (mid-1880s), Samuel Pannill Wilson Papers, UVA.

61. Kerr-Ritchie, *Freedpeople in the Tobacco South*, 165–171.

62. Advertisement, *Pittsylvania Tribune*, May 23, 1879, p. 2; and Cameron, *Sketch of the Tobacco Interests*, 112. For the importance of South Carolina phosphorus resources in the rise of southern fertilizer, see Helms, "Soil and Southern History," 752.

63. Tilley, *Bright-Tobacco Industry*, 168–178.

64. For selected examples of these endorsements, see "Danville as a Tobacco Centre," pamphlet, 1879, 9, 16, Duke; "Gilham's Tobacco Fertilizer," circular, March, 1871, in William Thomas Sutherlin Papers, Duke; Cameron, *Sketch of the Tobacco Interests*, 112; Robert Ragland, *Major Ragland's Instructions How to Grow and Cure Tobacco, Especially Fine Yellow* (Richmond, VA: Southern Fertilizing, 1885), 10, 12; fertilizer advertisements in the *Halifax Record*, April 23, 1886; and Tilley, *Bright-Tobacco Industry*, 158–159.

65. For the Reidsville Fertilizer Company's contest for farmers who used their Broad Leaf brand, see Advertisement, *Milton Gazette*, May 4, 1893, p. 3.

66. Receipts of Langhorn Scruggs, May 24, 1871 and May 30, 1871, both in Southside Virginia Family Papers, *RASP*, series E, part 3, reel 3.

67. William Sours to John Sours, May 6, 1877, Sours Family Papers, folder 21, SHC.

68. Charles Bruce, "Southside Virginia and the Agricultural Interest," *Southern Planter* 48, 5 (May 1, 1887): 249.

69. Otto, *Southern Agriculture*, 89–90, quote on 89.

70. G. A. Runner, *The So-Called Tobacco Wireworm in Virginia*, USDA Farmers' Bulletin No. 78 (Washington, DC: Government Printing Office, 1914), 9–10, 14; Billings, *Tobacco*, 437–439; American Tobacco Company (ATC), *Flue-Cured Tobacco: Diseases, Nutrient Deficiencies and Excesses, Injuries, Pests, Cured Tobacco* (Richmond, VA: American Tobacco, 1958), 19–20; and Garner, *Production of Tobacco*, 284–290.

71. Instead, most growers advised tolerating infestations, some sort of mechanical control (usually picking off worms or attracting their mature forms—moths—with flames), or, less often, applying early pesticides, including arsenic compounds. See John Ott, "The Outlook for Tobacco," *Pittsylvania Courier*, May 27, 1876, p. 2; J. M. Baker, "The Tobacco Horn-Worm," *American Farmer* 7, 5 (May 1878): 161; Robert L. Ragland, "Cultivation and Curing of Fine Yellow and Shipping Tobacco from the Plant-Bed to Market," *Southern Planter* 39, 5 (May 1878): 232–233; William B. Alwood, "Current Notes—No. III: The Horn Worm—Tobacco Wire Worm," *Southern Planter* 51, 8 (August 1, 1890): 369; and R. L. Ragland, "The Tobacco Horn-Worm: Some Practical Facts Concerning Its Life History Learned from Experience," *Southern Planter* 51, 9 (September 1, 1890): 406–407.

72. Ragland, "Tobacco Horn-Worm," 407.

73. Garner, *Production of Tobacco*, 244–246, 250–252, 262–264; ATC, *Flue-Cured Tobacco*, 8; Sharrer, *Kind of Fate*, 63–65; Geo. F. Atkinson, "Nematode Root-Galls," *Southern Cultivator* 48, 4 (April 1890): 156–158; George B. Lucas, *Diseases of Tobacco*, 3rd ed. (Raleigh, NC: Biological Consulting Associates, 1975), 22; Tilley, *Bright-Tobacco Industry*, 186–187; Bennett, "King Bacca's Throne," 79; and H. D. Shaw and G. B Lucas, *Compendium of Tobacco Diseases* (St. Paul, MN: American Phytopathological Society, 1991), 37–39. A Mecklenburg County farmer writing to the *Southern Planter* in 1895 illustrated the confusion that often accompanied tobacco diseases. J. D. Petty wrote that his crop had experienced "walloon" (likely mosaic virus) for the past three years, despite the fact that the land had produced quality yellow tobacco prior to the outbreak. Petty blamed poor fertilizer for the disease, rather than his continuous cropping practices. See J. D. Petty, "The Tobacco Crop. The Type to Make for 1895," *Southern Planter* 56, 2 (February 1895): 56.

74. Jane Dailey, *Before Jim Crow: The Politics of Race in Postemancipation Virginia* (Chapel Hill: University of North Carolina Press, 2000), 105, 119–125.

75. Ibid., 113; and Charles E. Wynes, *Race Relations in Virginia, 1870–1902* (Totowa, NJ: Rowman and Littlefield, 1971), 30.

76. *Coalition Rule in Danville* (Danville, VA: n.p., 1883), 1, 3.

77. Dailey, *Before Jim Crow*, 103–131, quote on 104. See also Jane Dailey, "Deference and Violence in the Postbellum Urban South: Manners and Massacres in Danville, Virginia," *Journal of Southern History* 63, 3 (August, 1997): 553–590. James Moore (*Two Paths to the New South*) describes the riot as a classic political struggle waged between Readjusters and Conservatives, with

race as leverage, an interpretation that follows that of Walter T. Calhoun, "The Danville Riot and Its Repercussions on the Virginia Election of 1883," in *Studies in the History of the South, 1875–1922* (Greenville, NC: Department of History, East Carolina College, 1966): 25–51.

78. *Coalition Rule in Danville,* 4. Italics in the original. For another account of tensions over tobacco marketing contributing to the riot, see Danville Tobacco Association, *100 Years of Progress,* 62.

79. Tilley, *Bright-Tobacco Industry,* 228.

80. Danville Committee of Forty, *Danville Riot, November 3, 1883: Report of Committee of Forty with Sworn Testimony of Thirty-Seven Witnesses, &c.* (Richmond, VA: John and Goolsby, 1883), 22.

81. Ibid., 18, 27–28, 42. For the dominance of black workers in 1880s Danville tobacco factories, see Robert E. King, *Robert Addison Schoolfield (1853–1931): A Biographical History of the Leader of Danville, Virginia's Textile Mills during Their First 50 Years* (Richmond, VA: William Byrd, 1979), 28; and William J. Erwin, *Dan River Mills: A Story of 75 Exciting Years in Textiles! (1882–1957)* (New York: Newcomen Society, 1957), 10.

82. Dailey, *Before Jim Crow,* 112, 117, 123.

83. Danville Committee of Forty, *Danville Riot,* 4–6, quote on 6.

84. Ibid., 6–7; "Meeting of Citizens," *Daily Register* (Danville), November 10, 1883, p. 3; and 1880 Census of Pittsylvania County.

85. Calhoun, "Danville Riot," 47. Moore claims a similar marked decline in total votes for Readjuster candidates in Danville. See Moore, *Two Paths to the New South,* 52.

86. Wynes, *Race Relations,* 33. Despite questionable Democratic Party tactics, less than ten years earlier black political activity at the Halifax polls had been much more assertive. See John Rilly to H. H. Hurt, November 17, 1874, Southside Virginia Family Papers, *RASP,* series E, part 3, reel 6.

87. "Election by Terrorism," *Southern Workman* 12, 12 (December 1883): 123.

88. Banner Warehouse circular, reprinted in Calhoun, "Danville Riot," 51.

89. "Lawlessness, North and South," *Independent* 36 (February 28, 1884): 5. Albion Tourgée made a similar observation in a contemporary editorial. He declared the riot the product of inherent white racism: "The people did not change their nature as the result of war or any accident of political existence. The people of the South are a product of development; the result of generations of formative events." See A. W. Tourgée, "Presidential Probabilities," *Continent; an Illustrated Weekly Magazine* 5, 110 (March 19, 1884): 377.

90. Charles Dudley Warner, "The Industrial South," *Harper's Weekly* 31 (January 29, 1887): 75.

91. No title, *Halifax Advertiser,* July 18, 1885, p. 2.

92. R. L. Ragland, "The Yellow Tobacco of North Carolina and Virginia," *American Farmer* 3, 12 (June 15, 1884): 178.

93. No title, *Halifax Advertiser,* December 3, 1887, p. 2. According to one purchasing power calculator, $3,300 in 1887 would equal $76,800 in 2009.

See Lawrence H. Officer and Samuel H. Williamson, "Purchasing Power of Money in the United States from 1774 to 2009," MeasuringWorth 2010, http://measuringworth.com/ppowerus/ (last accessed December 2, 2010).

94. "Humbuggery," *Milton Chronicle,* May 21, 1869, p. 2.

95. Thomas Pollard, "Letter from Dr. Pollard," *Southern Planter* 43, 12 (July 1882): 7.

CHAPTER 7. THE DECLINE OF THE BORDER

1. Gillespie, *Katherine and R. J. Reynolds,* chap. 3.

2. Garner, *Production of Tobacco,* 39; Daniel, *Breaking the Land,* 31–32; and Prince and Simpson, *Long Green,* 46–47.

3. Anthony J. Badger, *Prosperity Road: The New Deal, Tobacco, and North Carolina* (Chapel Hill: University of North Carolina Press, 1980), 3; Robert C. McAdams, "The Tobacco Culture of Wilson County, North Carolina" (Ph.D. diss., University of Tennessee, 1996), 1–2.

4. McAdams, "Tobacco Culture of Wilson County," 59–61.

5. Prince and Simpson, *Long Green,* 52–54, 57–58.

6. Cameron, *Sketch of the Tobacco Interests,* 28–29, 31.

7. E. M. Nix, *Production of Bright Tobacco* (Savannah, GA: Seaboard Air Line Railway, [1915?]), 6–7, 11; manuscript inventory notes of the Arthur M. Gignilliat Collection on Pineora (GA), Georgia Historical Society, Savannah, GA; Badger, *Prosperity Road,* 3; and Robert L. Ragland, "Nesbitt on Tobacco," *Southern Cultivator* 50, 1 (January 1892): 29. Ragland offered advice and seed to Georgia planters while W. H. Snow traveled through the region selling his modified curing barn system.

8. Robert L. Ragland, *Wholesale Catalogue of Reliable Tobacco Seeds* (Hyco, VA: R. L. Ragland Seed, 1894); and Robert L. Ragland, *Tobacco: How to Raise It and How to Make It Pay* (Hyco, VA: R. L. Ragland Seed, 1895), quotes on 4–5.

9. McAdams, "Tobacco Culture of Wilson County," 25–27.

10. Arnold, *History of the Tobacco Industry,* 24.

11. Walker and Seaton, *Report on the Productions of Agriculture,* 8–9, 302; Robert P. Porter and Carroll D. Wright, *Report on the Statistics of Agriculture in the United States at the Eleventh Census: 1890* (Washington, DC: Government Printing Office, 1895), 67, 445; and William R. Merriam, *Twelfth Census of the United States, Taken in the Year 1900: Agriculture, Part II, Crops and Irrigation* (Washington, DC: United States Census Office, 1902), 545, 570, 566.

12. In 1879, the county figures were: Pittsylvania, 12,271,533 pounds; Halifax, 7,553,842; and Caswell, 4,336,664. The corresponding 1899 figures were: Pittsylvania, 17,088,550 pounds; Halifax, 13,077,200 pounds; and Caswell, 5,633,900. The tobacco crop fluctuated rather dramatically from season to season depending on weather conditions and estimated market prices—the 1889 crop, for example, was slightly lower than that of 1879—but the general

trend in the last two decades of the nineteenth century was one of steadily increasing production (Walker and Seaton, *Report on the Productions of Agriculture,* 8–9, 302; Porter and Wright, *Report on the Statistics,* 67, 445; and Merriam, *Twelfth Census,* 545, 570, 566). The first postwar census crop, 1869, was markedly lower than the 1859 crop, but poor weather conditions were at least partly to blame. See Robert Somers, *The Southern States since the War, 1870–1* (New York: Macmillan, 1871), 26–27.

13. On Riverside Cotton Mills and its successor, Dan River Mills, see Julian R. Meade, *I Live in Virginia* (New York: Longmans, Green, 1935), 1–99, 245–310; King, *Robert Addison Schoolfield;* Erwin, *Dan River Mills;* and Robert Sidney Smith, *Mill on the Dan: A History of Dan River Mills, 1882–1950* (Durham, NC: Duke University Press, 1960). In the early twentieth century, textile production in Danville would rise as a serious rival to tobacco manufacturing, and Dan River Mills would for a time be the largest cotton mill in the world. During the nineteenth century, however, tobacco reigned supreme along the banks of the Dan.

14. William Sours to John Sours, February 16, 1885, Sours Family Papers, folder 21, SHC.

15. Gillespie, *Katherine and R. J. Reynolds,* 106.

16. Hahn, *Making Tobacco Bright;* and Hahn, "Paradox of Precision: Bright Tobacco as Technology Transfer, 1880–1937," *Agricultural History* 82, 2 (2008): 220–235.

17. John Estes, *Tobacco: Instructions for Its Cultivation and Curing,* USDA Farmers' Bulletin No. 6 (Washington, DC: Government Printing Office, 1892), passim; Otto Carl Butterweck, *The Culture of Tobacco,* USDA Farmers' Bulletin No. 82 (Washington, DC: Government Printing Office, 1898), 13; Frank D. Gardner, ed., *Traditional American Farming Techniques: A Ready Reference on All Phases of Agriculture for Farmers of the United States and Canada,* reprint of 1916 ed. (Guilford, CT: Lyons, 2001), 341–352; Mathewson, *Culture of Flue-Cured Tobacco,* 5, 18–21; and Trimble, *Man-Induced Soil Erosion,* 16.

18. Untitled article, *Milton Herald,* February 3, 1898, p. 3.

19. For examples, see Arnold, *History of the Tobacco Industry,* 24; Whitehead, ed., *Virginia,* 46–47; and Morrison, *Halifax County, Virginia,* 15–16.

20. For a typical article that ignored depressed prices and increasing competition, see "Tobacco Growing," *Southern Planter* 59, 4 (April 1898): 158–159.

21. Helms, "Soil and Southern History," 726, 751–752. For a remarkably similar process in an erosive Western landscape in the twentieth century, see Andrew P. Duffin, *Plowed Under: Agriculture and Environment in the Palouse* (Seattle: University of Washington Press, 2007), esp. chapter 6.

22. Butterweck, *Culture of Tobacco,* 17.

23. D. Wyatt Aiken, "The Patrons of Husbandry," *Southern Planter and Farmer* 9, 7 (July 1872): 404–406. This article was reprinted from the *Rural Carolinian.*

24. Kerr-Ritchie, *Freedpeople in the Tobacco South,* 144–145.

25. "Maj. William T. Sutherlin," 597; and Maddex, *Virginia Conservatives*, 172–173.

26. Tilley, *Bright-Tobacco Industry*, 397–401; Hairston, *Brief History*, 30; Pollock, *Illustrated Sketch Book*, 124–125; and Danville Tobacco Association, *100 Years of Progress*, 61–62.

27. William A. Link, "Cavaliers and Mudsills: The Farmers' Alliance and the Emergence of Virginia Populism" (master's thesis, University of Virginia, 1979), esp. 1–21.

28. "Alliance Notes—What We Want," *Caswell News and Alliance Advocate*, August 29, 1888, p. 2; "Alliance Notes," *Caswell News and Alliance Advocate*, July 18, 1888, p. 2; "Platform of the Farmer's Assembly," *Halifax Advertiser*, February 19, 1887, p. 1; and Tilley, *Bright-Tobacco Industry*, 411.

29. Susannah H. Jones, "Labor and Landownership in Halifax County, Virginia, 1865–1880" (master's thesis, University of Virginia, 1980), 14–15.

30. "Platform of the Farmers' Assembly," p. 1.

31. "Alliance Notes," *Caswell News and Alliance Advocate*, July 18, 1888, p. 2.

32. Link, "Cavaliers and Mudsills," 60–66.

33. Patrick G. Porter, "Origins of the American Tobacco Company," *Business History Review* 43, 1 (Spring 1969): 61, 63–70; Durden, *Bold Entrepreneur*, 16, 23–26; and Timothy M. Matthewson, "'Smoke Signals: Cigarettes, Advertising, and the American Way of Life' at the Valentine Museum, Richmond, Virginia," *Technology and Culture* 33, 3 (July 1992): 560–563. For the Alliance's fears of the ATC, see Kerr-Ritchie, *Freedpeople in the Tobacco South*, 195–204. For an early influential biography of Duke, see John K. Winkler, *Tobacco Tycoon: The Story of James Buchanan Duke* (New York: Random House, 1942).

34. Durden, *Bold Entrepreneur*, 36–37; Porter, "Origins of the American Tobacco Company," 59–60, 74–75; and Leslie Hannah, "The Whig Fable of American Tobacco, 1895–1913," *Journal of Economic History* 66, 1 (March 2006): 50.

35. Relli Shechter, "Selling Luxury: The Rise of the Egyptian Cigarette and the Transformation of the Egyptian Tobacco Market, 1850–1914," *International Journal of Middle East Studies* 35, 1 (February 2003): 58–60; Hannah, "Whig Fable," 44–45; and Durden, *Bold Entrepreneur*, 44. It was not until 1921 that cigarettes became the most common way that Americans consumed tobacco. See Herndon, *William Tatham*, 418.

36. Allan M. Brandt, "The Cigarette, Risk, and American Culture," *Daedalus* 119, 4 (Fall 1990): 157.

37. Sanborn Map Company, *Milton, Caswell Co., N.C., Sept. 1893*, North Carolina Collection at the University of North Carolina Chapel Hill, available online at http://dc.lib.unc.edu/cdm/compoundobject/collection/ncmaps/id/3844/rec/1 (last accessed September 26, 2012).

38. Shechter, "Selling Luxury," 60–61; Porter, "Origins of the American Tobacco Company," 62; Herndon, *William Tatham*, 430–431; and Goodman, *Tobacco in History*, 98.

39. Advertisement in Snow, *Snow's Modern Barn System*, 35.

40. Porter, "Origins of the American Tobacco Company," 62; Mathewson, *Culture of Flue-Cured Tobacco*, 2; Tilley, *Bright-Tobacco Industry*, 306–307; Herndon, *William Tatham*, 430–431; Gillespie, *Katherine and R. J. Reynolds*, 105, 113–115; "The Tobacco Market," *Southern Planter* 59, 11 (November 1898): 515; and Whitney, *Tobacco Soils of the United States*, 8.

41. Sanborn Map Company, Milton, *Caswell County, North Carolina, March 1908*, North Carolina Collection, University of North Carolina Chapel Hill, available online at http://dc.lib.unc.edu/cdm/compoundobject/collection/ncmaps/id/3842/rec/2 (last accessed September 26, 2012).

42. For the ATC's marketing tactics in the 1890s, see Arnold, *History of the Tobacco Industry*, 27–31, 50–52; Winkler, *Tobacco Tycoon*, 97–99; Hannah, "Whig Fable," 50; and Porter, "Origins of the American Tobacco Company," 59–60.

43. "Meeting of Caswell's Tobacco Growers," *Milton Herald*, January 11, 1900, p. 3.

44. A. K. Leake, "Anti-Monopoly," *Southern Planter* 53, 5 (May 1892): 292.

45. PCATU Ledger, August 16, 1890–May 6, 1892; PCATU Share Receipt Book, February–March 1891; and PCATU Share Receipt Book, March–June 1891, all in Records of the Pittsylvania Central Alliance Trade Union, UVA; Link, "Cavaliers and Mudsills," 34. Kerr-Ritchie also includes a good discussion of the PCATU in *Freedpeople in the Tobacco South*.

46. PCATU Ledger, 35; and PCATU Store Daybook, April 1–August 6, 1891, Records of the Pittsylvania Central Alliance Trade Union.

47. PCATU Freight Ledger, 1891–1892, Records of the Pittsylvania Central Alliance Trade Union. Between the above dates the cooperative store received 404,000 pounds of fertilizer and two railroad cars of undetermined weight.

48. PCATU Store Daybook, 1891.

49. PCATU Ledger, 24.

50. Ibid., 5.

51. PCATU, accounts of purchases (partial), Records of the Pittsylvania Central Alliance Trade Union. The Durham Fertilizer Company specialized in selling directly to Alliance chapters, going so far as to market a "North Carolina Farmer's Alliance Guano" brand. See Tilley, *Bright-Tobacco Industry*, 412.

52. Ibid.; and Link, "Cavaliers and Mudsills," 39. For the ongoing difficulties of the PCATU in collecting membership dues, see PCATU Ledger, 37, 41–43.

53. Peterson, *Historical Study of Prices Received*, 152, table 79; and Danville Tobacco Association, *100 Years of Progress*, 51.

54. The case of tobacco research support Elizabeth Sanders's argument that farmers were key instigators in the increasingly federalization of national and farm economics during the Progressive Era. See Sanders, *Roots of Reform: Farmers, Workers, and the American State, 1877–1917* (Chicago: University of Chicago Press, 1999).

55. The most popular regional agricultural journal, the *Southern Planter*, gave a free copy of Ragland's bright tobacco guide to all of its subscribers in 1896. See "Publisher's Notes," *Southern Planter* 57, 4 (April 1896): 182.

56. Estes, *Tobacco*, 7–8, quote on 7.

57. Butterweck, *Culture of Tobacco*, 5.

58. Richard V. Gaines, "A Tobacco Experiment Station and Model Farm," *Southern Planter* 61, 7 (July 1900): 405. The USDA eventually took over funding of the tobacco soils survey, though it seems clear that the initial impetus came from the state board. See *Yearbook of the United States Department of Agriculture* (Washington, DC: Government Printing Office, 1901), 56.

59. Whitney, *Tobacco Soils of the United States*, 15–16, 43, plate 9, quote on 8.

60. Milton Whitney, *Tobacco Soils*, USDA Farmers' Bulletin No. 83 (Washington, DC: Government Printing Office, 1898).

61. "The Tobacco Soils Report," *Milton Herald*, February 3, 1898, p. 4.

62. Whitney, *Tobacco Soils of the United States*, 3.

63. *Yearbook of the USDA*, 124–125, 128–130.

64. Tilley, *Bright-Tobacco Industry*, 189–190. North Carolina's eastern counties were the first to receive detailed county soil surveys, largely due to the survey's origination in the requests of aspiring bright tobacco growers.

65. Richard V. Gaines, "A Tobacco Experiment Station and Model Farm," *Southern Planter* 61, 7 (July 1900): 405.

66. Tilley, *Bright-Tobacco Industry*, 168–178.

67. Sanders, *Roots of Reform*, 391.

68. *Yearbook of the USDA*, 57.

69. Ibid., 57.

70. Ibid., 126.

71. Danville Tobacco Association, *100 Years of Progress*, 51. The ATC and the rising popularity of the cigarette cannot exclusively be blamed for this decline as the Panic of 1893 and high production throughout the decade undoubtedly contributed to depressed markets as well.

72. F. M. Rogers, Jr., "Top Your Tobacco Right," *Southern Planter* 52, 8 (August 1891): 414.

73. P. Lorillard Co., *Lorillard and Tobacco: 200th Anniversary P. Lorillard Company, 1760–1960* (New York: P. Lorillard, 1960), 28–32.

74. Ibid., 29; Tilley, *Bright-Tobacco Industry*, 550; Goodman, *Tobacco in History*, 101–102; and Winkler, *Tobacco Tycoon*, 49–51.

75. Goodman, *Tobacco in History*, 101–102; Gately, *Tobacco*, 205–206; Chandler, *Visible Hand*, 291–292.

76. R. L. Ragland, "Chemistry of Yellow Tobacco," *Southern Planter* 48, 11 (November 1, 1887): 576.

77. Edward M. Brecher et al., *Licit and Illicit Drugs: The Consumers Union Report on Narcotics, Stimulants, Depressants, Inhalants, Hallucinogens, and Marijuana—Including Caffeine, Nicotine, and Alcohol* (Boston: Little, Brown,

1972), 229; Goodman, *Tobacco in History*, 4–6, 98–99; Stanton A. Glantz, John Slade, Lisa A. Bero, Peter Hanauer, and Deborah E. Barnes, *The Cigarette Papers* (Berkeley: University of California Press, 1996), 65–67, 80; Brandt, *Cigarette Century*, 24; Gately, *Tobacco*, 210–211; E. Cuyler Hammond, "Inhalation in Relation to Type and Amount of Smoking," *Journal of the American Statistical Association* 54, 285 (March 1959): 48–49; C. M. Castleden, P. V. Cole, "Inhalation of Tobacco Smoke by Pipe and Cigar Smokers," *Lancet* 2, 7819 (July 1973): 21–22; L. A. Elson, T. E. Betts, and R. D. Passey, "The Sugar Content and the pH of the Smoke of Cigarette, Cigar and Pipe Tobaccos in Relation to Lung Cancer," *International Journal of Cancer* 9, 3 (May 1972): 666–675.

78. M. A. H. Russell, J. Peto, and U. A. Patel, "The Classification of Smoking by Factorial Structure of Motives," *Journal of the Royal Statistical Society* 137, 3 (1974): 315.

79. Hammond, "Inhalation in Relation," 35.

80. Brandt, *Cigarette Century*, 27–31; Patrick G. Porter, "Origins of the American Tobacco Company," *Business History Review* 43, 1 (Spring 1969): 61, 63–70; Durden, *Bold Entrepreneur*, 16, 23–26; Matthewson, "'Smoke Signals,'" 560–563; Hahn, *Making Tobacco Bright*, 92–94.

81. P. Lorillard Co., *Lorillard and Tobacco*, 28–32; Tilley, *Bright-Tobacco Industry*, 550; Goodman, *Tobacco in History*, 101–102; Winkler, *Tobacco Tycoon*, 49–51; Gately, *Tobacco*, 205–206.

82. Brandt, "Cigarette, Risk," 157; Relli Shechter, "Selling Luxury: The Rise of the Egyptian Cigarette and the Transformation of the Egyptian Tobacco Market, 1850–1914," *International Journal of Middle East Studies* 35, 1 (February 2003): 58–60; Hannah, "Whig Fable," 44–45; Durden, *Bold Entrepreneur*, 44. It was not until 1921 that cigarettes became the most common way that Americans consumed tobacco. See Herndon, *William Tatham*, 418.

83. Priscilla Murphy, "Affiliation Bias and Expert Disagreement in Framing the Nicotine Addiction Debate," *Science, Technology, and Human Values* 26, 3 (Summer 2001): 278–299. Of course much of the murkiness surrounding nicotine, addiction, and codependent variables comes from decades of intentional obstruction and misinformation on the part of the tobacco industry. For an illuminating, and disturbing, glimpse of what tobacco companies knew about addiction and when they knew it, see Glantz et al., *Cigarette Papers*. For the cover-up concerning polonium-210 in particular, see Rego, "Polonium Brief" and Muggli et al., "Waking a Sleeping Giant."

84. Caroline Jean Acker, "Addiction and the Laboratory: The Work of the National Research Council's Committee on Drug Addiction, 1928–1939," *Isis* 86, 2 (June 1995): 167–193.

85. Kerr-Ritchie, *Freedpeople in the Tobacco South*, 217, table 8.4. A 1906 auditor's report from Halifax estimated that black real and personal estates

made up less than 8 percent of the county total, despite the fact that African American were 52 percent of the population (Morrison, *Halifax County*, 29).

86. Calculated from HCDB.

87. Shifflett, *Patronage and Poverty*, 99.

88. C. S. Lewis, *The Abolition of Man* (New York: Macmillan, 1947), 35.

89. For the account of one family's movement into and out of the ranks of Pittsylvania landowners, see Jackson's poignant *My Father's Name*, chap. 8.

90. World War I temporarily brought strong prices as the army and navy provided soldiers with generous cigarette rations, but markets fell precipitously in the 1920s. One of the most ambitious early twentieth-century organization efforts was the Tri-State Tobacco Growers' Cooperative, founded in the early 1920s. This cooperative enrolled thousands of farmers, but like earlier efforts it collapsed after only a few years. Daniel, *Breaking the Land*, 35–36; Carrington, *History of Halifax County*, 58–59; Prince and Simpson, *Long Green*, 88–107; and Bennett, "King Bacca's Throne," 94–114.

EPILOGUE

1. Brandt, *Cigarette Century*, 38–42, 93–94.

2. Trimble, *Man-Induced Soil Erosion*, 2–3. In many cases this was the entire layer of sandy topsoil, leaving farms with nothing but exposed clay subsoil. See Rowalt, *Soil Defense in the Piedmont*, 2–4, 5–6.

3. F. F. Nickels, E. F. Goldston, and E. P. Deatrick, *Reconnaissance Erosion Survey of the State of Virginia* (Washington, DC: USDA, Soil Conservation Service, 1934); and G. L. Fuller, *Reconnaissance Erosion Survey of the State of North Carolina* (Washington, DC: USDA, Soil Conservation Service, 1934).

4. Brodell, *Cost of Producing Virginia Dark and Bright Tobacco*, quote on 44; Hearn and Drane, *Soil Survey of Caswell* (1910); Jurney et al., *Soil Survey of Halifax* (1938); and Kirk et al., *Soil Survey of Pittsylvania* (1922). For another account of the low quality of local nonagricultural land, see Ross O. Stevens, *Wildlife Conservation through Erosion Control in the Piedmont*, USDA Farmers' Bulletin No. 1788 (Washington, DC: Government Printing Office, 1937), 5–7.

5. P. F. Keil, "Two Centuries of Accruing Tragedy along the Dan River," *Soil Conservation* 1, 7 (February 1936): 1–5, quote on 3. *Soil Conservation* billed itself the "official organ of the Soil Conservation Service," and included the Secretary of Agriculture, Henry Wallace, and the head of the SCS, Hugh Hammond Bennett, on its masthead.

6. Ibid., 4.

7. Ibid.; for the SCS and other New Deal conservation agencies' association of poor land with poor people, see Sarah T. Phillips, *This Land, This Nation: Conservation, Rural America, and the New Deal* (New York: Cambridge University Press, 2007), esp. chap. 2.

8. Rowalt, *Soil Defense in the Piedmont*, 7. Virginia held four demonstration districts, North Carolina eight, South Carolina five, Georgia six, and Alabama three. Most districts were targeted at demonstrating proper cotton farming methods. A portion of the Dan River drainage in southwestern Pittsylvania composed another of Virginia's districts and may have operated in conjunction with the Bannister River district.

9. "Engineering Department," *Bannister River Banner* 1, 6 (January 1935): 3; and "Gully Control Work," *Bannister River Banner* 1, 7 (February 1935): 6.

10. "Forestry Department," *Bannister River Banner* 1, 6 (January 1935): 2; and "Forestry," *Bannister River Banner* 1, 10 (May 1935): 8.

11. "Kudzu," *Bannister River Banner* 1, 8 (March 1935): 6; "Diagram Showing a Three Year Cropping Plan for a Strip Rotated Field," *Bannister River Banner* 1, 7 (February 1935): 4; and "Forestry," 8. On the SCS's love affair with kudzu, see Mart A. Stewart, "Cultivating Kudzu: The Soil Conservation Service and the Kudzu Distribution Program," *Georgia Historical Quarterly* 81, 1 (Spring 1997): 151–167.

12. "Conservation in the Farm Woodlot," *Bannister River Banner* 1, 8 (March 1935): 3; "Erosion Department," *Bannister River Banner* 1, 12 (July 1935): 4; "Soils of the Bannister River Watershed," *Bannister River Banner* 1, 6 (January 1935): 6; and Rowalt, *Soil Defense in the Piedmont*, 45. For a contemporary observation that much of this advice was well-worn in the Upper South, see Hugh Hammond Bennett in Hall, *Early Erosion-Control Practices*, 1.

13. Charles E. Landon, "The Tobacco Growing Industry of North Carolina," *Economic Geography* 10, 3 (July 1934): 243.

14. Rowalt, *Soil Defense in the Piedmont*, 27–28; and Brodell, *Cost of Producing Virginia Dark and Bright Tobacco*, quote on 4.

15. Harold B. Rowe, *Tobacco under the AAA* (Washington, DC: Brookings Institution, 1935), 263–272, 282–300; Jack Temple Kirby, *Rural Worlds Lost: The American South, 1920–1960* (Baton Rouge: Louisiana State University Press, 1987), 60; Robert F. Hunter, "The AAA between Neighbors: Virginia, North Carolina, and the New Deal Farm Program," *Journal of Southern History* 44, 4 (November 1978): 537–570; Goodman, *Tobacco in History*, 198; Daniel, *Breaking the Land*, 110–133; Bennett, "King Bacca's Throne," chap. 4; Badger, *Prosperity Road*, 195–204; Benson, *Tobacco Capitalism*, 120–126.

16. Daniel, *Breaking the Land*, 117–120, quote on 119; Bennett, "King Bacca's Throne," 71–74; and Badger, *Prosperity Road*, 195–204.

17. *2007 Census of Agriculture: Virginia State and County Data* (Washington, DC: Government Printing Office, 2009), 519; and *2007 Census of Agriculture: North Carolina State and County Data* (Washington, DC: Government Printing Office, 2009), 524.

18. On the relative durability of small tobacco farms, see Petty, "Standing Their Ground," and on the emerging tobacco agribusiness model in North Carolina, see Benson, *Tobacco Capitalism*.

19. Pete Daniel has described this process of farm consolidation for eastern North Carolina, where mechanization and flatter terrain led to larger farms and fewer farmers at an earlier date. See Daniel, *Breaking the Land*, 256–270.

20. On increasing black landownership during this period and the subsequent out-migration of regional African Americans, see Kerr-Ritchie, *Freedpeople in the Tobacco South*, 223–244.

21. Morgan, *Slave Counterpoint*, 33.

22. This image, as well as similar ones, can be found online at the Library of Congress website's Prints and Photographs Online Catalogue: http://www.loc.gov/pictures/.

23. Sara M. Gregg, *Managing the Mountains: Land Use Planning, The New Deal, and the Creation of a Federal Landscape in Appalachia* (New Haven: Yale University Press, 2010), 196–197; Angela M. Thompson, "Ethics of Seeing and Politics of Place: FSA Photography and Literature of the American South" (Ph.D. diss., University of Oregon, 2006), esp. chap. 2; Nicholas Natanson, *The Black Image in the New Deal: The Politics of FSA Photography* (Knoxville: University of Tennessee Press, 1992), 3; Paul Hendrickson, *Looking for the Light: The Hidden Life and Art of Marion Post Wolcott* (New York: Alfred A. Knopf, 1992), xxv–xxvi, 44–45; F. Jack Hurley, *Marion Post Wolcott: A Photographic Journey* (Albuquerque: University of New Mexico Press, 1989), 27–28; James Curtis, *Mind's Eye, Mind's Truth: FSA Photography Reconsidered* (Philadelphia: Temple University Press, 1989), 6; Pete Daniel, Merry A. Foresta, Maren Stange, and Sally Stein, *Official Images: New Deal Photography* (Washington, DC: Smithsonian Institution Press, 1987), 1–4; Andrea Fisher, *Let Us Now Praise Famous Women: Women Photographers for the US Government, 1935–1944* (New York: Pandora, 1987), 5–6; Penelope Dixon, *Photographers of the Farm Security Administration: An Annotated Bibliography, 1930–1980* (New York: Garland, 1983), xvii; Sidney Baldwin, *Poverty and Politics: The Rise and Decline of the Farm Security Administration* (Chapel Hill: University of North Carolina Press, 1968).

24. On the prevalence and importance of the Historical Division's project photography, see Natanson, *Black Image*, 34, 59–60; Curtis, *Mind's Eye*, viii; Hurley, *Marion Post Wolcott*, 44–46, 103. On the widespread instruction in canning as self-improvement in the early twentieth century, see Elizabeth S. D. Engleheart, *A Mess of Greens: Southern Gender and Southern Food* (Athens: University of Georgia Press, 2011), 83–117.

25. F. H. Jeter in Landon, "Tobacco Growing Industry," 253.

26. Badger, *Prosperity Road*; Gilbert C. Fite, *Cotton Fields No More: Southern Agriculture, 1865–1980* (Lexington: University Press of Kentucky, 1984), esp. chaps. 7–9; Daniel, *Breaking the Land*; Jack Temple Kirby, *Rural Worlds Lost: The American South, 1920–1960* (Baton Rouge: Louisiana State University Press, 1987); Douglas Carl Abrams, *Conservative Constraints: North*

Carolina and the New Deal (Jackson: University Press of Mississippi, 1992), chap. 4; Phillips, *This Land, This Nation,* chap. 4; Benson, *Tobacco Capitalism,* 80–84.

27. C. Vann Woodward, *The Burden of Southern History,* 3rd ed. (Baton Rouge: Louisiana State University Press, 1993).

Index

addiction, 239–243
Agricultural Adjustment Act (AAA), 7, 251
agricultural reform, 82–83, 92–101, 250; bright tobacco in, 83, 106–118; criticisms of tobacco, 85–86, 101; failure of, 87–89; origins of, 84–85; publications, 89, 250; societies, 89–90, 91
Alamance County, NC, 176–177
American Farmer, 89
American Tobacco Company (ATC), 11, 216, 229–232, 242, 247
An Essay on Calcareous Manures, 85, 90
Arator, 84
auctioneers, 188

Bagby, George, 144
Bannister River Banner, 250
Barksdale, Elisha, 165–166
Barksdale, William, 166
bateaux, 19, 40–42
Bennett, Hugh Hammond, 88
Berry Hill plantation, 29
Blackwell, John, 148
Brown, Bedford, 174, 175

Bruce, Charles, 199
Bruce, James, 29
Bruce, James C., 29, 31, 86, 91, 93, 96, 99
Brumfield, Henry and Julia, 182, 183–184
Butterweck, Otto Carl, 235
Byrd, William, II, 16–18, 23–24, 26

Caswell County, NC. *See* Southside
Chappill, Allen, 164
charcoal, 57, 109
Chatham, VA, 26, 42, 147, 237
cigarettes, 128, 217–218, 229–232, 238–243, 247
Civil War, 118, 146, 150–151; effect on tobacco markets, 120–121, 123; impressment of slaves during, 135–136, 138; inflation, 139–140; taxes in kind, 133–134; tobacco consumption during, 129
Civilian Conservation Corps (CCC), 249
Clement, Maud Carter, 147
clover, 95, 96, 107, 117
Cobb, Griffin, 148
Cocke, John Hartwell, 86
Cohen, Benjamin, 112

Coleman, Charles, 101–106, 117
corn, 110, 130–132
country stores, 27
Craven, Avery, 88
Cronon, William, 189–190
crop rotation, 94–95, 107–108
cultivation, 35–36, 197

Dailey, Jane, 209
Dan River, 16, 18, 23, 248
Danville, VA, 5, 8, 26, 42, 120–121;
    auction system, 185, 186–190; Civil
    War refugees in, 119, 137–138, 141;
    industry in, 68, 74, 210, 222; prisons,
    137–139; racial politics in, 185–186,
    208–212; riot, 208–211, 212–213;
    tobacco market in, 65, 127–128, 222;
    Union occupation of, 141–142
Danville Circular, 209
Danville Committee of Forty, 211
Danville Tobacco Association, 227, 228
Davis, Jefferson, 141
deforestation, 108–109, 193–196,
    247–248
Dennett, John, 143, 178
Dew, Charles, 71
Dickerson, W. T., 163
Duke, James B., 229, 239

Earle, Carville, 88
Eden, VA, 26
Edmunds, John, 90, 95
emancipation, 121, 146, 152–153, 168
erosion, 107, 143, 196–199, 224, 247–250,
    253; control efforts, 93–94, 249–250
An Essay on Calcareous Manures, 85, 90
Estes, John, 235
Evans, Charles, 116, 129–130

Farm Security Administration (FSA),
    258–261
Farmers' Alliance, 217, 227–229
Farmers' Register, 85, 89
fertilizer, 201–206, 221, 225–226
Fiege, Mark, 118, 149
Flintoff, John, 155

flues, 55–57
forests, 22, 54, 196
Forrest, Nathan Bedford, 173
Freedmen's Bureau, 148, 153–155, 163–165,
    167–168, 169–170
freedpeople, 142, 155–156, 161; apprentice-
    ships, 166–168; landownership of,
    153–155, 179, 243–244, 252; white
    characterizations of, 9, 157, 162, 180,
    199–201, 255

Glasgow, Ellen, 112
Grange, 217, 226–227
Graves, William P., 210–211
guano, 34, 95–96, 117, 140, 202,
    204–205
gullies, 3, 107, 247, 250

Hahn, Barbara, 12, 47, 59, 223–224
Haines, Alanson, 142
Hairston, George, 29
Hairston, Peter, 116
Hairston, Ruth, 30, 93, 133
Hairston, Samuel, 29–30, 31
Halifax County, VA. See Southside
Halifax Courthouse, 42
Hall, Arthur, 88
Hatch Act, 237
Hatchett, William, 178
Hawk, P. J., 167
Henry, Patrick, 84
hogs, 99, 192
hogsheads, 39–40
Holden, William, 172, 173, 176–177
hornworms, 36–37, 206–207
Hundley, Elijah, 100, 179
Hunt, Gabe, 37, 61

inspection, 40, 186–188

Jeffreys, George, 62–64, 86–87,
    106–107, 110
Jennings, Polly, 165–166
Johnson, Ben, 172
Johnson, Jordan, 61
Jones, George, 127

Keil, P. F., 248
Kerr-Ritchie, Jeffrey, 161
King, James, 164
Kirk, George W., 176
Kirk-Holden War, 176
Ku Klux Klan, 171, 172–178

land surveys, 195–196
Lea, John G., 172, 175, 177
Leatherwood, VA, 124
Lederer, John, 24
Lewis, C. S., 244
Liedtke, F. W., 165
lime, 96
livestock, 132–133
Lotten, S. M., 171
Love brothers, 112, 196, 197

manure, 34–35, 97
Mathew, William, 88, 96
McFeely, William, 161
merino sheep, 98
Merritt, Daniel, 100–101, 151
Miller, N. C., 160
Millner, James, 125–126
Milton, NC, 26, 42, 230, 231

Nelson, Lynn, 88
Newlin, W. H., 142
nicotine, 53, 239–241

Occaneechi, 23–24
Olmsted, Frederick Law, 61,
    144
orchards, 100–101, 103
Ott, John, 162

Pace, Ed, 219, 220
Panic of 1837, 82, 187
Paxton, Jason, 133
peddlers, 27
Perry, Henrietta, 61
Petersburg, VA, 68
Piedmont. See Southside
pinhookers, 227
Pitt County, NC, 221

Pittsylvania Central Alliance Trade
    Union (PCATU), 232–234
Pittsylvania County, VA. See Southside
plant beds, 32–34
plaster, 96
plows, 60, 63, 92–93, 106–107
Pollard, Edward, 143
polonium-210, 203, 241
Post, Marion, 258–259

quota system, 251–252

R. J. Reynolds, 219, 223, 231
Ragland, Robert, 112, 196, 198, 205,
    214–215, 220–221
Readjuster Party, 208
Reconstruction, 149–150, 208;
    labor contracts, 157–165, 178–179;
    violence during, 147–148, 168–173,
    175, 177
Richmond, VA, 68, 78, 137
Richmond and Danville Railroad, 114, 115,
    119, 135, 138
Roanoke River, 23, 42, 43–44
Robert, Joseph, 71–72
Ruffin, Edmund, 85–86, 88, 111

Saura, 23–24
Schoolfield Reservoir, 248
Scots-Irish, 18–19
Seigel, Frederick, 8, 120
Self, Edward, 169
sharecropping, 158–159, 161–162, 178–179,
    191
Shelton, Samuel, 52, 57
Sims, William, 35–36, 132, 134, 136–137,
    152–153
Slade, Abisha, 46–49, 57, 112–113,
    115–116
Slade, Stephen, 46–49, 61, 112, 117
slavery, 25, 148, 150, 168; in agricultural
    reform, 91, 104–105, 118; expansion of,
    28–29
slaves, 28, 118, 134–136; factory work of,
    69–72; tobacco expertise of, 9, 48,
    60–62, 135, 200–201

Soil Conservation Service (SCS), 7, 247, 248–251
soil surveys, 217, 236–237
soil types, 21–22, 24–25, 52–54, 110
Sours, William, 123, 139, 145, 222
South Boston, VA, 130
Southern Planter, 89
Southside, 142, 146, 184, 191; environment of, 3, 14, 16–17, 20–23, 42–43, 248; internal improvements in, 114–115; lack of towns in, 26–27, 149; modern day, 252–253
Special Field Order 15, 155
Staunton River, 16, 18, 192
Staunton River Bridge, Battle of, 134, 135
Stephens, John W., 172, 173–176, 178
Stoll, Steven, 88, 97
Stryker, Roy, 258
Sutherlin, William, 100, 119, 145–146, 183, 184, 199; Civil War business of, 124–126, 128, 140
Sutherlin and Ferrell Company, 68–69, 72, 75–76

Taliaferro, Squire, 210
Tatham, William, 35, 37, 39
Taylor, John, 84–85
Thomas, James, 78, 137
threshing machines, 100
Tilley, Nannie May, 8, 47, 120
tobacco, bright, 3–4, 49–50, 149, 179–180, 214–215; appeal of, 11–12, 54–55, 60, 83–84, 109–118; creation myth, 46–49; curing of, 55–59, 108–109, 111–112, 240; definition of, 12–13; diseases of, 207–208; health effects of, 14, 203, 241, 252, 257; manufacturing, 67–79, 113–114, 190; marketing of, 75–78, 124–125; nutrient demands of, 53–54, 202; origins of, 50–60; pamphlets, 112; prices of, 64–67, 80–81, 116, 152, 215; problems

of, 80, 192, 194–195, 198; spread of, 217–218, 219–223; warehouses, 188, 190–191, 227, 228
tobacco, burley, 221
tobacco, dark, 17–18, 19, 78, 110, 130; Chesapeake, 19–20; curing of, 37–39, 109; global demand for, 25–26; introduction to Piedmont, 19–20, 26; routines of, 32–42, 94–95
tobacco barns, 38–39, 108, 192–193
tobacco seeds, 34, 51–52
topping, 36, 60
Tourgée, Albion, 171, 174, 177
Trimble, Stanley, 198
Trowbridge, John, 143–144
Tuck, Davis, 56

Union Agricultural Society, 91
Union League, 171, 173, 174, 177
United States Department of Agriculture (USDA), 235–238

Venable, George, 192–193
Virginia State Agricultural Society, 90

wheat, 20, 110, 130–132
Whitney, Milton, 236
Wilcox, J. F., 154, 167
Wiley, Bell Irvin, 129
Williams, Nancy, 37
Willis, Anderson, 127
Willis, Bobby, 257, 259–261
Wilson, James, 237–238
Wilson, Robert, 127, 160
Wilson, Samuel, 133–134, 160
Wilson County, NC, 221
Witcher, Vincent, 101
Withers, Robert, 141, 153
Womack, H. P., 115
Woodlawn Plantation, 102–105
Woodward, C. Vann, 261

Yanceyville, NC, 172